ENGINEERING: MODELING AND COMPUTATION

ENGINEERING: MODELING AND COMPUTATION

WALTER J. GAJDA, JR.
WILLIAM E. BILES
University of Notre Dame

HOUGHTON MIFFLIN COMPANY
BOSTON
Dallas Geneva, Illinois Hopewell, New Jersey
Palo Alto London

Printed in the U.S.A.

Library of Congress Catalog Card Number: 77-074378

ISBN: 0-395-25585-6

TO OUR PARENTS

AND THE PARENTS OF OUR WIVES

CONTENTS

PREFACE

Almost every new engineering student takes a course that introduces him or her to the study of engineering. Many textbooks have been written for this purpose and many different approaches to teaching introductory courses have been tried. Although these approaches can be categorized only in the most general terms, they can be said to be divided into two types.

The first type of introductory engineering course emphasizes a qualitative treatment of such topics as the history of engineering, professional standards and ethics, descriptions of the various fields within engineering, the types of jobs performed by engineers, and the approaches engineers take in designing and analyzing engineering entities. This first type of course describes *what* the engineer does.

The second type of introductory engineering course treats the quantitative tools engineers use in solving problems. Such a course requires the student to develop analytical techniques and exercise them in the solution of problems. This type of course describes *how* the engineer solves problems.

This textbook is designed for the second type of course. It stresses the development of models, and the mastering of computational tools and techniques necessary to manipulate these models to solve engineering problems.

This book has evolved from a decade of experience at the University of Notre Dame in teaching a course entitled Introduction to Engineering Concepts. For several years, this course examined the slide rule and the digital computer as computational devices needed in the study and practice of engineering, with a majority of the course devoted to the FORTRAN programming language. Problems from various engineering fields were examined, procedures for solving problems were described, and the students were required to develop the FORTRAN programs necessary for solving these problems. This approach afforded the freshmen engineering students exposure to typical problems from all engineering fields, and it imparted specific computational skills.

The engineering faculty observed, however, that students tended to regard the computer as an end in itself rather than the computational tool that it should be at this early stage of their education. There was also a strong opinion that FORTRAN programming did not justify an entire three-credit course. In response to these faculty reactions, we made an effort four years ago to broaden and systematize the engineering concepts to which the student was introduced, while retaining the essentials of FORTRAN programming. An initial attempt to use two textbooks simultaneously—one to introduce the desired engineering concepts and another to present FORTRAN programming—was unsatisfactory to faculty and students alike. It was apparent to us that we needed an introductory engineering textbook that would integrate fundamental engineering concepts with FORTRAN programming. We developed this textbook in response to that need.

This book is unusual among introductory engineering texts in that it presents to the reader such modern engineering topics as modeling, optimization, feedback and input/output analysis, probability and economic analysis within a framework of utilizing scientific and mathematical skills to solve engineering problems. We have tried to tread a path between oversimplification and excessive complexity, to demonstrate that engineering is an intellectually demanding endeavor, and to encourage and stimulate students to see engineering as an exciting and challenging career. The reactions of our students over the last three years have confirmed our belief that this textbook accomplishes those objectives.

This is not a book for marking time. It makes demands on both students and teachers. It is intended to motivate the student to master the science and mathematics that form the cornerstone of modern engineering education and that so often appear as hurdles to new students.

We have recognized engineering students' need to master computational tools and techniques early in their programs of study. We have observed the fascination that the seemingly boundless power of the digital computer holds for new students. Hence we have introduced FORTRAN programming early in the text (Chapter 3), and we have woven many examples and problems into succeeding chapters. We hope that engineering faculty and students will concur that computer programming can best be taught to engineering students by engineers in an engineering setting.

The book contains more than enough material for a one-semester course. The organization affords the opportunity to select course material in keeping with the specific future needs of a particular engineering program. For example, Chapter 8 treats probabilistic models in engineering. If a given engineering program requires a later course in engineering probability—as is the case at Notre Dame—the instructor might exclude Chapter 8 from an introductory engineering course. If no such course is required later, however, the instructor could devote as many as six or eight class periods to probabilistic models. The same is true of Chapters 9 and 10, which examine economic and feedback models, respectively. Even Chapter 1, which contains an extensive section on the electronic calculator, offers a choice in that the instructor can choose to allow the student to absorb that material in a self-study assignment. Chapters 2 through 7 represent fundamental engineering concepts that should be covered, at least in part, in any introductory engineering course.

The sequence of coverage makes flexibility feasible. If an instructor wants to expose students to such topics as modeling, units and dimensions, and optimization methods before giving any attention to computer programming, Chapter 3 could be covered after Chapter 5. Certain material in Chapter 5 requires a knowledge of differential calculus. If students are taking an introductory engineering course concurrently with a first calculus course, it may be necessary either to delay certain topics in Chapter 5 or to delete them entirely. Chapter 7 contains developments in both differential and integral calculus, but the instructor can treat these developments without relying on students having a rigorous understanding of calculus.

ACKNOWLEDGMENTS

We take pleasure in acknowledging the contributions of all those who helped in the development of this book. In particular, we thank the hundreds of students who have waded through poorly edited manuscripts. Some complained and most were patient, but all of them made valuable comments. We thank the many Notre Dame faculty members, past and present, who have taught the Introduction to Engineering Concepts course over the last decade. They have contributed engineering problems, comments, editing, corrections, and sometimes simply support and encouragement. Their influence is indelibly stamped on these pages and on the minds of the Notre Dame students who have taken this course. Professors Robert Betchov, Raymond Brach, Thomas Cullinane, Howard Demuth, Michael Doria, Vincent Goddard, James Kohn, R. Jeffrey Leake, James Lindell, Gary Long, John Lucey, James Massey, Stuart McComas, Andrew MacFarland, Albert Miller, Bruce Morgan, Edgar Morris, Arthur Quigley, Byron Roberts, Neil Schilmoeller, Thomas Smith, Adolph Strandhagen, Arvind Varma, and Leon Winslow have all contributed; we shall always be indebted to them. A special note of thanks is due to Professor Robert Eikenberry for his excellent review of the FORTRAN material in this book. We also extend our appreciation to our long-suffering secretaries—Janet Tillman, Valerie Flanagan, Kathy Simon, Ella Levee, and Shirley Wills—who have produced a decent product from relatively indecent feedstock. Our appreciation is also extended to our reviewers—W. George Devens, Virginia Polytechnic Institute and State University; Francis J. Jankowski, Wright State University; James M. McNeary, University of Wisconsin; Robert Stratton, Tulsa University; Joseph R. Troxler, Northern Arizona University; and Herbert W. Yankee, Worcester Polytechnic Institute—who have had measurable effects on the final version you hold in your hands. Finally, we would like to acknowledge the support of our wives and children in this endeavor.

TO THE STUDENT

We wish you well on your journey in education. You may decide that engineering is a career for which you are well suited, or you may discover that your talents lie in other fields. Or your involvement in this course may simply be an excursion—undertaken out of curiosity—from a non-engineering field of study. In any event, we hope that your moments spent in reading, studying, analyzing, and exercising the material in this book will be well spent and will afford you insights that will serve you well in the future.

W. J. G.
W. E. B.

CHAPTER ONE
ENGINEERING: THE APPLICATION OF SCIENTIFIC PRINCIPLES

The word engineer is abused by some people and ill understood by most. For example, trash collectors are often called "sanitary engineers," and some building superintendents are called "maintenance engineers." Further, many people—when they hear the word engineer—have a reflex thought and visualize the person who is in charge of guiding a railroad locomotive.

It is also true that the roles of the engineer and the engineering technician are often blurred. To many first-year students, engineering means specific technical jobs: drafting, fixing television sets, surveying land to establish a building site, repairing automobile air conditioners, and so forth. These kinds of jobs are properly seen as belonging to engineering technicians in that performing them successfully does not involve, to any significant degree, those abilities that characterize the engineering profession. The skills that differentiate the engineer from the technician are well defined. In general, the engineer has a much deeper understanding of mathematics and the physical sciences. The career of a technician does not ordinarily require more than two years of education beyond high school, while most engineers have had at least four years of education beyond high school.

In addition, the jobs assigned to technicians tend to be, with some exceptions, primarily of a routine nature. For example, assembling electronic circuits according to another individual's specifications requires gathering the necessary components and tools as well as following a standard procedure. Repairing machinery involves following an established set of diagnostic procedures which isolate those components that are at fault (although the insight gained from this experience is often invaluable).

Our intent in this introduction is not to demean in any way the importance of the roles of engineering technicians. Without these highly skilled individuals, the efforts of a large number of engineers would have little value. Our intent is to point out those functions that define the engineering profession.

Engineers have to have a stronger math and science background than technicians because they have to be able to meet the many challenges of system design. The word *design* means the developing of a physical system (some collection of things) that performs a required function. For example, a civil engineer who designs a system for a municipal water supply must specify all the elements of such a system: the performance requirements of the pumps, the sizes of pipes to be used in various segments of the system, and the required valves and interconnections. Similarly, an aeronautical engineer must design the wings (among other necessary parts) to be used on a given aircraft. This design must include statements of the materials to be used, the arrangement of supporting structural members, the number of rivets needed to fasten the skin to the supports, and the means of attaching the wings to the fuselage. Figure 1–1 shows aluminum bulkhead fittings that are important structural elements in aircraft.

These two examples raise many questions that must be understood and answered if the proposed designs are to have a chance of yielding end products that operate satisfactorily.

To design municipal water-supply systems, civil engineers must know how much water will flow through pipes of different sizes, and must be able to estimate the

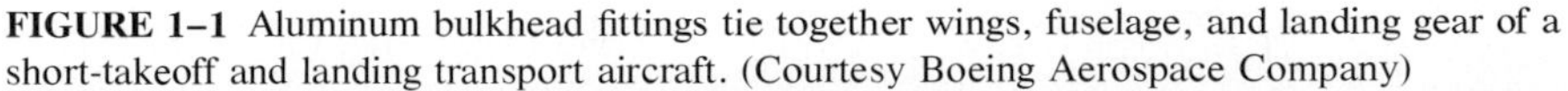

FIGURE 1–1 Aluminum bulkhead fittings tie together wings, fuselage, and landing gear of a short-takeoff and landing transport aircraft. (Courtesy Boeing Aerospace Company)

pressure drops that will occur along the pipeline. (If the pressure drop in a given location is too great, families living at elevated locations in the city will not receive an adequate flow of water.) They must also have knowledge of material behavior in order to be able to choose the right pipe. In the same vein, aeronautical engineers must know how much mechanical stress a plane's wings will have to withstand, and they must understand the relative strengths of materials as well as the load-bearing capabilities of different arrangements of supporting members.

Both problems clearly require, in addition to knowledge of physical phenomena, the ability to perform mathematical calculations. Figure 1–2 shows one commonly used calculating instrument. Any civil engineer who has to build three of four water-supply systems in order to find the one that works best will have an exceptionally short career. The engineer must choose a design on the basis of accurate mathematical analysis of the system's performance before the system is built.

We have discussed only two specific problems, so far, but these problems demonstrate that people need a wide range of engineering skills if they are to design systems that work well. We already see that engineers need to understand the flow and pressure variations for fluids such as water. They also need an understanding of the mechanical strength and corrosion resistance of various materials. They need to

FIGURE 1–2 A compact electronic calculator facilitates the work of both engineer and engineering student. (Courtesy Hewlett–Packard)

have a good grasp on engineering mechanics. They must also be able to write equations describing physical systems and use these equations to mathematically predict the performance of these systems.

Additional examples from other fields of engineering would reveal the necessity of understanding physical phenomena involving electrical, thermal, chemical, metallurgical, and other concepts. The one common thread is the ability to write equations and perform calculations.

We can summarize this discussion in the following statements. Engineering is the application of scientific principles to meet human needs. The language of engineering is mathematics. Thus undergraduate engineering students must have significant exposure to mathematics and the physical sciences. Further, they must be adequately exposed to the important methods that have been developed to apply scientific principles to solve problems. To a large extent, courses in the physical sciences, mathematics, and their applications constitute the undergraduate experience. Finally, to be successful, an engineer must be able to abstract and to model.

1–1 ABSTRACTION AND MODELING

The stated aim of this text is to provide an introduction to engineering. Yet we have just used terms that have not been carefully defined. What does it mean to abstract and to model?

This is a difficult question, and no single brief answer will suffice. Philosophers have written books about these two ideas. Indeed, the science of philosophy is one of abstraction. To begin a discussion of this topic, consider an algebraic expression:

$$y = x + z \tag{1–1}$$

What does this seemingly simple expression mean? It is not, at least in form, very different from the following:

$$4 = 2 + 2 \tag{1–2}$$

In fact, we can carry this process one step further and arrive at

$$,,,, = ,, + ,, \tag{1–3}$$

or even

$$,,,, = ,,,, \tag{1–4}$$

This sequence of four equations progresses from the quite general (Equation 1–1) to the very specific (Equation 1–4). Let us review this sequence step-by-step. The first equation simply states that one symbol (y) can be defined (this is the meaning of the equals sign) as the algebraic sum (plus sign) of the other two symbols (x, z). We did not attempt in this equation to indicate either the physical or the numerical meaning of the symbols. The second equation made this relationship more specific by restricting the entities involved to certain integers within the real number system. The third equation is even more specific in that it indicates the physical quantities (a set of commas) that are involved on both sides of the equals sign. In other situations, for example, the physical quantities could be beads or pieces of paper or newtons of force or flows of water or many other things. Equation (1–3) still retains a minor element of abstraction because a symbol (the plus sign) denotes the process of grouping things together. The last equation in the series is self-evident and appears almost nonsensical—four commas are indeed equal to four commas. This illustrates that increasing the level of abstraction broadens the scope of applicability of the equations or statements used. Equation (1–1) is much more general than (1–4) and represents a higher level of abstraction.

It is not possible to state a useful application of the last equation. However, Equation (1–1) *is* useful in a number of situations. For example, an electrical engineer who has to specify the size of a copper wire needed to supply x amperes to an air conditioner and z amperes to a refrigerator could use this equation to calculate y, the total number of amperes required. This is a very simple equation and a correspondingly simple example.

Learning to abstract is essential in engineering education. Rather than writing down the specific equations that govern each different example of chemical reactions, a chemical engineering student can study and master those general (more abstract) equations that govern all chemical reactions in a given classification. Rather than examining the equations applicable to each specific electronic circuit, an electrical engineering student concentrates on a general set of equations that hold for all circuits. Engineering education involves the exposition of general sets of equations, although there are many examples in which these equations are applied to specific detailed problems.

This need to develop engineering principles in abstracted general form becomes clearer if we compare the varied demands on the talents of an engineer over an entire career with the scientific principles that can be communicated in a four-year educational program. Faculties simply cannot hope to cover enough specific topics to enable their students to know a detailed answer to each problem an engineer is likely to encounter over the span of a whole career. They can teach only a broad range of general scientific principles and their applications. When engineers encounter a specific engineering problem on the job, they must generalize it to perceive the important applicable principles. Successful abstraction enables engineers to apply a relatively small amount of knowledge to a wide range of problems. For example, the interstate highway network may appear to be quite different from the controls that guide an aircraft, but when they are abstracted both can be seen in terms of engineering systems.

Now let us consider what the word modeling means. Modeling denotes a theme that will be explicitly obvious throughout the following chapters. Modeling can be easily explained, but only practicing engineers who have gained quite a lot of experience can fully appreciate its power.

Any event involving physical entities is extraordinarily complex, even when it appears simple on the surface. For example, when the compressor in a refrigerator runs, the billions of atoms that make up the refrigerant gas are placed in organized motion. Even with high-speed computers, we cannot calculate the detailed paths atoms travel as they move through the refrigerator's cooling system. There are simply too many atoms, and their motions are too complex. The paths of the atoms depend on the compressor itself and on the details of the tubing, valves, and other elements used to construct the total system.

In exactly the same fashion, calculating the flow of water in a municipal water-supply system is impossible if the calculations are based on the motions of all the individual water molecules. An aeronautical engineer cannot begin to analyze the stresses a given wing can withstand by writing the equations governing the response of each individual atom to the applied stress.

There are two ways to react to problems of such complexity. The first is simply to walk away from the problem in despair. Such a response will not help solve the problem. Without the efforts of engineers to understand and design these systems, our standards of living would be much lower. Without refrigeration systems, municipal water supplies, and aircraft, for example, the problem of getting food would be unimaginably complicated.

The other possible response is to model the problem. When we model a complicated problem, we throw away all the nonessential elements that complicate matters and retain the essential physical processes.

Definition. A model is a simplified description of some real-world phenomenon or device. Its purpose is to aid analysis, understanding, and design ability.

Models of refrigeration systems have been successfully formulated and analyzed. The key concept underlying these models is simple. It involves developing the equations of motion for an "average" small volume of the refrigerant gas. With this model, we can accurately calculate the response of the billions of atoms in the actual refrigerant gas by considering the motion of this "average" volume. This is true, but certainly not obvious. Many engineers have had to do much theoretical and experimental work to demonstrate the model's validity. We can make analogous remarks for municipal water supplies and aircraft wings.

Abstraction and modeling are the cornerstones for much of this book, as well as for subsequent engineering courses. You probably don't yet feel comfortable with these ideas, for this introduction was written only to get you to think in some new directions and to appreciate the complexities involved in physical processes that you may now take for granted. Stop and think about the problems involved in setting up an electrical power distribution system or in designing highways. How does an engineer decide whether a given plot of land could support a 70-story building? What factors determine which materials should be used to construct an internal combustion engine? We could think of literally thousands of such questions. Education and experience enables engineers to answer them.

1–2 SYSTEMS, MATERIALS, AND ENERGY

In addition to abstraction and modeling, a number of major themes recur throughout engineering. Engineers often find it convenient to approach a physical problem in terms of its major elements. For example, they often discuss a specific problem in terms of three key concepts: systems, materials, and energy.

A system is any collection of interacting physical objects. Almost anything that serves a function is a system.

For example, the distance meter shown in Figure 1–3 provides information about distances and directions useful in construction surveys. Similarly, a digital computer (Figure 1–4) takes questions—as long as they're asked in the proper way—processes them, and returns answers.

An engineer must also consider the properties of the *materials* from which the elements of a given system are assembled. The performance of the total system depends strongly on these properties in a number of obvious and not-so-obvious ways. For example, in a home electrical system, the properties of certain materials are obviously important. The material used for the wire core must be a good conductor of electricity. Further, a suitable insulating material must surround this core, to minimize the chance of short circuits, and resulting fires, inside walls. It is a simple matter to go to a handbook of physical properties and observe that gold, silver, copper, and aluminum are the best electrical conductors. The first two can be immediately eliminated on the basis of cost. Very few people could afford homes

FIGURE 1–3 This distance meter is an example of a system. Its energy source is the battery pack on the ground. (Courtesy Hewlett–Packard)

wired with gold or silver. Copper and aluminum are both excellent conductors of electricity. Copper is somewhat better, but aluminum is less expensive. Thus it might appear that aluminum's excellent electrical performance and lower cost make it the better choice. In fact, aluminum tends to corrode slightly because a chemical reaction takes place between the surface of the aluminum wire and oxygen in the atmosphere. The resultant compound is an insulator. Experience has shown that carelessly installed connections to aluminum wire tend to corrode and generate electrical heat that is a fire hazard. Copper is thus the best choice, even though it costs more.

Similar considerations of materials apply to all engineering designs. A consumer has the right to expect that a product will perform well not only the day it is purchased but for a reasonable time in the future. Therefore engineers must understand the long-term performance of the materials used. Will a rivet used in the wing of an aircraft fail after it has undergone vibration for a long time? Will the pipes used in a municipal water-supply system corrode? Will the electrical components in a television set fail after the set has been turned off and on only a few times? To avoid early system failure, the engineer must choose materials wisely.

The last key concept mentioned above was *energy*. The word energy has been mentioned almost daily by the news media over the past several years, but in the

FIGURE 1–4 A compact computer system. The input device is a typewriter-like keyboard with an associated display screen. Output may be obtained on the printer to the left of the operator. (Courtesy Hewlett–Packard)

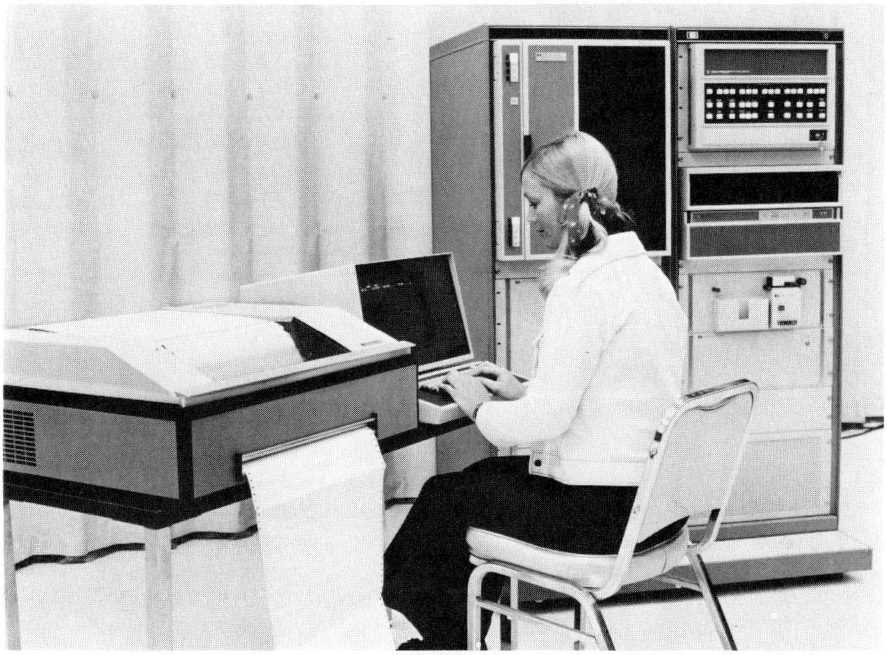

present context, energy simply means the *motive power within a system*. In simple terms, what makes a system operate? The answer to this question is the energy source.

Not all systems change with time. Some are static. Such systems do not have an identifiable source of energy. A bridge is a good example. It can absorb energy, as it must when a vehicle passes over it. The motion of the vehicle generates vibration in the structure, but if the bridge was properly designed, the vibration is damped as energy is absorbed and dissipated within the bridge itself.

Almost any physical entity that serves a function may be considered an engineering system. For example, we can consider the Earth as a system. The definition of system implies that we must be able to identify a set of inputs and a set of outputs, because we could not observe that a given system was performing any function unless we could observe inputs and outputs. In what ways can we consider this planet on which we live as a system?

First let's consider the Earth as a system viewed on the scale of the solar system. Figure 1–5(a) shows an appropriate sketch. The Earth intercepts electromagnetic radiation (visible light, heat, x-rays) that falls on its surface from many directions

FIGURE 1–5 Two views of the world as a system: (a) A sketch of the basic elements. (b) A block diagram showing inputs and outputs.

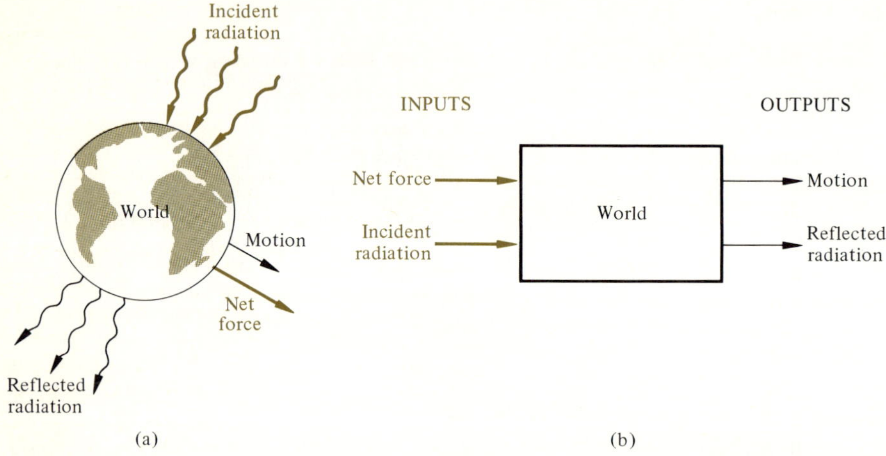

and sources. The most visible source is the sun, although the stars also contribute to this radiation field. The Earth interacts with this radiation and scatters some of it back into space. The reflection of incident energy is critical because this is what determines the average temperature at the Earth's surface. The incident radiation is an *input* to the system, while the scattered radiation is an *output* from it.

The Earth also interacts with its surroundings on a mechanical level via the forces involved in its motion. Figure 1–5 shows a net force acting on the Earth and the motion that results from that force. From the systems point of view, the forces are inputs and the subsequent motion is a resulting output. Figure 1–5(b) shows the Earth in a schematic block diagram. Inputs are represented by arrows entering the box from the left and outputs as arrows leaving to the right. The Earth, when viewed on this vast scale, is quite simple. This should be somewhat surprising, in view of the complexities we see in our day-to-day lives!

Exercises

1–1 Examine the Earth as a system from a much smaller point of view: your home town. Treating everything other than the town as a system, list the inputs and outputs as you perceive them. This exercise has no wrong answer. It is intended to encourage you to begin to apply systems concepts to familiar situations.

1–2 What are the inputs and outputs associated with the system that constitutes the school you attend?

1–3 What properties would be important in selecting materials with which to build a child's swing set?

1–4 Identify the sources of energy used in this nation's electrical system.

1–5 Identify the inputs and outputs associated with an automobile.

Choosing materials for the various parts of any system is complex because a wide range of materials is in common use and many properties are of interest. An automobile is a system with definable inputs and outputs that contains significant quantities of many different materials: steel, aluminum, copper, plastics, rubber, and glass. Each material was chosen to perform a particular function efficiently on the basis of its physical properties. Examples: Material used to make a window must be transparent to visible light. Glass is transparent, but it has the dangerous property of shattering into sharp fragments when it breaks. The solution was safety glass—a hybrid sandwich of materials in which plastic and glass sheets are laminated together. Copper is used in the electrical wiring of an automobile for the same reasons it is used in homes. Its resistance to the flow of electrical charge is low, and it is relatively inexpensive.

Each of these materials was selected because its properties match a given application. Engineers have developed many materials with properties tailored to meet specific sets of requirements. The entire area of materials science is aimed at understanding such properties. In this context, the word *properties* implies chemical, electrical, mechanical, optical, and thermal behavior.

In designing and fabricating a complete system, we cannot simply collect a number of parts with regard only to the properties of their materials. We must also consider the source or sources of energy that will act within the system. In some familiar cases, the energy source is obvious. Mechanical forces power an automobile; these forces are produced when the chemical energy stored within gasoline is converted, by combustion, into heat and mechanical force. The energies responsible for the operation of a municipal sewage system are less obvious. When we take a shower, the water flows downward because it minimizes gravitational potential energy by seeking the lowest level available. All unpumped drainage systems operate on this principle, and sewage pipe is installed with a downward pitch to ensure that water will flow under the influence of gravity.

This concept of an engineering system, along with the associated considerations of materials and energy, provides a framework for understanding the engineering profession. It is not enough for an engineer to be familiar with this collection of terms and definitions, since these are inadequate for finding solutions to actual problems. The next logical question is: What tools does a typical engineer have available? This question is analogous to the proverbial question: How long is a piece of string? Both questions have many correct answers. Different engineers have different skills, and we cannot enumerate all the combinations of skills displayed by all practicing engineers. But we can seek a common thread: the general set of ideas and methods common to all engineers.

Any engineer worthy of the title must have gained familiarity with many scientific principles because, in essence, engineering is the application of these principles to solve problems and meet human needs. Engineers must master these ideas before they can apply them. Naturally, a chemical engineer and a nuclear engineer will master different sets of principles. Regardless of the field, scientific principles are basic to the practice of engineering.

There is no simple, short definition of the word engineer. Many definitions have been formulated, but in light of the vast diversity of engineers and engineering problems, they fail to cover the scope of the word. In the following two sections we shall attempt to operationally define engineer and engineering by two complementary approaches. First we shall examine the roles of engineers within an economic system, and then we shall briefly describe the fields within engineering. This discussion will include both the traditional disciplines and the newer interdisciplinary areas.

1–3 THE ECONOMIC ROLES OF THE ENGINEER

Any discussion of the economic roles of engineers will necessarily be controversial. Reasonable people can often reasonably disagree in their interpretations of the same economic data. We recognize that the following discussion contains certain personal interpretations, and you should recognize this also.

To begin, consider Figure 1–6. It presents a simplified view, in block-diagram form, of the workings of a private company. The economic process begins with a need. Someone or some group needs something. The need can range from the fundamental, such as a source of pure water for drinking, to the somewhat frivolous, such as a dog food with a higher nutritional content. The company itself may have induced the need, perhaps through a clever advertising campaign. A company may even have made a product and then induced a need for it.

In any event, once the need (or potential need) is apparent, the company's management must estimate the potential size of the market, the cost of manufacturing a proposed product to satisfy the need, and the eventual selling price. Then they must answer the question: Is it economically feasible? If the answer is no, the process stops, as indicated in Figure 1–6. This company, at least at this time, has concluded that it cannot satisfy the need profitably. Profit is essential if the company is to grow and provide jobs to support the economy. If the answer to the question is yes, the company begins a process that may ultimately lead to a new product. (We say "may" because many projects designed to develop new products fail for a variety of reasons.) And this is where a large company's research and development department (usually shortened to R and D) enters the picture.

Once the company decides to proceed, they assign a research team (in some cases the team has only one member) the task of determining whether scientific principles needed to develop the proposed product are known. The research team searches the scientific and engineering literature as well as the patent registry. If the scientific principles are either unknown or inadequately understood, the company must

FIGURE 1–6 Decision making in the economy—a simplified view.

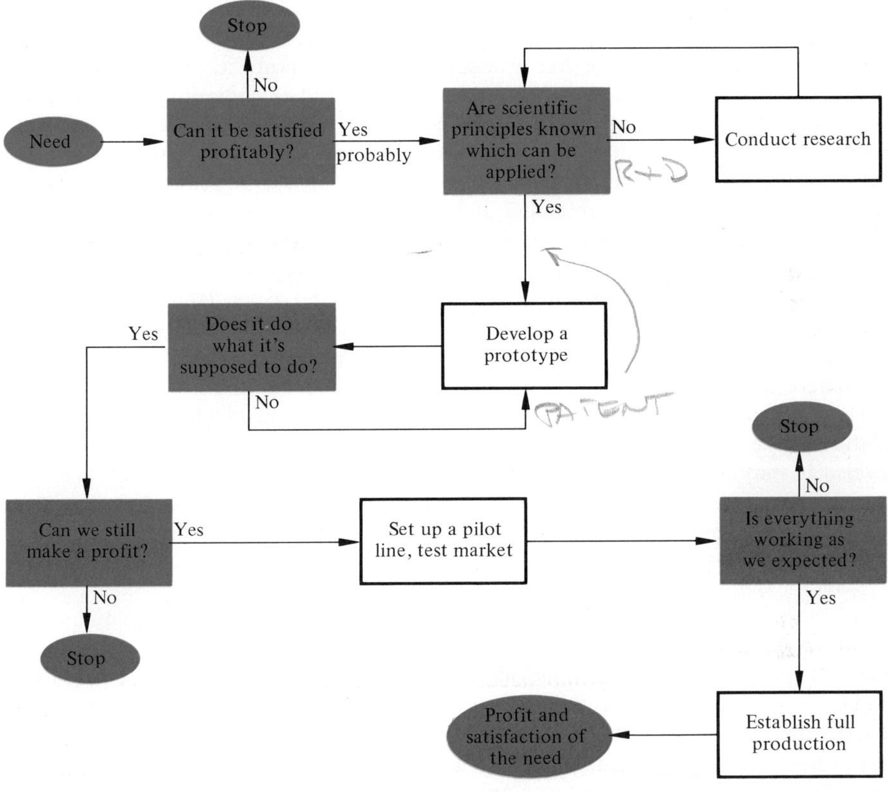

perform research to discover new scientific principles. It is here that an engineer can first be involved in the overall process. Once the engineer finds a principle that has no serious drawbacks or flaws, the economic process can proceed to the next phase. The company normally seeks a patent. We should point out that research costs money, and sometimes when a company does significant work without making progress toward a solution, the company's management decides not to spend any more money on the project. Figure 1–6 does not explicitly show this possibility, but it is an important consideration at each step in the process.

The next phase is development. A prototype of the final product is assembled and tested to see if it meets all the necessary conditions. These include performing as required, as well as satisfying all applicable government regulations and laws. If a given prototype is inadequate, another must be developed and tested. Again, if this process is repeated too many times without progress, the project may be terminated. Once a successful prototype has been developed, the company's marketing personnel again investigates and assesses the profit potential of the product. This reassessment is essential if the research and development phases have consumed significant amounts of time.

The company's management next normally establishes a pilot production line, produces limited amounts of the product, and test markets it. If these tests are positive, they commit money to the construction of a full-scale production plant and sell the product nationally or even internationally. Periodically, they review the performance of each product line to ensure that it remains profitable.

Engineers fit into this economic flow at a number of points. Research engineers search for new scientific principles and more complete understanding of these principles. Development engineers exploit these principles to produce prototypes. Product design is also a basic task for the engineer. In setting up a pilot production line, engineers must specify the necessary equipment and space. Plant engineers see that the design and operation of a new production facility are successful. In addition, in companies in which the product is primarily technical in nature, graduate engineers work with sales and marketing personnel.

This brief examination of the roles of engineers is based on a highly idealized view of the private economy. We have not dealt with the effects of governments at all levels. It may surprise you that the roles of engineers in government are not significantly different from their roles in private industry. Of course, whereas the profit motive governs private industry, different motives are present in government.

1–4 THE MANY FIELDS WITHIN ENGINEERING

Now let us briefly describe some identifiable segments of the highly diversified engineering profession, and indicate some of the different physical principles and systems that are of immediate concern to engineers in each.

The *aerospace engineer* is concerned primarily with the motion of solid objects through air and space. The aeronautical engineer is interested in the scientific principles governing flight within the atmosphere, while the astronautical engineer is concerned with flight beyond the atmosphere. These two titles are commonly lumped together in the one title—aerospace engineer. Such engineers are concerned with everything required for successful flight. They have to answer important questions such as the following: What propulsion systems should be used? Of what materials should the systems be composed? What physical laws govern the trajectories of bodies in flight? How does the shape of a wing influence the amount of lift it produces? What techniques can the pilot use to control the engines and movable surfaces of vehicles? Can the nation's air transportation network be made more efficient and safe? These questions delineate areas in which aerospace engineers work, and will continue to work.

The *architectural engineer* designs and supervises the construction of buildings. He or she is necessarily concerned with the harmony between buildings, their environments, and their usage. The interior must provide levels of lighting and background noise and temperatures that allow humans to live and work comfortably. The exterior must be visually appealing and harmonious with its surroundings.

Architectural engineering thus requires a significant degree of artistry. In addition, to design buildings, architectural engineers must know certain scientific principles. They must be aware of the mechanisms by which materials withstand mechanical loads imposed during use. They must take into account the suitability of the building site chosen for construction. They also need a knowledge of acoustics if they are to be able to control noise levels within buildings.

The field of *biomedical engineering* did not even exist a decade ago. The biomedical engineer attempts to combine the methods and techniques of engineering with the biological and medical sciences. Current areas of activity within this field include developing more reliable, longer-lasting heart pacemakers to keep a damaged heart beating regularly, establishing noninvasive techniques to obtain information about the interior of a human body without using potentially hazardous forms of radiation, understanding the chemical and electrical processes by which the brain communicates with other parts of the body, and developing better models of the brain itself to allow further insight into problems involving brain damage. The biomedical engineer must have strong training in biology and medicine as well as in engineering.

The *chemical engineer* is concerned primarily with products and processes that result from the science of chemistry, the interaction of elements in forming compounds under various conditions of temperature, pressure, radiation, and concentrations. Chemical engineering concerns processes in which material is treated to effect a change in state, energy content, or composition. Refining crude oil to get different useful products is an example of chemical engineering at work. Areas of active concern include the development of improved fuels, plastics, textiles, and medicines. The chemical engineer attempts to understand, control, and optimize the processes whereby some chemicals are converted into others through reactions. This field requires a thorough understanding of the basic principles of chemistry. Modeling chemical reactions requires emphasis on the equations governing mass and heat flows in physical systems. Only through these principles can complex systems be analyzed and controlled. Figure 1–7 shows a chemical engineer.

Civil engineering includes a wide array of activities. The civil engineer is responsible for the design and construction of structures, water-supply and waste-disposal systems, air and water pollution-control systems, flood control and transportation systems. Typical tasks include understanding and measuring the properties of soils to determine the suitability of building sites, analyzing the properties of building materials, developing techniques for handling and purifying wastes, designing and supervising the construction of buildings, tunnels, and bridges, and modeling transportation systems. Civil engineers address the following kinds of questions: Under what conditions is it necessary to add reinforcing iron bars to concrete walls and slabs? How thick should the bars be? What equations govern the vibration induced in a bridge by a heavily loaded truck crossing at 55 miles per hour with a crosswind of 50 miles per hour? By what geometric configuration can girders be interconnected to produce the most efficient load-bearing structures? Can better methods be developed for the construction of tunnels? What innovations will improve public transportation systems? How can air and water pollutants be measured more

FIGURE 1–7 A chemical engineer at work. (Courtesy Boeing Aerospace Company)

simply? The civil engineer normally concentrates in one of these areas. Some require a strong background in mechanics and strength of materials; others need a sound mastery of chemical principles. You can see that civil engineering offers a really diverse set of challenges.

The processing of energy and information in electrical form is the concern of *electrical engineers*. In establishing the national power grid, they have developed techniques to generate electrical energy, transmit it over large distances, and distribute it to the ultimate users. This work necessarily involves an understanding of the generators that convert mechanical energy to electrical energy, plus a knowledge of how electrical signals are transmitted along power lines. Electrical engineers are also involved in handling of all information in electrical form. Examples: the telephone system, radio and television communications, and communication satellites. To satisfy society's needs, electrical engineers must develop better devices to carry out these functions. Electrical engineers working in microelectronics brought about a revolution when they developed rugged miniature integrated circuits that now replace large fragile vacuum tubes (the size of light bulbs) in a wide array of electronic equipment. Electrical engineers focus their technical education on the understanding of electronic circuits and the materials used to produce electrical devices.

The digital computer—a vast field all by itself—is a large electronic circuit, although it may use magnetic materials for storing information. In some colleges, computer engineering is a part of electrical engineering, while in others it is a separate academic discipline. *Computer engineering* is divided into two parts: hardware and software. Engineers concerned with the design and realization of computer hardware are truly electrical engineers, in that they design circuits. Software engineers tend to concentrate on understanding and improving the organization and languages of a computer system.

The *industrial engineer* applies engineering methods and management principles to the design, improvement, and installation of systems of personnel, material, and facilities. Examples: An industrial engineer may design a refrigerated warehouse for the efficient handling and distribution of frozen foods. Or model the criminal justice system to reveal ways to make it operate with greater dispatch. Or arrange work stations on an assembly line to give the most trouble-free operation. Such problems often involve statistical questions. Thus the industrial engineer must have a sound mathematical foundation in probability and statistics. Engineering economics is an important aspect of the industrial engineer's approach to problems.

Mechanical engineers design and produce systems composed of machines and related apparatus that use energy to augment human effort in accomplishing a desired result. Two major areas concern the mechanical engineer: *mechanical* systems and *thermal* systems. In mechanical systems, the generation, transmission, and use of energy in mechanical form are of primary interest. In thermal systems, it is, naturally, thermal forms of energy that are of interest. In addition, there are some mechanical engineers who examine the interaction between energy in these two forms. For example, the design of air conditioning systems involves both. The compressor in such systems is used to convert mechanical work to a thermal form; this conversion removes heat from some region of space which subsequently cools. This area of thermal engineering will become more important in the future as the world's need to conserve energy and use it more efficiently grows. Students of mechanical engineering study thermodynamics, fluid mechanics, mechanics (both statics and dynamics), and mechanical design. Mechanical engineers attempt to answer the following kinds of questions: Can a refrigeration cycle be made more efficient? What techniques can be used to reduce the loss of heat through the walls and roofs of buildings? How can a mechanical system be designed to minimize friction and vibration? What can be done to improve the efficiency of a turbogenerator?

We have already discussed the work of the *materials engineer* in earlier sections of this chapter. The materials engineer focuses on understanding and modifying the useful properties of materials and on specifying materials for various applications. Therefore materials engineers are interested in mechanical, electrical, magnetic, and optical properties of materials. There is no limit, other than their imagination, to the range of properties and materials that can interest them. Figure 1-8 shows a materials engineer using an electron microscope to examine a part of an integrated circuit.

Nuclear engineering encompasses the theory and utilization of basic nuclear reactions. In particular, nuclear engineering involves understanding and designing

FIGURE 1–8 The scanning electron microscope uses electrons rather than visible light to make possible greatly magnified views of material structures and integrated circuits. (Courtesy General Electric Company)

nuclear-*fission* reactors to produce the heat and then the steam required to drive electrical generators. In the next several decades, nuclear *fusion,* under controlled conditions, may also become a reality, and the scope of the nuclear engineer will increase as a consequence. These engineers must thoroughly understand the physics that determines nuclear structure and nuclear reactions. Materials are very important to the safe design of nuclear reactors. Questions for nuclear engineers include the following: What physical factors control the diffusion and moderation of neutrons? What equipment is available for detecting radiation, and how accurate is it? How do properties of materials change when the materials are bombarded by neutrons?

Within the last ten years, *ocean engineering* has become an identifiable branch of the profession. Engineers try to understand the physical processes that give rise to the observable behavior of the oceans. They are also trying to find techniques for harvesting the mineral riches of the oceans. They are actively pursuing projects such as deep-submersible vehicles and ocean mining.

Although we know of no formal programs in *plant engineering,* this area employs large numbers of engineers. Any operating company needs individuals whose primary responsibilities are the various systems necessary for the successful operation of the plant: systems for providing electrical service, supplying water, ensuring safety, and handling materials. Plant engineers are responsible for the installation and continuing operation of products and service machines. The typical plant engineer has a background in one of the traditional areas of engineering, as well as much on-the-job training. Plant engineers often labor without widespread professional recognition, but they are essential to the efficient operation of the nation's productive capability. The dividing line between the industrial engineer and the plant engineer is fuzzy, since both share a number of concerns.

For purposes of our discussion, we may conveniently gather the traditional areas of *agricultural, forest, and mineral engineering* under the single title of *resource engineering*. Some colleges have a separate program for each of these areas. Engineers in these fields are concerned with the efficient use of natural resources, including forests, productive crop lands, and mineral deposits. For most of recorded history, development of resources involved primarily the application of processes that had been successful in the past. In this century, engineers have attempted to systematize this knowledge to make the best use of our planet's resources. Engineers are today developing techniques by which we can find mineral resources using observations from satellites; this will minimize the need for extensive excavation and test drilling. Techniques for improving crop yields and fertilization are also under intensive investigation.

Engineers must also be aware of the effects of their technologies on social systems. This need has given rise to *social-technological engineers*. These engineers are concerned with the impact of new technologies on society at large, so they must understand not only technology, but the political, economic, and social sciences as well. These engineers will increasingly serve to unite the disciplines of the social and physical sciences.

In these paragraphs, we have attempted to sketch the major concerns of identifiable branches of the engineering profession. Although, in order to do so, we have had to categorize engineers, we realize that engineers, in the course of their careers, tend to work productively in many different areas. The important common threads are (1) the language of mathematics, (2) the knowledge and application of physical principles, and (3) the intelligent use of abstraction and modeling.

1–5 FORMULATING A MODEL

An engineering model is essential to any successful mathematical analysis of a complex physical problem. In approaching a problem, an engineer must either formulate a model or use one that was developed by someone else. Once an engineer has defined the problem and asked the important questions, the next step is to formulate a model. In general, unless suitable models are available, no numerical analysis is possible.

An appropriate model helps us understand the relationships between the important parameters of a given system and its behavior. For example, consider the design of a bridge that must withstand certain maximum mechanical loads. For a given bridge design, the load-carrying capability depends on the size of the structural components and the properties of the materials of which they are made. To evaluate a given design requires a set of mathematical formulas that accurately relate the total supporting capability of the bridge to the size, material, and arrangement of the supporting girders. For large complex structures, these formulas involve many mathematically complex terms. To calculate the load capacity of a specific design, a civil engineer must substitute numerical values for constants and sizes of materials and must specify the geometric arrangement of these elements. With this information, a civil engineer can analyze the performance of several possible bridge designs and choose the best design on the basis of satisfactory performance, in all important characteristics, and minimum cost. In this way, the engineer is making optimum use of available resources.

The design process depends critically on the engineer's ability to model a physical system and perform an accurate mathematical analysis to predict the system's behavior. The model enables the engineer to evaluate many possible designs without incurring the prohibitive expense of building each. The solution of nearly all realistic engineering problems involves this stage of modeling and mathematical analysis. The engineer must have an accurate set of models, often equations, into which numerical data can be substituted. The model then generates a set of predictions regarding the expected behavior of the specified system.

The knowledge of suitable models—as well as experience in applying them to realistic problems—constitutes a significant part of the engineer's stock in trade. To reiterate an earlier theme, these models depend on valid scientific principles.

1–6 THE TOOLS OF CALCULATION

The modeling and mathematical analysis that engineers must do to solve problems means that they frequently need to carry out calculations. It is no accident that for decades the slide rule was a visible symbol of the engineering profession. To be sure, the revolution in electronics made possible by the integrated circuit has brought about a modern equivalent of the slide rule, the electronic calculator. This device, despite its sophistication, performs essentially the same function as a slide rule. It enables a person to rapidly evaluate complex mathematical expressions. In addition, it is a great deal more accurate than a slide rule. Figure 1–9 shows a slide rule and electronic calculators.

The slide rule, which served the engineering profession for so many years, makes possible multiplication and division, plus logarithmic and trigonometric calculations accurate to three figures. Figure 1–10 shows the different scales on a slide rule. The slide rule is reasonably compact, about four to ten inches long, and extremely portable, since it does not depend on a source of electrical power. The major drawback of the slide rule is accuracy. It is not adequate for calculations requiring more than three significant figures.

Slide rules have, as far as we can see, only two advantages over electronic calculators. They never run out of charge in the middle of a problem, and—unlike

FIGURE 1–9 Electronic calculators are more compact, more accurate, and easier to use than slide rules.

FIGURE 1–10 Each horizontal scale on a slide rule serves a different computational function.

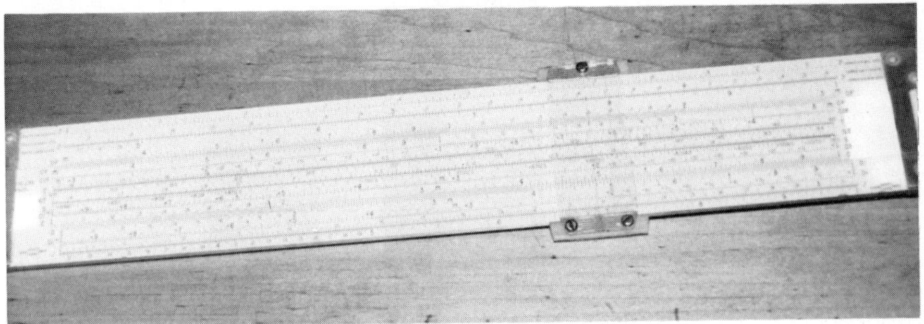

calculators—they do not induce static on nearby radios when in use. The electronic device is more accurate, faster, and capable of carrying out a wider range of mathematical operations. Although the full impact of these instruments on the engineering profession has not yet been studied in depth, it is clear that they greatly increase the productivity of engineers.

So, although the slide rule has long been a fixture in engineering courses, we shall not consider it further here, because the electronic calculator has almost entirely displaced it in the engineering profession. Electronic calculators that perform all slide-rule functions to 10 significant figures are now available with quite modest price tags. Scientific calculators now cost on the order of $20. Since these electronic devices are more accurate and can carry out more functions than slide rules, we see no need to include an exposition of the principles and use of slide rules.

CALCULATORS WITH DIFFERENT LOGIC SYSTEMS

Figure 1–11 shows the four major elements of a scientific calculator. The input device is a collection of keys, which the user depresses in proper sequences to enter numbers and operations. The output device is an array of light-emitting diodes which display answers. The details of the electronic circuits that perform the memory and logic processes are beyond the scope of this course. Some circuits store numbers in binary (zeros and ones) form. Other circuits carry out mathematical operations on these numbers. The last major element provides the electrical energy necessary to run the calculator; this is commonly a rechargeable nickel–cadmium battery.

Several companies produce a wide variety of electronic calculators with different features. It is not possible to discuss all the devices now available. Instead, we shall focus on two important features and differences. The first is the logic system that a calculator uses, and the second is whether the calculator is programmable.

One logic system widely used today is called *algebraic notation* (AN). Another is

FIGURE 1–11 The four basic elements of an electronic calculator: keyboard, processor, battery, and display.

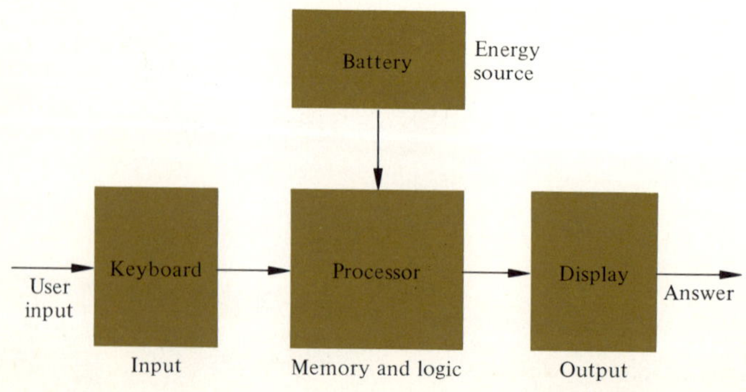

FIGURE 1–12 Two programmable calculators. The machine on the right has a motor and magnetic system to allow the use of magnetic cards for permanent storage and convenient reuse of keystroke sequences. The machine on the left retains a program until it is turned off.

called *reverse Polish notation* (RPN). Texas Instruments calculators that are now available use the AN system, while Hewlett–Parkard devices use RPN. (We mention two specific manufacturers not as an endorsement, but as a reflection of the kinds of electronic calculators now available.) Figure 1–12 shows one AN calculator and one RPN calculator.

Algebraic notation You can immediately tell an electronic calculator using AN when you see a key labeled with an equals sign. Such a key is lacking on an instrument using RPN. In general, problems are entered into algebraic instruments in the same way they are written. For example, to solve

$$5 + 3 = ?$$

you depress the following keys in sequence.

[5] [+] [3] [=]

The answer displayed is 8.

[8.]

We shall use rectangles containing numbers to represent the displays and squares to denote keys.

This example indicates that the user need not enter the decimal point after whole numbers. All electronic calculators assume that the decimal point follows the last digit entered before an operation (+, = in this example). The decimal point does appear explicitly in the answer.

Subtraction, multiplication, and division are performed in a similar fashion.

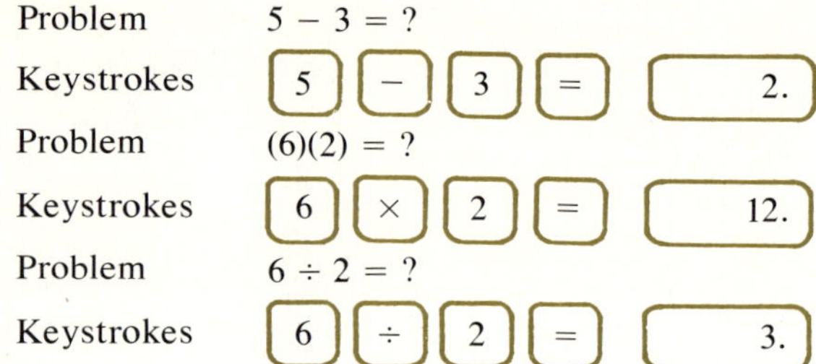

Multiple additions and subtractions are also carried out directly, as are successive multiplications and divisions.

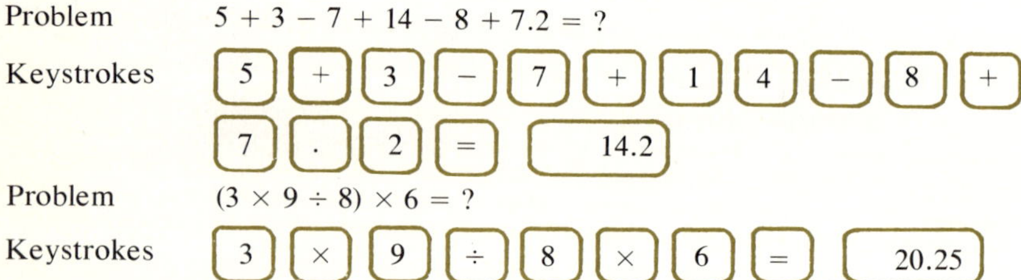

The keystroke sequence becomes more complex when different operators and numbers are mixed together in a chain calculation. The manufacturers of AN machines have not settled on an industry-wide standard for the order of operations. Some calculators carry out the operations in the order in which the user enters them. Other calculators execute some operations (usually multiplication and division) before others (usually addition and subtraction). We shall try to examine the impact of these differences by working some examples. Not all these exercises will work as indicated on all algebraic calculators.

When chain operations mix addition and/or subtraction with multiplication and/or division, the specific calculator's order of operations becomes important. For example, look at this.

Problem $(5 \times 3) - 2 = ?$

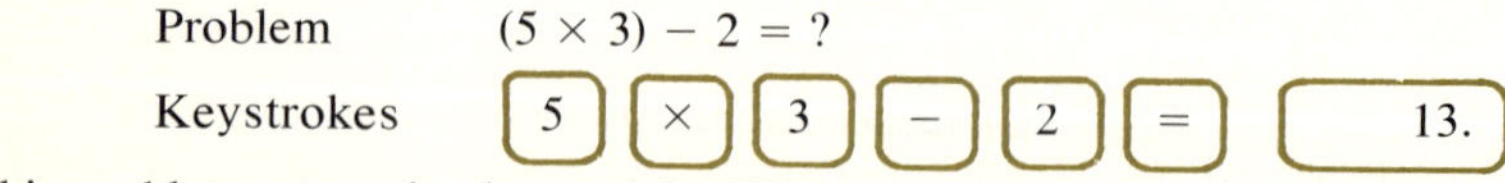

This problem was solved correctly. However:

Problem $(5 - 2) \times 3 = ?$

Keystrokes [5] [−] [2] [×] [3] [=] [−1.]

See what happened? The last sequence of keystrokes gave an incorrect answer (−1 instead of 9) because the problem was executed on an algebraic calculator that performed multiplication and division before it performed addition and subtraction.

In this example, the calculator executed the series of commands in the following order:

$$
\begin{aligned}
5 - 2 \times 3 &= \\
5 - 6 &= \quad \text{(multiplication before subtraction)} \\
&\ -1 \text{ (carrying out subtraction)}
\end{aligned}
$$

When a calculator performs multiplications and divisions before additions and subtractions, we say that it assigns these operations a higher *priority*.

As we shall point out in Chapter 3, this concept of operator priority is present in FORTRAN (the language we shall use to communicate with digital computers). Consequently, this idea of a hierarchy of mathematical operators will become quite familiar in later chapters. We hasten to add that not all algebraic machines display this feature. Some carry out operations as they are entered. For example:

Keystrokes [5] [−] [2] [×] [3] [=] [9.]

If you have an algebraic machine, you can check to see whether it has an operator priority by entering the above keystroke sequence and observing whether you get 9 or −1.

Exercises

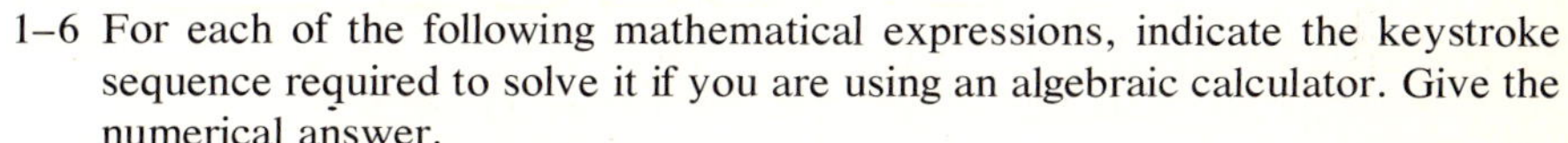

1–6 For each of the following mathematical expressions, indicate the keystroke sequence required to solve it if you are using an algebraic calculator. Give the numerical answer.

(a) $5 - 3 - 7 + 8 - 9 =$
(b) $(((5 \times 3) \div 4) \times 7) \div 8 =$
(c) $1.1 + 2.2 - 3.7 + 17.4 =$
(d) $2.4 \times 2.8 \div 7 \times 8 =$
(e) $5 \times 4 \times 3 \div 6 =$

1–7 If you are using an algebraic calculator *without* an operator priority, what number will be displayed at the end of each of the following keystroke sequences?

(a) [5] [×] [3] [−] [4] [=]
(b) [2] [+] [9] [÷] [4] [=]
(c) [3] [×] [2] [÷] [6] [+] [1] [=]
(d) [7] [−] [5] [÷] [4] [=]
(e) [2] [+] [2] [×] [3] [−] [7] [÷] [2] [=]

1–8 Repeat Exercise 1–7, assuming that you are using an algebraic calculator that executes multiplications and divisions before additions and subtractions.

We summarize the discussion of operator priorities briefly because they are important in mastering an algebraic calculator. Instruments without operator priorities carry out mathematical operations in the order in which you enter them. The following example of this shows the contents of the calculator display at intermediate points.

Problem	$3 + 2 \times 4 = ?$	
Keystrokes	[3] [+]	3.
	[2] [×]	5.
	[4]	4.
	[=]	20.

A calculator with an operator priority also carries out operations as entered, but it defers any additions and subtractions until all multiplications and divisions are completed. The following example illustrates this procedure.

Problem	$3 + 2 \times 4 = ?$	
Keystrokes	[3] [+]	3.
	[2] [×]	2.
	[4]	4.
	[=]	11.

Owners of calculators using algebraic notation should carefully study the differences between these two examples.

Reverse Polish notation Calculators that use reverse Polish notation (RPN) lack a key labeled with an equals sign. In its place, there is a key with one of the following labels:

The basic idea underlying RPN is illustrated by the following two keystroke sequences.

AN	RPN
[1] [+] [3] [=]	[1] [↑] [3] [+]

Both sequences display the same value (4.). With the RPN calculator, the user pushes the operators (+, −, ×, ÷, etc.) *after* the two numbers on which they operate. The following additional examples will clarify this procedure.

AN					RPN				
2	−	5	=	−3.	2	↑	5	−	−3.
3	×	4	=	12.	3	↑	4	×	12.
8	÷	2	=	4.	8	↑	2	÷	4.

An RPN calculator uses a "stack," a set of storage locations in which numbers may be placed. Calculators have different-sized stacks, but four levels are common. We denote this stack simply as a set of four horizontal lines.

____
____
____
____

These four lines represent the four locations within the calculator in which numbers may be stored. The first location, the one at the bottom of the stack, is the number in the calculator display. For example, when you depress the key labeled 1, this number shows up in the display and is also present on the first line of the stack.

____
____
____
___________________1

Pressing the ENTER or ↑ key pushes the number in the display into the second level. This number remains stored on the first level, but pressing the ENTER key conditions the first level to receive a different number directly from the calculator keyboard.

____
____
___________________1
___________________1_c

We have attached a c to the first level to indicate its conditioned status. If you next depress the key labeled 3, this number is written into the first level and the stack takes the form

____
____
___________________1
___________________3

Next you enter the desired mathematical operation by pressing the + key. The calculator carries out the indicated operation on the numbers currently stored in the first and second levels of the stack and places the result in the first level.

Before depressing +	After depressing +
1	
3	4_c

The presence of the c on the first level of the stack in the illustration on the right indicates that the level is conditioned to accept a keyboard entry. At first glance, there appears to be little difference between the algebraic and reverse Polish notations other than in the labeling of the keys used and the position of the operators. After all, on both calculators, you have to use the same number of keystrokes to carry out a simple operation (addition, subtraction, multiplication, or division) between two numbers. However, the approaches differ significantly when you are carrying out chained calculations.

Chained calculations Calculators using AN may come with or without an operator priority. In addition, such calculators may or may not provide parenthesis keys. The lack of such keys on an algebraic calculator is a severe handicap, and you are cautioned against buying such an instrument. For example, consider the following problem executed on an algebraic machine that doesn't have a parenthesis key, but does have an operator hierarchy.

Problem $\frac{5+3}{6-2} = ?$

Keystrokes [5] [+] [3] [÷] [6] [−] [2] [=] [3.5]

The displayed answer is incorrect because the calculator divides before it adds. To remedy this, the user of such a calculator has to take advantage of the available addressable memory (or memories) in the following way.

Problem $\frac{5+3}{6-2} = ?$

Keystrokes [5] [+] [3] [=] [STO] [6] [−] [2] [=]
[÷] [RCL] [=] [1/x] [2.]

The key labeled STO stores the displayed result in a separate memory location. The key labeled RCL recalls that number into the display. The last keystroke in this sequence (1/x) gives the reciprocal of the displayed number and is necessary to obtain the correct solution to this problem when this keystroke sequence is used. We could have avoided this additional step by modifying the keystroke sequence in the following way.

Problem $\frac{5 + 3}{6 - 2} = ?$

Using a temporary storage location by means of the STO and RCL keys is analogous to using scratch paper to save intermediate results in a lengthy calculation. Most calculators provide more than one such temporary location. These locations are normally addressed by more complex keystrokes, as the following execution of the above problem illustrates.

Problem $\frac{5 + 3}{6 - 2} = ?$

Keystrokes [6] [−] [2] [=] [STO] [0] [0] [5] [+] [3] [=]
[÷] [RCL] [0] [0] [=] [2.]

Solving this simple problem requires a total of 16 keystrokes! Using an algebraic calculator with parentheses can reduce the number of keystrokes by *pending* mathematical operations. [*Note:* To *pend* means to defer execution of material that is in parentheses.] We can show this most conveniently by rewriting this problem using parentheses and writing the necessary keystroke sequence.

Problem $(5 + 3) \div (6 - 2) = ?$

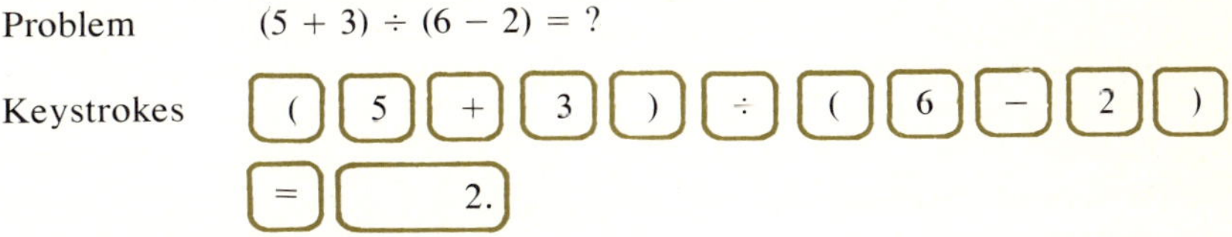

Parenthesis keys are used in a keystroke sequence exactly as you would write them in mathematical expressions. By writing expressions carefully, a person using an algebraic machine with parentheses can work through the expressions in a natural left-to-right order and obtain the correct answer. On some machines, depressing the equals key automatically adds any missing closing parentheses. This mechanism simplifies the solution to the above problem.

Problem $(5 + 3) \div (6 - 2) = ?$

Keystrokes

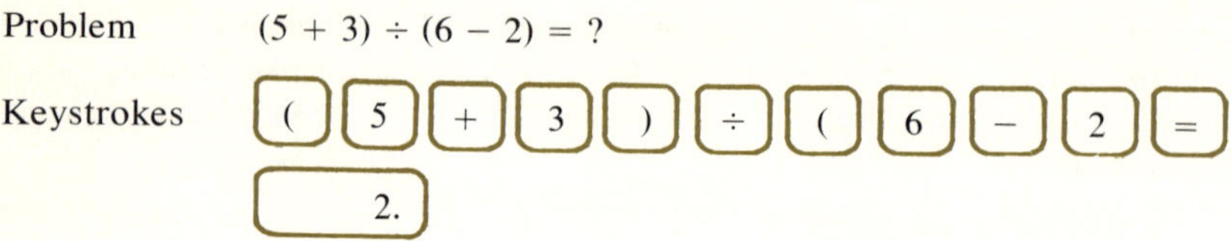

The following problem further illustrates the power of parentheses. It will execute as shown only on calculators providing at least six levels of parentheses.

Problem
$$\frac{6 + 2}{4 - \dfrac{2}{4 + \dfrac{7}{3 - 5(7 - 5)}}} = ?$$

Rewritten as $(6 + 2) \div (4 - 2 \div (4 + (7 \div (3 - 5 \times (7 - 5))))) = ?$

Keystrokes
(6 + 2) ÷ (4 − 2 ÷
(4 + (7 (3 − 5 × (
7 − 5 = 2.4

The use of parentheses to pend mathematical operations greatly simplifies the evaluation of this expression. Anyone using an algebraic machine without this feature would be forced to store intermediate results in memory locations and recall them as necessary for additional computations. We feel strongly that a person should not buy an algebraic machine without parentheses for engineering use. The number of parentheses available on a given calculator differs from model to model. You have to consult the owner's manual to find out how many levels of parentheses are available on a particular machine. The larger the number, the better and more useful the machine is. However, as a practical matter, a total of ten parentheses is sufficient for nearly all problems.

The hallmark of an algebraic calculator is the execution of mathematical expressions in the exact order in which they are written. Indeed, this feature is one of the strongest points made by the manufacturers of these machines. Users can evaluate complex expressions from left to right as written, provided they have used parentheses to write the expressions on one line, as we did in our example.

The RPN calculator offers an alternative method of evaluating such complex expressions. The method involves no rewriting, but does require the user to evaluate the expression from the "inside" to the "outside." Before considering a complex example, let's evaluate the quotient of the sums of two sets of numbers. (Here the e means empty.)

Problem $\frac{5 + 3}{6 - 2} = ?$

Keystrokes	Stack	Display
	e e 5 5_c	
[5] [↑]	e e 5 5_c	
[3]	e e 5 3	
[+]	e e e 8_c	
[6] [↑]	e 8 6 6_c	
[2]	e 8 6 2	
[−]	e e 8 4	
[÷]	e e e 2	[2.]

For clarity and ease of composition, we have omitted the horizontal lines and just shown the stack as a set of four symbols arranged vertically. The first two keystrokes push the number 5 onto the second level, and the third keystroke enters the number 3 on the first level. The fourth keystroke is the addition operator. It acts, as always in an RPN calculator, on the numbers in the second and first levels; the result is stored on the first level. At this point, the number 8 is displayed and the upper three levels of the stack are empty, as denoted by the e.

The next two keystrokes push the number 8 into the third level and the number 6 into the second level. The number 2 is then entered and the subtraction operator depressed. The calculator performs the operation on the numbers in the first and second levels; it stores the number 4 in the first level, and slides the number 8 down from the third to the second level. Finally, when the division operator is pushed, it acts on the 8 in the second level and the 4 in the first level. The result (2) is stored in the first level and simultaneously displayed. The four-level stack is used to store intermediate results and does away with the need for separate memory locations or parentheses.

It is important that you understand this example. Each time you invoke an operation, it acts on the numbers in the first and second levels of the stack. The result is stored in the first level, and the rest of the stack slides down one level. Whenever you push the ENTER key, it raises the numbers in the stack one level, and any number stored in the fourth level is lost. When the stack is collapsed by means of a mathematical operation, the number stored in the fourth level slides to the third level and is electronically duplicated in the fourth level. The significant point here is worth repeating: The stack replaces intermediate scratch-pad memories as well as parentheses. However, the stack is limited to four levels, and you can see that this limits the complexity of the expressions that you can evaluate directly using the stack.

To show the problems this limitation can cause, let us next consider the evaluation of a more complex expression on an RPN calculator.

Problem $$\frac{6+2}{4-\dfrac{2}{4+\dfrac{7}{3-5(7-5)}}} = ?$$

Keystrokes	Stack
[6] [↑] [2] [+] [4] [↑] [2] [↑] [4]	8 4 2 4
[↑] [7]	4 2 4 7

The above set of keystrokes may appear obscure at first glance, but let's trace through each step of the procedure. In evaluating the above expression, we first enter the numbers 6 and 2 and then sum them by pressing the addition key. We next enter the number 4, and this pushes the 8 (the result of the addition operation) into the second level. When we depress the ENTER key, this pushes the 8 into the third level, while the 4 follows into the second level. This prepares the first level for the number 2, which we then enter. We now have the numerator of our mathematical expression stored in the third level of the stack, while the first number (4) in the denominator is in the second level and the numerator of the remaining expression (2) is in the first level.

The next step is to depress the ENTER key, which pushes the 8 into the fourth level (the 4 goes to the third level and the 2 to the second level). The 4 is then entered in the display, and the stack is as shown (8, 4, 2, 4). We still have not completed the expression and cannot carry out any mathematical operations. The next step is to depress the ENTER key to make room for the number 7. When we have entered the 7, the stack is as shown, and the important number 8 (the numerator of the final expression) is lost from the stack. Proceeding with calculation in this manner will give an incorrect answer.

The basic problem is that we are trying to store more intermediate results than the stack has room for. As a result we have lost information and obtained incorrect results.

To get around this problem, the user of an RPN calculator should reduce a complex mathematical expression from the "inside out" instead of trying to work through it from left to right. With RPN instruments, it is helpful, although not necessary, to rewrite the expression on a single line, using parentheses, as we did earlier with the algebraic calculator.

$$(6 + 2) \div (4 - 2 \div (4 + (7 \div (3 - 5 \times (7 - 5))))) = ?$$

To work "inside out," begin the mathematical solution with the expression within the innermost parentheses, in this example, (7 − 5). Then proceed through the successive parentheses.

Keystrokes	Stack
	e e 7 5
7 ↑ 5	
−	e e e 2_c
5	e e 2 5
× CHS	e e e -10_c
3	e e −10 3

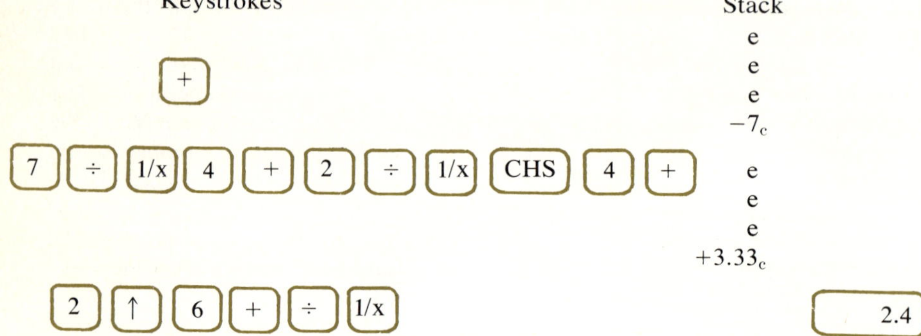

The CHS key *changes the sign of the number* in the first level of the stack. This is a complex set of keystrokes, but by studying this example carefully, you can understand the basic idea involved in working from the inside to the outside of a complicated expression.

Evaluating the expression on the RPN calculator required 26 keystrokes; doing so on the AN calculator required 27. In general, we have found that for most problems, RPN machines require fewer keystrokes. However, the difference is normally insignificant, and either logic structure (RPN or AN) will serve the user well. Both calculators are widely used by engineers and scientists; each has its strong supporters. An individual can become familiar with either and use it confidently. Although manufacturers spend many advertising dollars asserting the superiority of their method over those of their competitors, each solves problems efficiently.

In summary, the RPN method does away with the need to rewrite expressions and parentheses. The AN approach enables the user to evaluate complex expressions from left to right, although it often requires the user to rewrite the expressions on a single horizontal line with the required parentheses.

Shortcuts The keystroke sequences used to evaluate the complex expression in these examples are not the most efficient for either the algebraic or reverse Polish calculators. You can use various operating tricks to eliminate steps. On RPN instruments, these tricks use the stack more efficiently to eliminate, for example, the use of the reciprocal key in some situations. In algebraic calculators, the operator hierarchy can be used to eliminate some parentheses. An example illustrates these shortcuts. First, consider evaluating the expression

$$\frac{1 + 2}{4 - 3/(6 + 7)} = ?$$

on an algebraic machine. A solution is to do this.

Keystrokes [( ] [1] [+] [2] [ )] [÷] [( ] [4] [−] [( ] [3] [÷]
[( ] [6] [+] [7] [=] [.796]

The equals sign automatically supplies the three missing right-facing (or closing) parentheses.

The first parenthesis can be eliminated (at least on some machines). The simplified solution is as follows.

Keystrokes [1] [+] [2] [)] [÷] [(] [4] [−] [(] [3] [÷]
[(] [6] [+] [7] [=] [.796]

On an RPN calculator, a possible sequence to evaluate the above expression is

Keystrokes [4] [↑] [3] [↑] [6] [↑] [7] [+] [÷] [−] [1] [↑]

[2] [+] [÷] [1/x]
E F

[.796]

The contents of the four levels of the stack are shown for the six points labeled A, B, . . . , F above.

A	B	C	D	E	F
4	4	4	4	4	4
3	4	4	4	3.769	4
6	3	4	4	1	4
7	13_c	$.231_c$	3.769_c	2	1.256_c

The key to reducing the number of keystrokes is to fully use the stack. You can evaluate the denominator of the original expression using three successive operations (addition, division, and subtraction) after you have entered the proper numbers into the stack (A above). To use this strategy successfully, you must remember what numbers are stored where in the stack and in what order the operations must be carried out.

We do not recommend that novices use such shortcuts. When you first begin to work with an electronic calculator of either type, it is important to use as many parentheses as necessary on algebraic machines and always to work from the inside out on RPN calculators.

Exercises

1–9 Evaluate the following expressions on an electronic calculator.

(a) $\dfrac{1 + (2 \times 3)}{4 - 6/7} = ?$

(b) $1 + (3 \times 6) - 4/5 = ?$

(c) $\dfrac{(1 + 7.4)/(3 - 7.8)}{4 + 5} = ?$

(d) $6 - \dfrac{3}{4} + \dfrac{(7 \times 5)}{33} - \dfrac{6 \times 5 \times 4}{(5.8)(19)} = ?$

(e) $-3 \times 7 + \dfrac{28}{1 - (3)(0.5)} = ?$

1–10 Write the keystroke sequence required to evaluate each of the following expressions on an AN calculator.

(a) $\left(\dfrac{2 + 3}{4 + 6}\right) \times \left(\dfrac{4 - 2}{16 - 4}\right)$

(b) $1 + 2 \times \left(\dfrac{3 - 6}{4 + \dfrac{7}{2}}\right)$

(c) $\dfrac{2 \times 5}{3 \times 4} + \dfrac{6 \times 7}{5 \times 11} + \dfrac{3}{1 + \dfrac{7}{4}}$

(d) $5 + (7)(3) - (4)(-6)$

1–11 Repeat Exercise 1–10 for an RPN calculator.

Trigonometric and logarithmic functions A calculator suitable for engineers or engineering students provides full trigonometric and logarithmic capability denoted by the following keys.

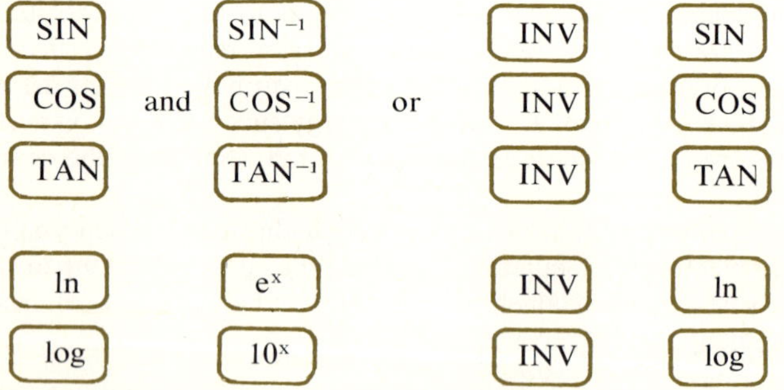

It will also provide a way to designate that quantities used as arguments of trig functions are in degrees or radians. Some calculators have a separate switch that can be placed in either the radian or degree position. The following equation gives the conversion between angular measure in radians and degrees.

$$\text{Angle in degrees} = \left(\frac{180}{\pi}\right)(\text{Angle in radians})$$

If the calculator's switch has been set to degrees, typical trigonometric evaluations are performed in the same way on both algebraic and reverse Polish calculators.

$\sin 30° = ?$

Keystrokes [3] [0] [SIN] → 0.5

$\cos 80° = ?$

Keystrokes [8] [0] [COS] → 0.174

$\tan (45°/2) = ?$

Keystrokes [4] [5] [÷] [2] [=] [TAN] → 0.414 (AN)

[4] [5] [↑] [2] [÷] [TAN] → 0.414 (RPN)

$\sin^{-1} 0.5 = ?$

Keystrokes [.] [5] [SIN^{-1}] → 30.

The electronic calculator also enables you to evaluate logarithms and exponentials using simple keystroke sequences. (We shall cover this in detail in Chapter 7.)

$\log 5 = ?$

Keystrokes [5] [log] → 0.699

$\ln 5 = ?$

Keystrokes [5] [ln] → 1.61

$10^{0.2} = ?$

Keystrokes [.] [2] [10^x] → 1.58

$e^{-2} = ?$

Keystrokes [2] [CHS] [e^x] → 0.135

Or (on some machines): [2] [CHS] [INV] [ln] → 0.135

A key labeled y^x is also available to raise one number to a power. The implementation is slightly different on the two logic systems, as the following example illustrates.

Problem $2^3 = ?$

Keystrokes [2] [y^x] [3] [=] → 8. (AN)

[2] [↑] [3] [y^x] → 8. (RPN)

Exercises

1–12 Evaluate each of the following trigonometric expressions.

(a) $\sin(15°) = ?$

(b) $\sin(0.2 \text{ rad}) = ?$

(c) $\cos(21.5°) = ?$

(d) $\dfrac{\sin(2\theta)}{\sin\theta} = ?$ where $\theta = 9°$

(e) $\tan^{-1}(-10) = ?$

(f) $\cos^{-1}(\sin 15°) = ?$

(g) $\sin\dfrac{\theta - \phi}{\theta + 2\phi} = ?$ where $\theta = 0.5$ rad, $\phi = 0.1$ rad

(h) $\dfrac{(\sin 15°)(\cos 4°)}{\tan(30°)} = ?$

1–13 Evaluate the following logarithmic and exponential expressions.

(a) $\log 375 = ?$

(b) $\ln 27.3 = ?$

(c) $10^{3.4} = ?$

(d) $10^{7/2} = ?$

(e) $e^{(3 + 7/4)} = ?$

$e^{-\sin 30°} = ?$

(g) $\log(e^3) = ?$

(h) $\ln(10^{-2.4}) = ?$

1–14 Evaluate each of the following expressions.

(a) $[\sin(32°)]^3 = ?$

(b) $[\cos(30°)]^{\tan 50°} = ?$

(c) $(14)^{1.8} = ?$

(d) $(7)^{-2} = ?$

(e) $(4)^{\cos 80°} = ?$

1–15 Evaluate each of the following expressions.

(a) $(e^2)^{0.2} = ?$

(b) $(\log 4)^{\ln 4} = ?$

(c) $(3)^{\ln 1.9} = ?$

(d) $(\ln 36)^{(3 - 2.7)} = ?$

(e) $7^{(\log 10/\log 7)} = ?$

Scientific notation The largest and smallest (in magnitude) numbers that a calculator can display are fixed by the number of individual numerical elements in the output device. On most scientific instruments, the displays have ten positions. This implies that the greatest possible number is 9,999,999,999, while the smallest in magnitude (closest to zero) is 0.0000000001.

Although this is a large range, the instruments would be unable to manipulate, for example, the magnitude of the electrical charge associated with an electron.

$$1.6(10^{-19}) \text{ coulomb}$$

To circumvent this limitation, calculators normally provide two display positions for an exponent that is always a power of 10. The following show the display representations of various numbers written in this so-called *scientific* notation.

Number	Display
$1.5(10^{10})$	1.5 10
$-7.4(10^{-36})$	−7.4 −36
$8.7(10^{-98})$	8.7 −98

The following examples show the introduction and manipulation of such numbers.

Problem $(3(10^{-19})) \times (4(10^{7})) = ?$

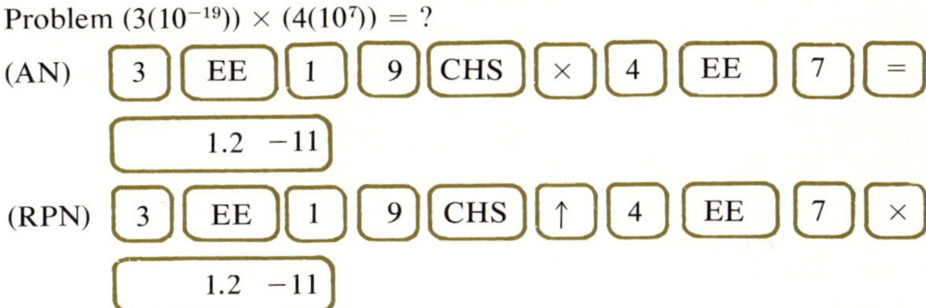

The key labeled EE means *Enter Exponent* and enables you to enter the power of 10 directly into the display. The CHS key, when pressed after the exponent is entered, changes only the sign of the exponent. If you wanted to place the number $-5(10^{-20})$ into the display, the necessary keystroke sequence would be

5 CHS EE 2 0 CHS   −5 −20

This brings us to the end of our brief discussion of the basic elements of nonprogrammable calculators. In addition to the ones we have described, there are different models that have various other keys. Examples include x^2 to enable you to square a number, a key to extract the xth root of y, a % key, and keys to clear the display or the stack (if any). There are also calculators with addressable memories. The best source of information about the special keys available on a given instrument is the manual provided by the manufacturer.

PROGRAMMABLE CALCULATORS

A programmable calculator (one that can "remember" a sequence of keystrokes) is now available for $35. This price will continue to decrease as more manufacturers introduce such products and as they are sold in greater volumes. To see the exceptional value of this feature, consider the task of evaluating the function

$$f(x) = 6x - 7$$

for the following values of the argument:

$$x = 1, 2, 3, 4, 5$$

One way to proceed, on an algebraic calculator, is repetitive.

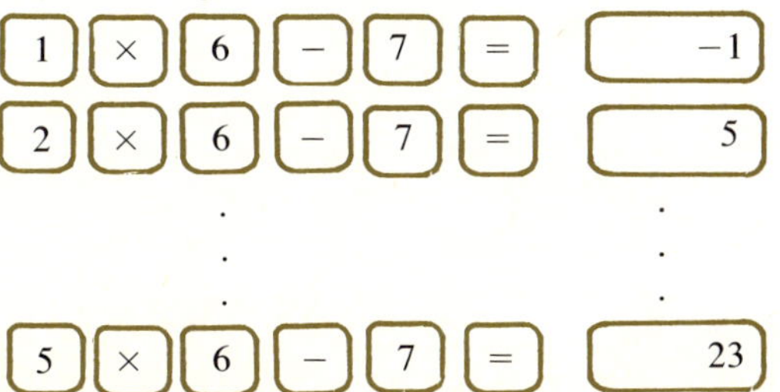

Evaluating the function *f(x)* requires six keystrokes for each separate value of *x*. However, five of the six are the same for each evaluation; only the first keystroke changes. With a programmable calculator, you do not have to repeatedly enter keystroke sequences that are used again and again. The following set of keystrokes are used to "teach" a Texas Instrument SR 52 the desired algorithm in this example. [*Note:* An algorithm is a procedure for solving a problem.]

[LRN] [HLT] [×] [6] [−] [7] [=] [rset] [LRN]

The first (LRN) signals the calculator to remember or *learn* the keystrokes following. The second (HLT) instructs the system to *halt*. The next five keys are the set of common instructions used to evaluate the function *f(x)* for each of the five values of the argument *x*. These are followed by a *reset* (rset), which causes the calculator to branch back to the beginning of the set of instructions it just learned. Finally, the second learn (LRN) designates the end of the set of keystrokes to be "memorized." Once you have entered these nine instructions, you can evaluate the quantity $f(1)$ directly.

[rset] [RUN] [1] [RUN] [−1]

It is even easier to find subsequent values.

[2] [RUN] [5]

[3] [RUN] [11]

[4] [RUN] [17]

[5] [RUN] [23]

After you depress the LRN for the second time, the instruction "rset" causes the calculator to return to the first memorized instruction (HLT). The key RUN causes this instruction to be executed. Since this is a halt, the calculator stops immediately and does nothing. You then enter the number 1 and depress RUN again. The calculator multiplies this entered quantity by 6, subtracts 7, and displays the result (−1). It also executes the reset instruction and finally reaches the halt, where it stops to enable you to record the value $f(1)$.

You next enter the number 2 and depress RUN. This causes the number 2 to be multiplied by 6, 7 to be subtracted from the product, and the result (5) to be displayed. The machine again halts. You can find the values of the function for the other arguments in exactly the same fashion.

Although we have used a very simple function in this example, the amount of operator effort saved in evaluating the function several times is impressive. With some electronic calculators, the set of keystrokes "memorized" can even be permanently stored on small magnetic cards and later reused. These features enable many engineers and engineering students to have the equivalent of a digital computer immediately available at their work station. In addition, printers are becoming available. These enable users to record calculations and answers on paper tape so that they don't have to copy answers from the calculator display. Although beginning engineering students do not need such sophisticated equipment, advanced students and practicing engineers may find its advantages significant. For undergraduates, a simple scientific calculator is entirely adequate.

COMPUTERS

Calculations that would be much too long and tedious to perform on electronic calculators require a digital computer. The techniques needed to use such machines efficiently are not obvious; to master them normally requires much study. A number of languages serve as common ground between a user and the digital computer. The language that is most widely used by engineers is FORTRAN. In Chapters 3 and 6 we shall examine the use of digital computers and FORTRAN.

It is worth recalling the motivations that have led us to consider digital computers: (1) Engineers must do calculations to obtain numerical values of physical performance, design parameters, and cost, so that they can make meaningful comparisons between different designs. (2) Most realistic engineering problems, at some point, generate the need to perform quite a lot of computation. (3) As physical systems have grown more complex and interconnected, the calculations required to understand their behavior have increased dramatically in difficulty. (4) The digital computer provides the only practical means of carrying out these calculations.

Let's look at an example that will illustrate the magnitude of the mathematical complexities: The design and construction of radar systems are major concerns of some engineers in the United States Air Force. We can appreciate their problems by reviewing the difficulty involved in solving simultaneous sets of equations with increasing numbers of unknowns. For example, the problem of solving two equations in two unknowns is straightforward. Many of these problems can be solved in about a minute, using a calculator. Three equations and three unknowns require several minutes, but still cause little procedural difficulty. When problems involve four or five unknowns, solving them on a calculator becomes quite tedious. The time involved can grow to about one hour. In designing radar systems, engineers have to solve systems of 1000 equations in 1000 unknowns! Even with an electronic calculator, an engineer could not solve such a problem manually within a lifetime. Only a modern high-speed digital computer can solve such complex problems. And it requires several hours to complete the necessary processing of even these sophisticated devices.

1–7 SUMMARY

In a very real sense, this chapter has set forth a model of the engineering profession. It has presented a simplified description of the engineering profession. Like any model, this description is designed to help you analyze and understand the subject.

Two basic approaches can be used to provide a deeper understanding of the engineering profession. The first is to greatly expand the job descriptions in Sections 3 and 4 of this chapter in an attempt to describe the precise challenges engineers face. The second is to focus attention on the methods and tools used by engineers. This is our approach in this book. The chapters that follow will present detailed discussions of basic engineering concepts, ideas that are the foundations of engineering. Examples, exercises, and problems will be used to convey information about the challenges that arise in the different fields within the profession.

If a model is to be useful, those working with it must constantly remember its assumptions and its limitations. In the following chapters, we shall continually point out these assumptions and limitations, because any model can mislead if it is not understood or properly used.

Chapter 2 provides a careful overview of the different kinds of models commonly used, and develops the idea of a flowchart. This overview will set the stage for an organized approach to problem solving using digital computers. Chapter 3 will introduce FORTRAN, the computer language that we shall use exclusively.

PROBLEMS

P1–1 Engineering involves the creative application of scientific principles to problem solving. The following is a list of such scientific principles. Describe one engineering application of each.
(a) Water flows to the lowest level attainable.
(b) Metal "whiskers" glow when sufficient electrical current is passed through them.
(c) A mixture of gasoline and air burns and produces a mechanical force when properly compressed and ignited.
(d) When a compressed gas is allowed to expand, it absorbs heat from its surroundings.

P1–2 Consider an operating television set as a system.
(a) Identify the inputs and outputs.
(b) What are the important material considerations involved in the design of a television receiver?

P1–3 What are the inputs and outputs associated with an electronic calculator?

P1–4 What are the inputs and outputs associated with a candy vending machine? First, answer this question from the perspective of a buyer, and then answer it again from the perspective of a seller.

P1–5 What are the scientific principles involved in the operation of your home heating system?

P1–6 List one engineering application of each of the following scientific principles.
(a) A pitched blade revolving in a fluid exerts a force on the fluid.
(b) A column of mercury expands when heated and contracts when cooled.
(c) Bone strongly absorbs x rays, but tissue weakly absorbs them.

P1–7 What material considerations are important for a heart pacemaker that is to be implanted in a patient's body? What kind of energy source should be used?

P1–8 Classify each of the following activities as either research or development. Justify your answers.
(a) Investigating atomic vibrations in solids held at a temperature of −270°C
(b) Designing a new water pump
(c) Building a new instant-picture camera
(d) Analyzing the motion of electrons in thin hydrocarbon films

P1–9 What material considerations are important in each of the following?
(a) Roofing shingles (b) Tennis racquets (c) Automobile bumpers
(d) Contact lens

P1–10 Discuss, in one or two paragraphs, the role you presently see for yourself in engineering after you leave college. What are your motivations?

P1–11 The volume V of a sphere of radius R is given by the equation

$$V = \frac{4}{3}\pi R^3$$

Calculate V for spheres with the following radii.

$R = 1$ meter	$R = 1.04$ meters
$R = 7.4$ meters	$R = 5(10^6)$ centimeters
$R = 3(10^{-4})$ meters	

P1–12 In many engineering applications, the hyperbolic functions are important. Two of these are defined by the equations.

$$\sinh x = \frac{e^x - e^{-x}}{2}, \qquad \cosh x = \frac{e^x + e^{-x}}{2}$$

Evaluate each of these functions for the following values of x.

$x = 0.0$	$x = -0.2$
$x = 0.2$	$x - -3.5$
$x = 2.1$	$x = -6.2$
$x = 5.7$	$x = -20.0$
$x = 20.0$	

P1–13 Figure P1–1 shows a pair of axes and a point P which has the coordinates x, y. Any point may be specified by a value x and a value y. This is the rectangular representation of a position on the plane shown. In addition, the location of the point P can be uniquely specified through a distance R and an angle A. These quantities are also indicated in the figure. The equations linking x and y to R and A are found by applying simple trigonometry.

FIGURE P1–1 Rectangular and cylindrical coordinate systems are both commonly used in engineering disciplines.

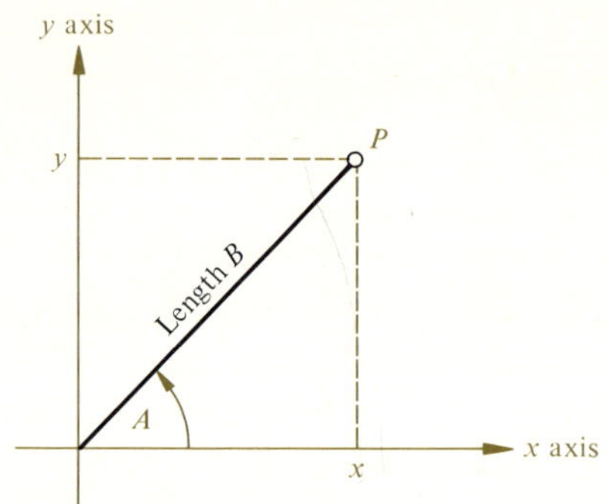

$$R^2 = x^2 + y^2 \quad \text{or} \quad R = \sqrt{x^2 + y^2}$$

$$\tan A = \frac{y}{x} \quad \text{or} \quad A = \tan^{-1}\frac{y}{x}$$

Evaluate R and A for each of the following points.

$x = 3$	$x = 3$	$x = -3$	$x = -3$
$y = 4$	$y = -4$	$y = 4$	$y = -4$
$x = 2$	$x = 2$	$x = 2$	$x = 2$
$y = 5$	$y = 7$	$y = 9$	$y = 11$

P1–14 The equations developed in Problem 1–13 are useful when values of x and y are known. They allow the immediate calculation of R and A. However, when R and A are known and x and y are to be found, the above equations are not helpful. Again refer to Figure P1–1 and use basic trigonometry to write the following equations.

$$\sin A = \frac{y}{R} \quad \text{or} \quad y = R \sin A$$

$$\cos A = \frac{x}{R} \quad \text{or} \quad x = R \cos A$$

Use the two equations on the right and find the values of x and y corresponding to the given values of R and A.

R	5	R	5	R	10
A	53.13 degrees	A	0.9273 radian	A	10 degrees
R	8	R	7	R	8
A	115 degrees	A	205 degrees	A	285 degrees

P1–15 Problems 1–13 and 1–14 introduced the concepts of rectangular (x, y) and polar (R, A) representations of points on a two-dimensional plane. The equations developed are the rectangular–polar conversions. If your calculator has a built-in capability to carry out such conversions, list the keystrokes necessary to carry out the following two conversions. Also, in answering this question, identify the manufacturer and model of your calculator.

$$\begin{array}{lcl} R = 1 & & x = ? \\ A = 10^\circ & \rightarrow & y = ? \\ x = 3 & & R = ? \\ y = 4 & \rightarrow & A = ? \end{array}$$

If your calculator does not have such a provision, list the keystrokes necessary to do the conversions using the formulas developed in the last two problems.

P1–16 Einstein's theory of relativity states that the mass m of a body depends on its velocity v according to the equation

$$m = \frac{m_0}{\sqrt{1 - (v/c)^2}}$$

where m_0 is the mass of the body at rest ($v = 0$) and c is the speed of light. Given that

$$m_0 = 1 \text{ kilogram}, \qquad c = 3(10^8) \text{ meters/sec}$$

calculate the mass m for the following values of velocity v (in meters/sec); 0, 10, 10^3, 10^7, 10^8, $2(10^8)$, $2.9(10^8)$.

CHAPTER TWO
MODELING AND ENGINEERING PROBLEMS

An engineer approaches the problems of designing, constructing, and controlling systems composed of people, materials, and machines by formulating and manipulating models of such systems. As we stated in Chapter 1, a *model* is a simplified description of some real-world entity or phenomenon. Its purpose is to aid analysis, understanding, and design. An accurate model enables the engineer to understand a real entity or phenomenon, or to predict its behavior, without actually constructing or operating the real thing. In this chapter, we shall expand this definition of a model, study several types of models, and examine some of the ways engineers use models.

2–1 MODELS AND THE DESIGN PROCESS

Our definition of a model contains several terms that we must define carefully in order to fully understand the concept of a model. The first is "entity or phenomenon." These words are not synonymous; they distinguish two different states of nature that models may represent. The word "entity" refers to a thing or an object. It is analogous to a noun in grammar. The word "phenomenon" refers to an activity. It can occur within a single entity or among several entities, and it is analogous to a verb. For example, a pipe and a quantity of water are two different entities. The flow of water within a pipe is a phenomenon because the water is moving within the pipe. Other examples are the rotation (phenomenon) of a flywheel (entity), the pH (phenomenon) of a reactant (entity), and the flow (phenomenon) of electrical charge (entity) in a wire (entity). These entities and associated phenomena form systems that engineers seek to model. Distinguishing entities from phenomena will not only enable you to establish a basic framework for modeling engineering systems, but will also provide an analytical mechanism that you will find useful in your advanced courses in science and engineering.

The second phrase that we need to understand is "to aid analysis, understanding, and design." In fact, there are two major ideas embedded in this term. A model may be either *descriptive*, if its purpose is to explain or to aid understanding, or *prescriptive*, if its purpose is to predict or duplicate the behavior of a real system. A prescriptive model might also be descriptive, but the reverse is usually not true. A prescriptive model is used primarily in design, but at the same time it may explain the behavior of a system. A descriptive model enables us to understand a real-world system, entity, or phenomenon. It may even serve as a device to communicate ideas. But a descriptive model has little utility in the design process itself. The engineer or engineering student who is choosing a model should always be aware of its purpose—whether it is to describe or predict the system's behavior. We shall examine these ideas later in this chapter.

A third phrase needing amplification is the "design process," an activity in which many engineers are engaged. Enumerating the steps in design will further show the importance of models in engineering. We can describe design by the following steps.

Step 1. Define the *problem*. State the *objective* of the problem. Identify the criteria and constraints by which the possible solutions will be evaluated. A *criterion* is a measure by which one decides that a given solution is better than another. The single solution that is better than all others is the best or optimum solution under that criterion. For instance, when one is comparing several different engineering designs with respect to a cost criterion, the least costly design is preferred. In contrast to a criterion, which provides for relative comparisons, a *constraint* is a condition that must be met before a solution is acceptable. If an engineering design must attain a certain level of performance (e.g., top speed for an aircraft), we eliminate candidate designs that do not achieve this level. The process by which we identify the best solution is called *optimization*. (We discuss optimization in detail in Chapter 5.)

Step 2. Determine the *output*. What results are necessary in order to properly evaluate solutions in terms of the stated criteria and constraints?

Step 3. Determine the *input*. What data and information are necessary in order to properly formulate and manipulate the model?

Step 4. Choose or develop the relations (equations, graphs, or computer models) that describe the functioning of the elements and their interaction. This is basically the *model*.

Step 5. Perform the analysis required.

Step 6. Compare the result with the requirements and criteria.

Step 7. If necessary, repeat this process. Repeating a stepwise approach to the final product is known as *iteration*.

The fourth step is actually formulating the model. Steps 1–3 help us select or develop the model; steps 5–7 tell us how to use the model. In the remainder of this chapter, we will develop the idea and formulation of models.

2–2 CLASSIFICATION OF MODELS

Besides classifying models as descriptive or prescriptive, we can use several other schemes for classifying them. Each scheme serves a particular purpose in the study or practice of engineering. Some of these means of classifying models are as follows.

1. Static–dynamic
2. Deterministic–probabilistic
3. Iconic–analog–symbolic

When we use a static–dynamic classification, we consider the effects of time on the system of interest. If we consider a system which doesn't change with time, we are examining its *static* behavior. For example, it is often convenient to examine the characteristics of a bridge in terms of its dimensions, its materials, and the forces acting on it. In fact, the study of forces acting on a body at rest is called *statics*, while the study of motion (the change of position with time) of a body, due to the forces imposed on it, is termed *dynamics*. Similarly, the study of fluids (liquids and gases) at rest is called fluid statics, while the principles governing fluids in motion are called fluid dynamics. Together these topics form *fluid mechanics*, an extremely important subject in aerospace, chemical, civil, and mechanical engineering. Whether a system is static or dynamic is a fundamental concern in many branches of engineering. Thus the static–dynamic classification of models is a basic one.

The deterministic-probabilistic classification distinguishes models that tell us what *will* happen from those that tell us what is *likely* to happen. Essentially a *deterministic* model describes or predicts the behavior of a system in which the outcome of an experiment under a specified set of conditions occurs with certainty. For example, in a simple electrical circuit, the relationship between the voltage E (the electrical "pressure"), the resistance R (a measure of the difficulty with which charge moves through a part of a circuit), and the current I (the rate at which electrical charge moves) is expressed by *Ohm's law*, which is written:

$$E = RI \tag{2–1}$$

When we know the current I and the resistance R, the voltage is also precisely fixed. That is, we know the voltage with certainty. Ohm's law, as expressed in the above equation, is a deterministic model.

In many experiments, however, the outcome is not known with certainty. For example, when a three-stage missile is launched to place a satellite in orbit, each stage must ignite at a specific point in flight. With some past launchings, one of the stages has failed to function properly. Thus the outcome of a given launch is uncertain. Usually we are able to say (after studying many such experiments) that a given outcome occurs in some fraction of the total number of trials conducted. We call such an experiment *probabilistic*, and we can only describe it accurately by using suitable probabilistic models. An engineer must be prepared to deal with both deterministic and probabilistic models of real-world systems.

The third scheme used to classify engineering models rests on the iconic, analog, or symbolic nature of a given representation of the physical system, entity, or phenomenon. *Iconic models* are those which look like the real thing. Examples are a miniature mock-up of an aircraft, a full-scale photograph of an electronic calculator, or an enlarged model of a molecule. *Analog models* are those which behave like the real system. *Symbolic models* are abstractions of the real system, entity, or phenomenon in terms of symbols, letters, and numbers, as in mathematical formulas. Example: Ohm's law (Equation 2–1) is a symbolic model. These three categories span the abstraction spectrum from the least abstract iconic models to the most abstract symbolic models. Let us now examine each of these classes more carefully.

Iconic models look like the real thing. The word iconic comes from the Greek word *eikon*, which means an image or likeness. Such models are used primarily to describe the static characteristics of the system, and they are usually used for entities rather than phenomena. Iconic models are visual geometric equivalents of real entities, either miniatures (scaled down), enlargements (scaled up), or duplicates (same scale). These geometric representations can be two dimensional (photographs, drawings, or maps) or three dimensional (an enlargement of a chemical

FIGURE 2–1 Examples of two-dimensional iconic models: (a) Engineers preparing a 500× enlargement of integrated circuit. (Courtesy Intel) (b) Three views of a machined metal casting.

(a)

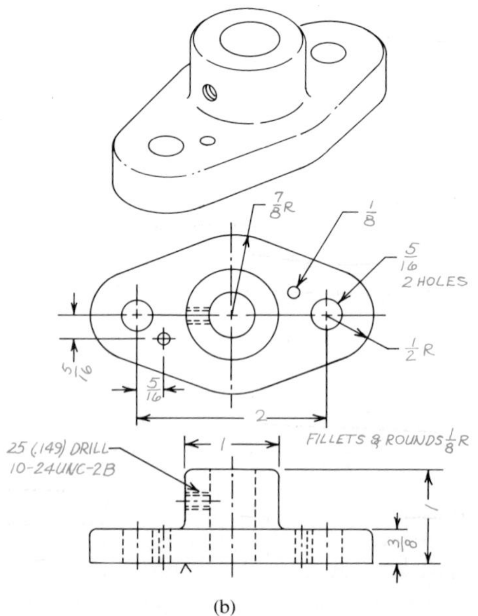

(b)

molecule in the form of plastic spheres and rods or a scaled-down balsa-wood model of a chemical process plant).

The first class of iconic models is the two-dimensional representation. Figure 2–1(a) shows one such representation, an integrated circuit enlarged 500 times. Figure 2–1(b) is a mechanical drawing showing three views of a machined part. Engineers have traditionally received extensive training in engineering graphics in order to develop skills useful in preparing such iconic models. Indeed, some engineers find that a major component of their work is preparing drawings and diagrams that specify the details of the manufacture and assembly of products or the construction of bridges, dams, and other structures. Figure 2–2 gives simple examples of drawings prepared by engineers.

FIGURE 2–2 Drawings and sketches as two-dimensional iconic models: (a) Exploded view of an assembly. (b) Detailed drawings of a caster assembly.

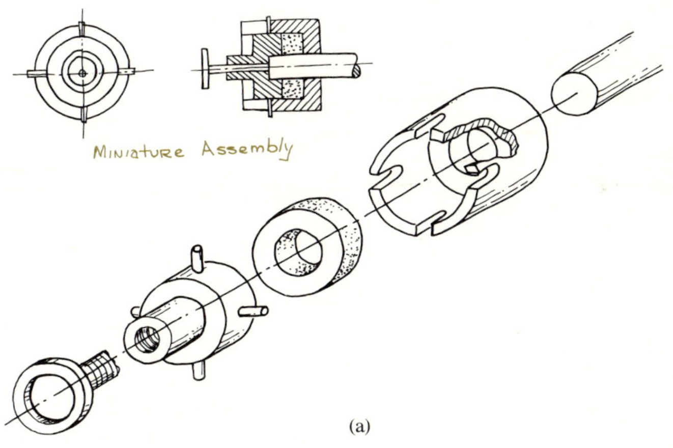

(a)

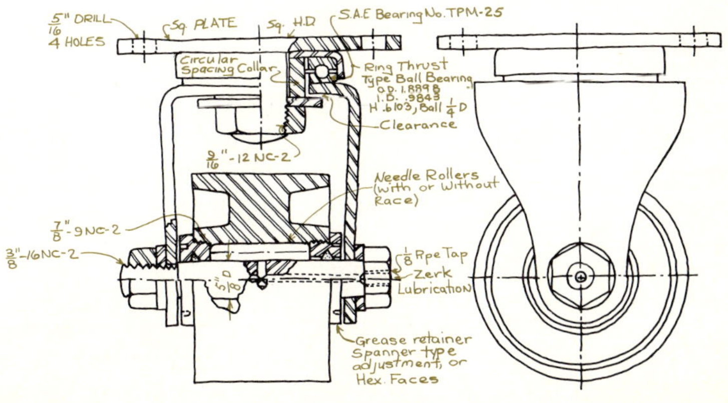

(b)

The second class of iconic models includes three-dimensional physical replicas: a miniature layout of a manufacturing facility, a balsa-wood-and-paper model airplane, an enlarged model of a molecule, a full-size wood-and-plaster representation of a proposed design for an automobile body, or a three-dimensional relief map of a sector of terrain. Engineers should be able to fabricate useful three-dimensional iconic models of systems they are analyzing. Formal courses in plant layout and design, as well as certain courses in structural design, emphasize this ability. Engineers frequently develop three-dimensional models to help them analyze and design a given system. Figure 2–3 shows some three-dimensional iconic models prepared by engineers.

These models duplicate the geometric configuration and spatial relationships of the components of a system. Iconic models enable engineers to perfect their concept of a system and eliminate errors in designing, constructing, or improving a system. They also help engineers communicate their ideas to managers, customers, and other engineers. In addition, engineers may also use these models for dynamic tests in wind tunnels or model basins.

Analog models behave like the physical objects or systems they represent. They are called analog models because they use quantities analogous to corresponding quantities in the physical entity or system of interest. In contrast to iconic models, analog models need look nothing like the real thing. Rather, an analog model either obeys the same principles as a physical system or simulates the behavior of that system. Thus engineers are likely to use analog models for phenomena and iconic models for entities.

The slide rule (Figure 2–4) is an example of an analog model. The C and D scales of the typical slide rule use length or distance to represent the logarithms of numbers. The person using a slide rule multiplies and divides by adding and subtracting, respectively, the lengths (and thus the logarithms) of the numbers of interest. For instance, to multiply the quantities x and y, one locates x on the D scale, places the index of the C scale over this value of x, slides the hairline to the value y on the C scale, and notes the product xy on the D scale. Figure 2–5(a) illustrates that multiplying is adding lengths, since

$$\log (xy) = \log (x) + \log (y)$$

Similarly, to divide x by y, one places the value of y on the C scale directly over the value of x on the D scale and observes the quotient x/y on the D scale under the index of the C scale. Figure 2–5(b) shows that dividing is subtracting lengths, since

$$\log \left(\frac{x}{y}\right) = \log (x) - \log (y)$$

The length corresponding to the logarithm of y is subtracted from the length corresponding to the logarithm of x, and this results in a length proportional to the log of the quotient x/y. This discussion and Figure 2–5 illustrate that lengths on the C and D scales of a slide rule behave like the logarithms of the quantities being manipulated. The slide rule is an analog computer, since its primary function is computation.

A graph, such as the one sketched in Figure 2–6, is another example of an analog model. A *graph* is simply a drawing representing the relationship between two or more physical quantities. A graph is, in fact, an analog model because distance represents the magnitude of each of the different physical quantities. A graph reflects the real functional relationship existing between these quantities. We shall treat graphical models in greater detail later in this chapter, as well as in subsequent chapters, because they are important to engineers.

The electronic analog computer shown in Figure 2–7 is another useful analog model. This device differs from a slide rule in that it uses electrical quantities rather than lengths to represent the magnitudes of physical entities; voltage is used as an analog for physical quantities. To use this device, the user must choose scale factors to relate the electrical voltages in the analog computer circuits to the variables in the problem being modeled. For example, 1 volt is equivalent to 60 feet or 10 volts is equivalent to 1 pound of mass.

The gasoline gauge in an automobile uses both the electrical analog and the length analog. The gauge itself uses a physical length to represent the quantity of fuel in the tank at any given time. The gauge is driven by an electrical voltage that is produced by a combination of an electrical circuit and a float riding in the gas tank. The float falls as the gasoline is consumed and the electrical voltage produced by the circuit decreases. This overall system represents a two-stage analog model of the volume of fuel in the tank.

Electrical analogs are often used in systems in which physical information has to be analyzed, transmitted, processed, or displayed. They are useful for these systems because a wide range of electrical circuits is available to process the information. Chapter 10 covers these ideas in much more detail. However, to cite just one example, consider the needs of a chemical engineer responsible for the proper operation of a processing plant. In one part of the operation, five storage tanks are used to hold an essential chemical. Each tank has a float that measures the liquid's level and an electrical circuit that produces a voltage analogous to that level. At any time, the engineer has available five voltages,

$$e_1, \quad e_2, \quad e_3, \quad e_4, \quad e_5$$

where e_1 is proportional to the volume of chemical in the first tank, e_2 is proportional to the volume in the second tank, and so forth. Further, the electronic circuits have been calibrated so that the following equation gives the relation between tank volume and electrical voltage:

$$V \text{ (gallons)} = 1000e \text{ (volts)}$$

A voltage of 1 volt implies that a tank holds 1000 gallons.

The engineer is responsible for ensuring that the total volume of chemical in the five tanks is adequate to meet the processing-plant requirements. It would be convenient to have an electrical circuit with five inputs, each connected to a different tank voltage, and one output equal to the sum of these five voltages. This setup would enable the engineer to accurately and continuously monitor the entire volume of chemicals in all the tanks. Fortunately, such electrical circuits do exist. We shall briefly analyze one here.

FIGURE 2–3 Examples of three-dimensional iconic models: (a) Enlarged models of crystalline structures. (b) Subscale model of a chemical process. (c) Anthropometric model of man before low-speed collision test. (d) Vehicle and anthropometric model after low-speed collision test. (Courtesy Chrysler Corporation)

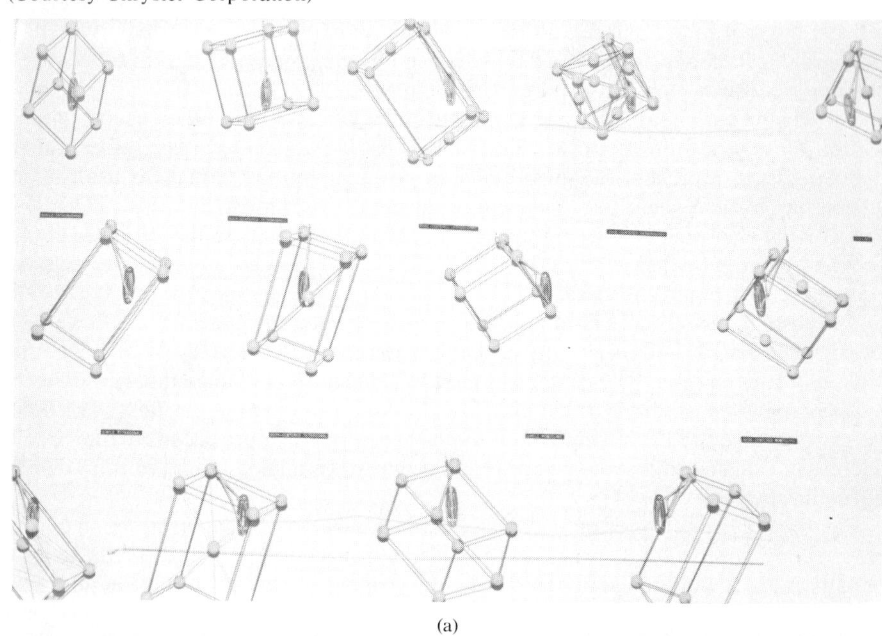

(a)

(b)

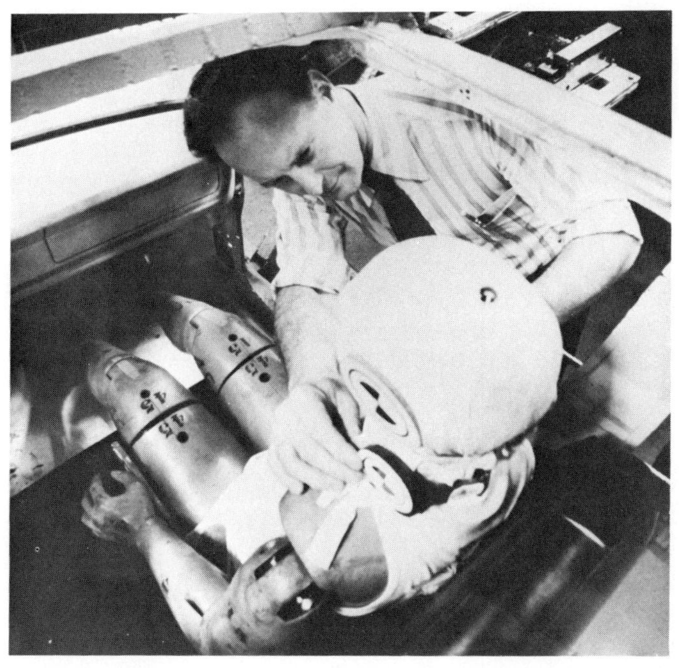

(c)

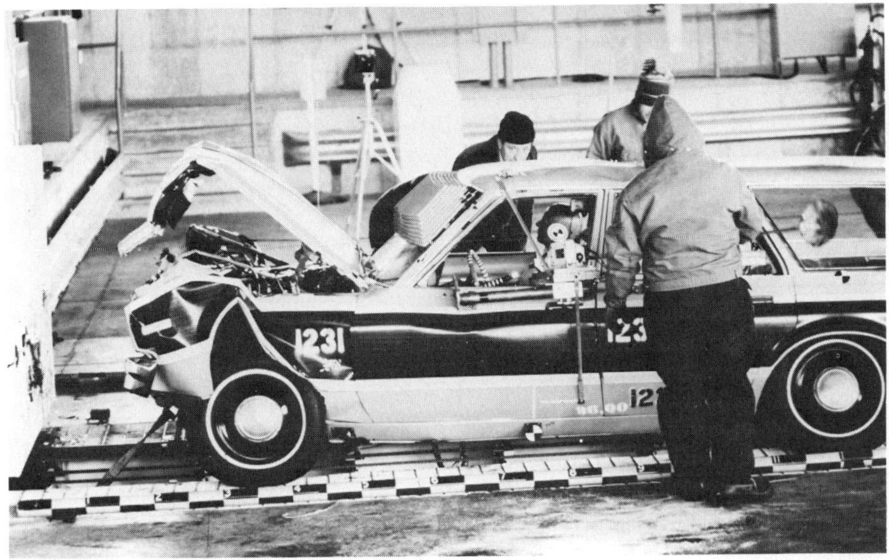

(d)

FIGURE 2–4 The slide rule as an analog model

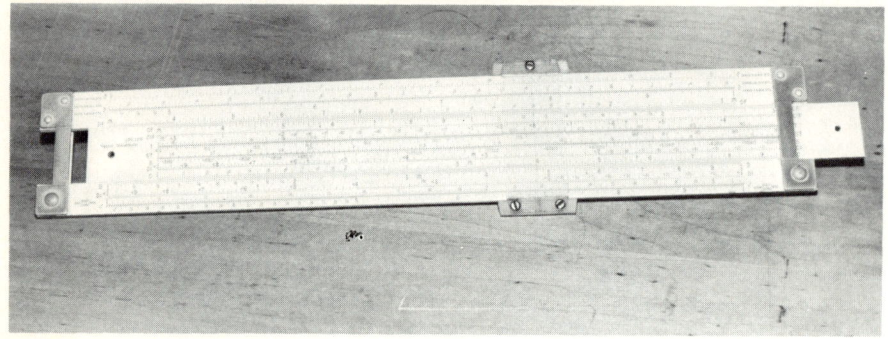

FIGURE 2–5 Slide-rule operations—length as an analog: (a) The slide rule operation for multiplication. (b) The slide rule operation for division.

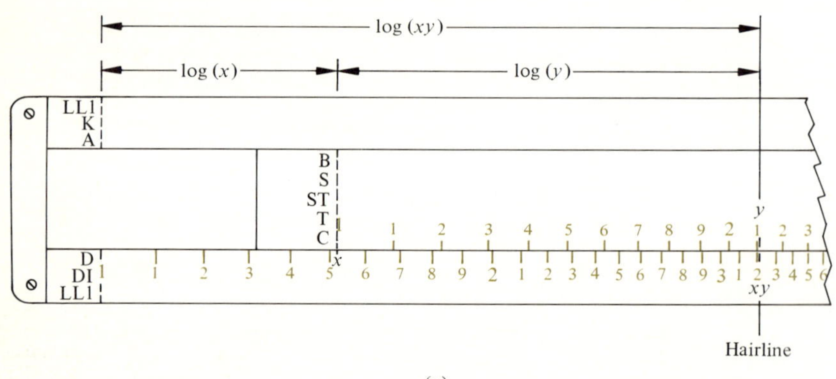

(a)

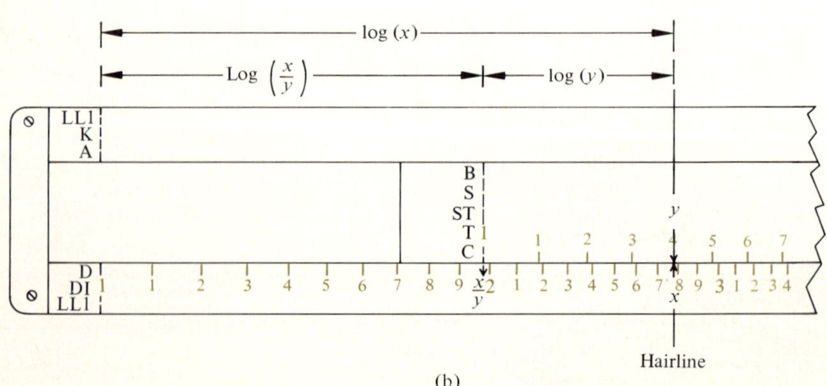

(b)

FIGURE 2–6 The graph as an analog model

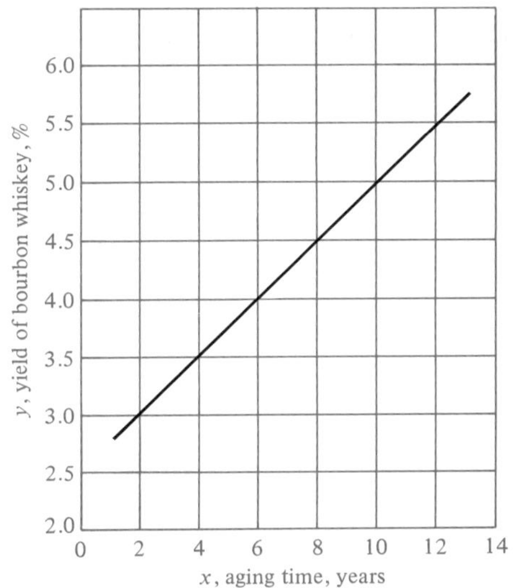

FIGURE 2–7 Engineers using an electrical analog computer (Courtesy General Electric)

Figure 2–8(a) shows a sketch of this problem. Figure 2–8(b) shows the electronic circuit used to sum the five voltages. The operational amplifier A is a complex device, but for this example, we only need to know that it holds the voltage at the point SJ to zero and forces the current i_A to zero also.

The electrical current flowing into the point SJ must equal the current flowing away. In symbols, this statement is

$$i_1 + i_2 + i_3 + i_4 + i_5 = i_f \tag{2–2}$$

Ohm's law (Equation 2–1) relates the currents to the respective voltages by the following equations:

$$i_1 = \frac{e_1}{R}, \qquad i_2 = \frac{e_2}{R}, \qquad i_3 = \frac{e_3}{R}, \qquad i_4 = \frac{e_4}{R}, \qquad i_5 = \frac{e_5}{R}$$

$$i_f = -\frac{e_o}{R}$$

Note that the voltage at point SJ is zero, as is the current i_A. These conditions are forced by the amplifier A used.

We can rewrite Equation (2–2) as

$$-\frac{e_o}{R} = \frac{e_1}{R} + \frac{e_2}{R} + \frac{e_3}{R} + \frac{e_4}{R} + \frac{e_5}{R}$$

After multiplying both sides by R, we obtain

$$e_o = -(e_1 + e_2 + e_3 + e_4 + e_5)$$

Thus the output voltage e_o is equal to the negative of the sum of the other five voltages. By monitoring only this one voltage, the engineer can see exactly how much of the chemical is available.

FIGURE 2–8 Summing operation in an electrical analog computer: (a) Adding volumes of stored chemicals. (b) Equivalent circuit in an electrical analog computer.

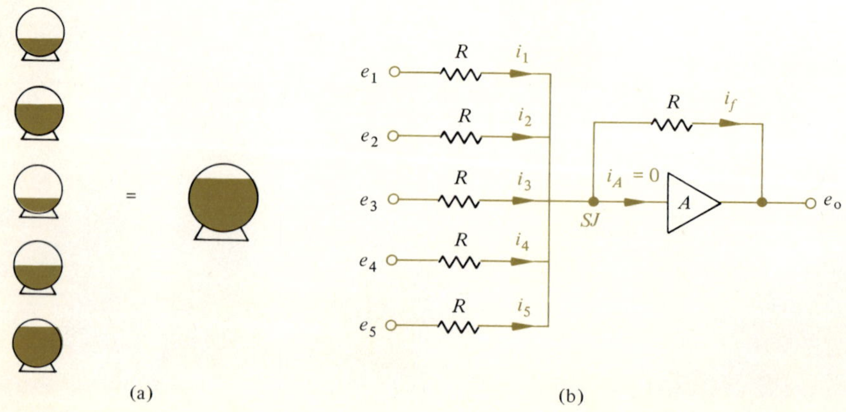

We can carry out other mathematical operations by using electrical analogs along with suitable circuits. In fact, a general analog computer provides the circuitry necessary to carry out basic operations such as addition of voltages, subtraction, multiplication, sign changes, differentiation, and integration. We can observe the behavior of a complex engineering system by viewing the time-varying electrical response of a suitable analog circuit.

Flow charts constitute an important class of analog models. At first glance, a flow chart may appear to be a two-dimensional iconic model. But if you look at it more closely, you will see that a flow chart behaves like a physical system or a sequence of steps in a logical procedure. There are basically two types of flow charts: process flow charts and program flow charts.

Process flow charts represent the movement of entities or the evolution of activities in a physical system. For example, people in a service facility such as a cafeteria or an outpatient clinic are entities, as are materials in any manufacturing process. The activities related to these entities are the services provided in a cafeteria or clinic and the operations performed in the various manufacturing steps. Process flow charts can represent both the entities and the phenomena in a physical system, as Figure 2–9 illustrates. The blocks in this diagram represent the activities (phenomena) in which the entities are engaged. For example, the block labeled "raw materials storage" represents the waiting (activity) of the materials (entities) at that step of the process. The arrows represent the flows from one activity to

FIGURE 2–9 Process flow chart of a manufacturing process for bedding and seating

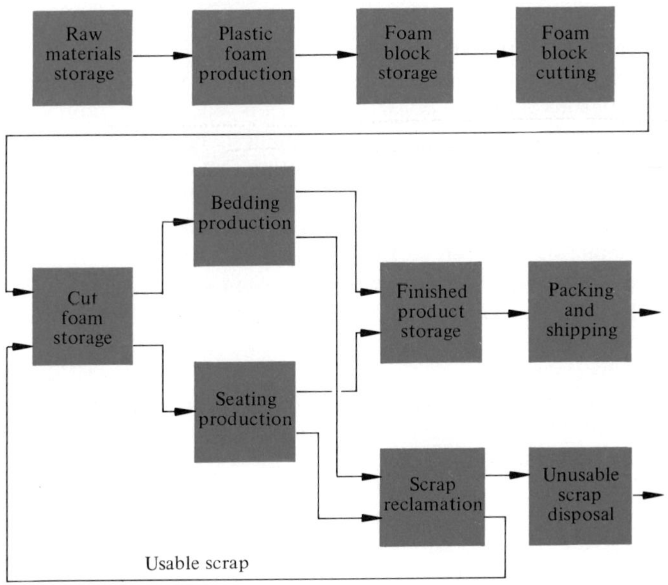

another, including the branching and feedback flows. Feedback flows proceed back to some block that had previously been encountered in the physical process. Chapter 10 examines these concepts in more detail.

Program flow charts indicate the succession of logical steps in an algorithm. They are especially useful for preparing detailed computer programs. As Figure 2–10 illustrates, the blocks in the flow chart represent the individual steps in the algorithm. This figure shows the steps required to calculate N factorial, usually symbolized $N! = (1)(2) \cdots (N-1)(N)$. The flow chart is valid for any value of N. It illustrates the first use of a decision block, represented by the diamond. At this block, the next step is to take one of three available branches, depending on the outcome of the decision.

Symbolic models are abbreviated abstractions of the important and quantifiable (that is, representable by numbers) components of a physical system. The simple addition operation

$$2 + 3 = 5$$

uses the numerals of the set

$$(2, 3, 5)$$

and the symbols

$$(+, =)$$

to indicate that anyone having two units of some entity or phenomenon and acquiring three more such units will have a total of five. Similarly the operation

$$a + b = c$$

generalizes this concept through the use of alphabetic characters to represent variable quantities. Implicit in such a model is the understanding that the physical units of the quantities involved are *consistent*. That is, each term (number or letter) in the model presents the same entity or phenomenon. For example, one can add two apples and three oranges only if the objective is to acquire pieces of fruit. Similarly it is meaningless to add two feet and three minutes. The importance of respecting certain conventions concerning units and dimensions in formulating and manipulating symbolic models will be treated in Chapter 4.

Engineers routinely use symbolic models to represent such concepts as geometric relationships and physical laws. For example, the model

$$A = \pi r^2$$

establishes that the area (A) of a circle is equal to (=) the product of the quantity pi (π) and the square of the radius (r). Likewise, Ohm's law

$$E = RI$$

states that the voltage (E) in a simple electrical circuit is equal to (=) the product of the resistance (R) and the current (I). In both models, alphabetic characters represent physical quantities, and a widely accepted symbol (=) indicates equality.

FIGURE 2–10 Program flow chart for calculating N factorial

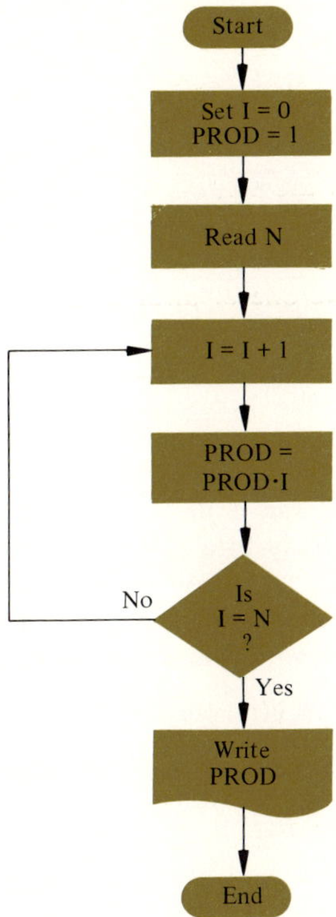

The adjacency of the variables implies the operation of multiplication. Other universally used representations of multiplication include

$$E = (R)(I) \qquad \text{and} \qquad E = R \cdot I$$

Other aspects of symbolic models frequently encountered in engineering are set notation and functional representation. A *set* identifies a collection of items having some common attribute. It is usually represented as

$$S = \{a_1, a_2, a_3, a_4\}$$

for example. This notation indicates that the set S has four distinct elements a_1, a_2, a_3, and a_4. An example of *functional notation* is

$$F(x, y) = 3x^2 - 9xy + y^2$$

This model indicates that some function F depends on the variables x and y and has the form

$$3x^2 - 9xy + y^2$$

FIGURE 2–11 Functional notation for motion in a plane as a function of time

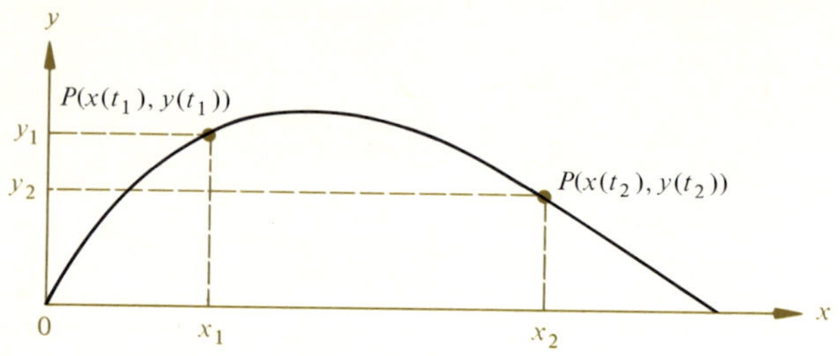

The representation $F(x, y)$ does not mean that the quantity F is multiplied by x and y. This symbolic model may be entirely new to you, but it arises early in math and science courses. For example, in a first course in physics, a basic problem is expressing the position of a body in a plane as a function of time. One possible representation is the function $P(t)$, where P is the two-dimensional vector (x, y). The symbol x denotes the location along the x axis, while y gives the location along the y axis. Since x and y depend on time t, we could represent the position as

$$P[x(t), y(t)] = P(x,y)$$

Figure 2–11 illustrates this concept in a graphical model, the topic of the next section.

Exercises

2–1 Classify each of the following models as iconic, analog, or symbolic.
- (a) A satellite photograph of cloud formations
- (b) A plot of the equation $x + y = 10$
- (c) The equation for the electric power p in a simple circuit as a function of the current i and voltage e, $p = ei$
- (d) A graph of gasoline consumption as a function of the highway speed of an automobile
- (e) A flow chart for computing the quantity sin (x) from the approximation

$$\sin x = x - \frac{x^3}{3!} + \frac{x^5}{5!} - \frac{x^7}{7!} + \frac{x^9}{9!}$$

- (f) A sketch of a new design for a hair dryer
- (g) A subscale model of a supersonic aircraft
- (h) A free-body diagram showing the forces acting on a bridge

2–2 Which of the following are iconic models? Give a reason for each that is not.
- (a) A photograph of an electronic calculator
- (b) The total number of individual keys on an electronic calculator

(c) An x-ray photo of an electronic calculator
(d) A sketch of the arrangement of the electronic components in a calculator

2–3 Consider a simple electronic calculator that can perform the operations of addition, subtraction, multiplication, division, and exponentiation, and has one memory location for storing numbers. Design and sketch the face of the calculator, showing keys and registers. The calculator is to be 2.5 inches wide, 4 inches long, and 0.75 inch thick. What type of model does this sketch represent? Discuss the information that such a sketch communicates.

2–4 Draw a program flow chart of the algorithm for calculating sin (x) from the polynomial approximation

$$\sin x = x - \frac{x^3}{3!} + \frac{x^5}{5!} - \frac{x^7}{7!} + \frac{x^9}{9!}$$

2–5 For a thermocouple, the following data relates voltage in millivolts (y_i) and various temperatures (x_i) expressed in degrees Celsius. Graph this relationship and discuss how it could be used.

i	1	2	3	4	5	6	7	8	9	10
x_i	0	20	40	60	80	100	120	140	160	180
y_i	0.11	0.22	0.34	0.48	0.61	0.77	0.94	1.11	1.25	1.41

2–6 State the model that gives the volume V of a closed cylindrical tank with hemispherical ends. Define any symbols you use. Sketch two views of such a tank. Discuss the models your answers represent.

2–7 Develop a circuit diagram for an electrical analog that will add four quantities and produce the result

$$-e_0 = e_1 + e_2 + e_3 + e_4$$

2–3 GRAPHICAL MODELS

Because graphs are important in both the study and practice of engineering, they deserve a separate section in this chapter. They are a valuable aid in analyzing data and in presenting the results of engineering studies. Professors routinely employ graphs to illustrate important relationships, and students find that graphs help them understand difficult new concepts.

Graphs can be classified into two categories: *descriptive*, those used for the general presentation of information, and *prescriptive*, those employed in engineering design and analysis. Examples of descriptive graphs are the *bar graph* and the *circle graph*, shown in Figures 2–12 and 2–13, respectively.

FIGURE 2–12 Bar graph showing growth in computer storage of data acquired by satellites

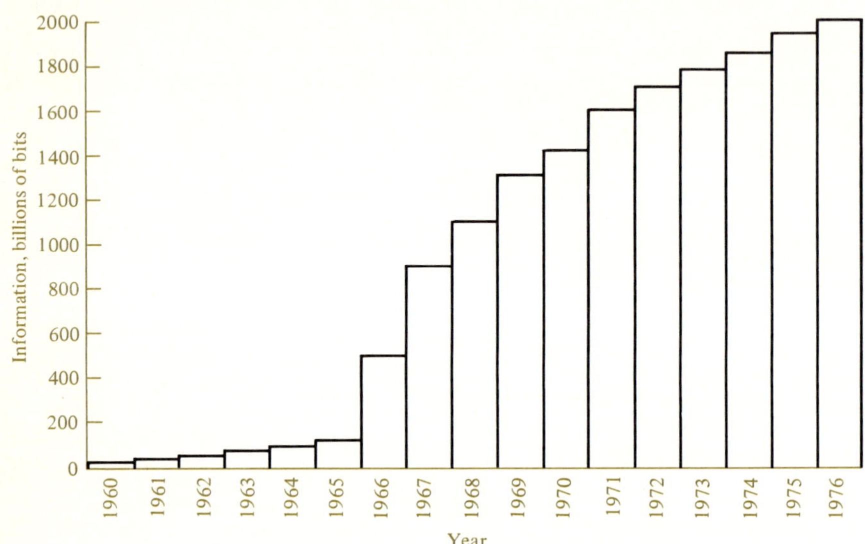

The bar graph in Figure 2–12 shows the growth in the estimated quantity of information acquired by satellites and stored in computerized data banks from 1960 to 1976. The graph is clearly descriptive because it presents information. The increasing height of the bars dramatizes the surge in the requirements for storage capacity in computers as aerospace technology flourished.

The circle graph (Figure 2–13) presents a breakdown of estimated U.S. coal reserves according to sulfur content. This type of graph is especially useful for showing data on certain aspects as parts of a whole situation. We are all familiar, for instance, with presentations of the federal budget broken into such categories as amounts spent for defense, debt service, welfare, health, education, transportation, and so forth. These provide information in an easily understood pictorial form.

With such additions as color and cross-hatching, bar and circle graphs are valuable aids for communicating technical and financial information. The same information could be presented in words or in tables, but it would be much harder to comprehend.

One of the most important graphical models in engineering design and analysis is the line graph, shown in Figure 2–14. A *line graph* expresses a relation between two variables. The *dependent variable* is typically plotted along the *ordinate,* the vertical axis, while the *independent variable* is shown along the *abscissa,* the horizontal axis. The relationship between the variables is shown as a solid line. If several of these lines appear on the same graph, dashed lines can be used to distinguish between them. Figure 2–15 shows the use of such lines.

Figure 2–16 is a line graph used to display data with rectangular coordinates. Small circles represent the data points; a solid line connects these points to form the

FIGURE 2–13 Circle graph showing estimated breakdown of U.S. coal reserves by sulfur content

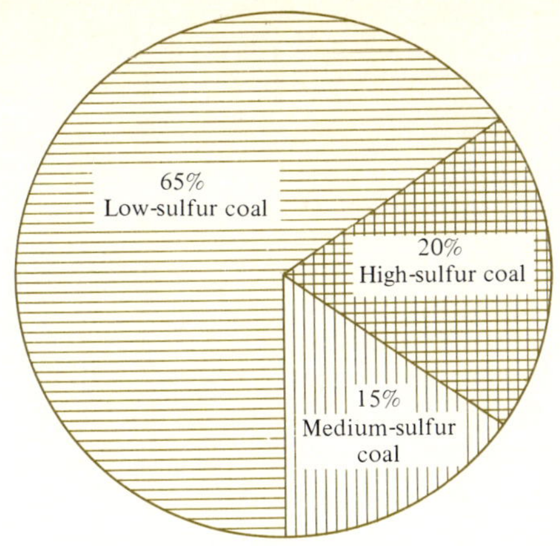

FIGURE 2–14 Example of a line graph

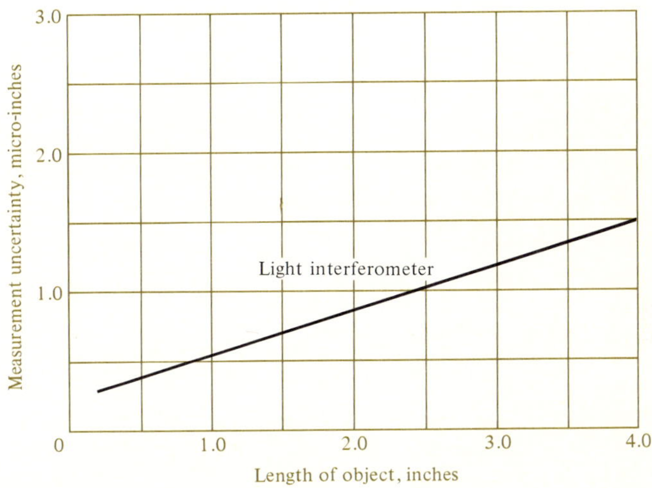

FIGURE 2–15 Example of multiple line graphs

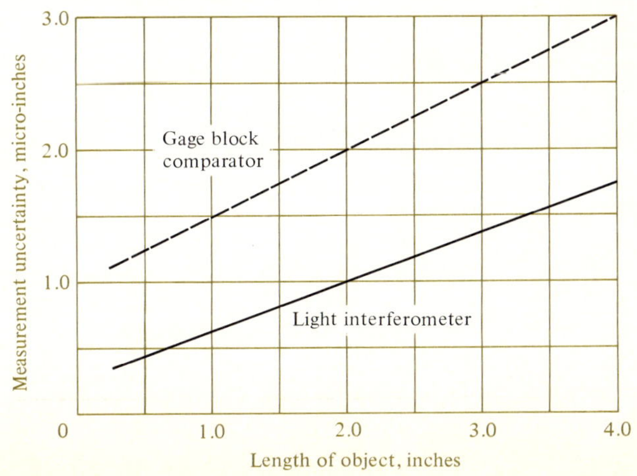

FIGURE 2–16 A line graph with a specific data set

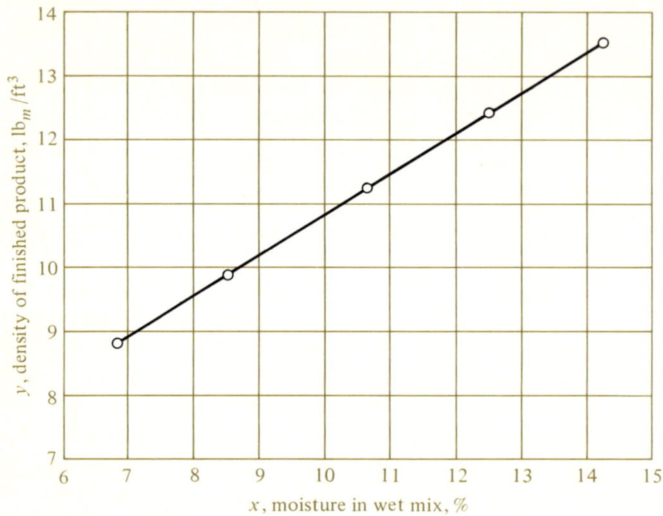

FIGURE 2–17 Multiple line graphs with different data sets

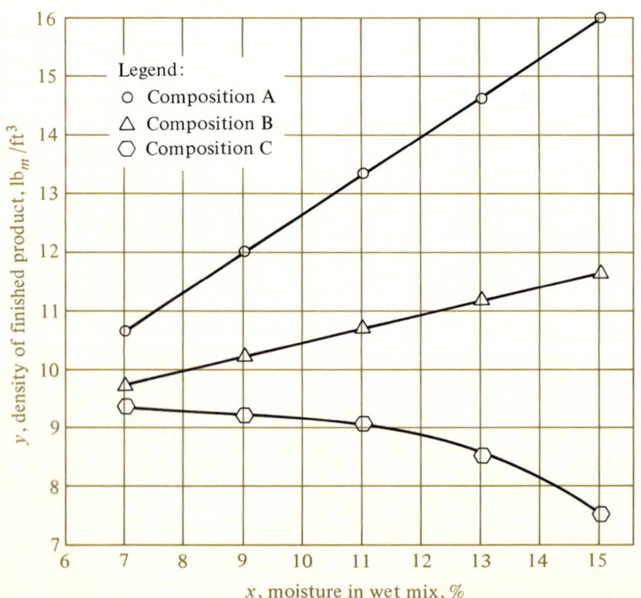

line. When several sets of data are plotted on the same coordinates, different symbols are used to represent the different sets, as shown in Figure 2–17.

It is often desirable to use the logarithms of experimental data as the variables of a model. (We shall explain this in Chapter 7.) If only one of the two variables is logarithmic, the graph is plotted on semilog paper. With this paper, one can locate the value of the variable on the logarithmic scale without performing the logarithmic transformation; the scale is itself proportional to the logarithm (base 10) of the value plotted. As Figure 2–18 shows, only one coordinate of the semilog paper has logarithmic intervals; the other displays the normal linear scale, as in Figure 2–16. When the logarithms of both variables are required, logarithmic (or log–log) graph paper provides the most convenient format, as Figure 2–19 shows.

Figure 2–20 illustrates polar coordinates. In this system, one of the two variables is represented as an angle, given in either degrees or radians. The second variable is plotted as a radial distance from the center of the graph. This variable can be plotted either on a linear or a logarithmic scale. An important application of polar graphs is the circular chart, used on a recording instrument where time is the independent variable, as shown in Figure 2–21. The 360 degrees of the circle are divided into 24 hours. The angle represents time, the first variable. The radius represents the

FIGURE 2–18 A semilogarithmic graph of the time required to assemble each of 60 hypothetical missiles

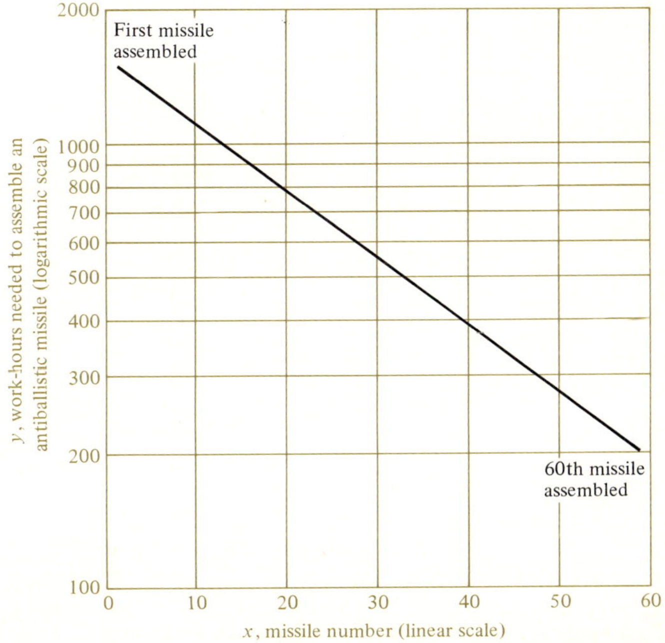

FIGURE 2–19 A logarithmic graph (also log–log graph) of the amount of reactant remaining after a given reaction time

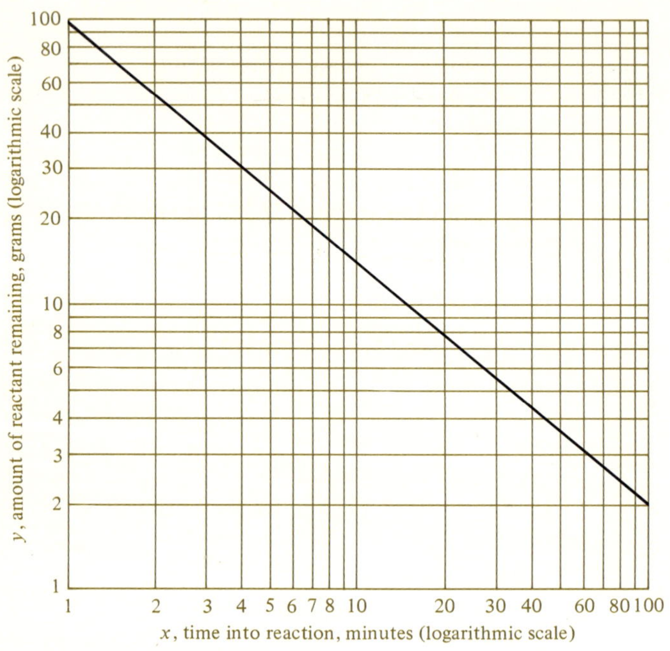

FIGURE 2–20 Polar coordinate graph of the function $y = \cos\theta$

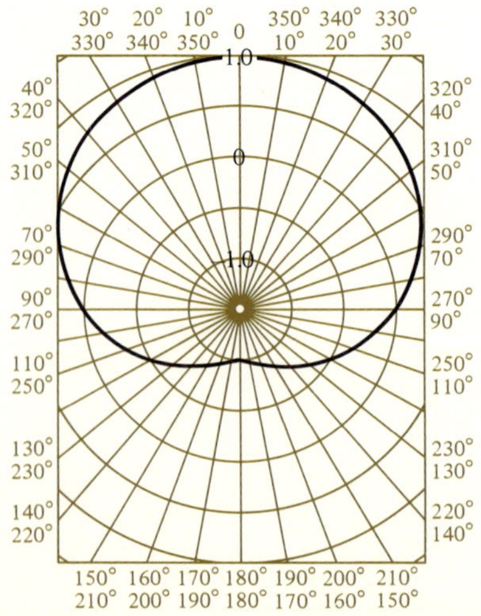

FIGURE 2–21 Circular graph of pH in a chemical reactor ~ (0 ≤ pH ≤ 10)

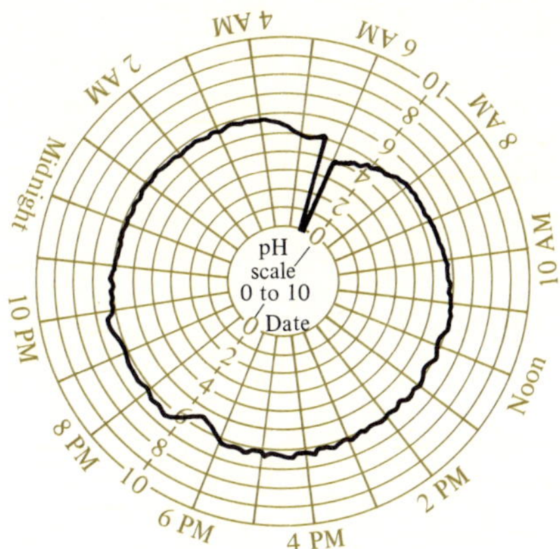

second variable, in this case the pH of the contents of a chemical reactor. The ink recorder plots the dependent variable (pH) as the chart rotates one revolution each day. Such graphs can be easily removed and filed to provide a permanent record of the data of interest.

Exercises

2–8 Develop a bar graph for the following data; they give the relationship between weight and particle size for a fine powder.

Particle size, micromillimeters	*Weight, %*
0–5	8.7
5–10	16.8
10–15	16.8
15–20	13.9
20–25	9.2
25–30	8.7
30–35	7.5
35–40	5.8
40–45	4.6
45–50	3.5
50–55	2.3
55–60	2.2

2–9 Prepare a circle graph using the data from Exercise 2–8. Is this graph an effective way to present the data? Why?

2–10 Prepare a circle graph showing the composition of clean, dry air at sea level. Use the following data, and discuss the visual effect of this graph.

Component	*Volume (%)*
Nitrogen	78.08
Oxygen	20.95
Argon	0.93
Carbon dioxide	0.03
Others	0.01

2–11 Use the following data to prepare a line graph of the green strength of molding sand in pounds force per square inch (psi) as a function of mulling time in minutes. Mulling is a process by which molding sand is mechanically stirred and kneaded with a binder to distribute sand particles uniformly throughout the mix.

Mulling time, minutes	*Green strength, psi*
0.5	4.0
1.0	9.2
1.5	11.2
2.0	12.2
3.0	13.0
4.0	13.3
5.0	13.6
6.0	13.6

2–12 Graph the following relationships on rectangular coordinate graph paper; use values of x between -6.0 and $+6.0$ in increments of 0.5.

$$y = 1.3^x, \qquad y = 0.8^x$$

2–13 Plot suitable semilogarithmic graphs of the two relationships given in Exercise 2–12.

2–14 Following are the semimajor axes of the elliptical orbits of the planets in our solar system, along with the times required for each planet to make one complete orbit. Prepare a log–log graph of these data.

Planet	Semimajor axis (10^6 miles)	Planet year (days)
Mercury	43.36	88.0
Venus	67.65	224.7
Earth	94.45	365.3
Mars	154.8	687.0
Jupiter	506.7	4333
Saturn	935.6	10,759
Uranus	1867	30,686
Neptune	2817	60,188
Pluto	4600	90,737

2–4 SYMBOLIC MODELS

Symbolic models are mathematical representations of the relationships among the entities and phenomena in engineering systems. The major advantage of these models is that they can be manipulated by mathematical rules or numerical procedures. A symbolic model usually expresses a quantitative relationship. For example, the simple model

$$y = r + s + t$$

relates the *dependent* variable y to the *independent* variables r, s, and t. We say that the quantity y is a *function* of the quantities r, s, and t. In a generalized fashion, we can state that y is a function of these three independent variables, as follows:

$$y = f(r, s, t)$$

This symbolism implies a functional relation, as we discussed in Section 2 of this chapter.

Symbolic models fall into two categories: theoretical and empirical. *Theoretical models* represent established and universally accepted principles or laws. These include physical laws and mensuration formulas. *Empirical models* are based on experience and experiment; they are relationships for which theoretical principles or general laws are unknown. We can illustrate this distinction by considering a few examples.

The *perfect gas law,* which describes the behavior of ideal gases, is a theoretical model. This law states that, for n moles of an ideal gas held at temperature T, the volume V occupied by the gas is inversely proportional to the pressure P to which the gas is subjected. The model representing this relationship is

$$PV = nRT$$

where R is a constant of proportionality that depends on the units associated with p, V, and T.

The engineer encounters many such theoretical models in chemistry, physics, mechanics, fluid flow, heat transfer, and many other areas.

An example of an empirical model is the equation used to predict the heat transfer from a pipe carrying a fluid (e.g., the Alaskan pipeline) to the surroundings:

$$h = 0.026\, G^{0.6}/D^{0.4}$$

where h = heat transfer coefficient, Btu/hr-ft^2-°F
G = flow rate of fluid, lb/hr-ft^2
D = outside diameter of the pipe, ft

This model does not represent an exact relationship among the three variables h, G, and D, nor is it based on fundamental physical principles. It is an approximate model based on experimental results and the technical judgment of engineers.

2–5 COMPUTER MODELS

Digital computer models are increasingly being used to solve complex or repetitive problems. Such a model consists of a set of instructions, called a *program,* written in a language a computer can interpret. FORTRAN, COBOL, BASIC, APL, and PL1 are some of the more popular, general-purpose computer programming languages. Each has a unique structure, but they all store data, make logical decisions, control logical flow, and perform arithmetic operations as well as input and output functions.

Since FORTRAN is now the most widely used language in engineering modeling and computation, we have chosen it for developing computer models in this text. Chapter 3 introduces the basic concepts of FORTRAN, and Chapter 6 develops the more advanced FORTRAN techniques that you will need. Each of the chapters dealing with specific classes of engineering models also presents related FORTRAN models. One of our fundamental goals is to intertwine a computer language with some important engineering concepts. We believe that this combined approach is the optimum technique for learning the fundamentals of engineering.

The recommended approach to developing computer models of engineering systems parallels the engineering design process described earlier in this chapter. The first three steps are identical, while the remaining steps detail the process by which an engineer develops a computer model. This detailed method is the following.

Step 1. Define the problem. State the objectives and identify the criteria and constraints.
Step 2. Determine the output. What results are required?
Step 3. Determine the input. What data and information are needed?
Step 4. Determine the algorithm. What steps must be carried out to transform input into related output?

Step 5. Develop a program flow chart that outlines the essential steps in the logical operation: introduce input, execute the desired algorithm, and produce the necessary output.

Step 6. Write the computer program. Prepare statements, in the selected computer programming language, to execute the steps in these input, algorithm, and output procedures. Remember that the computer program itself is a model.

Step 7. Verify the program. Execute the program to ensure that it performs as intended and that it produces the desired results.

Step 8. Validate the model. Make certain that the model truly represents the problem or system being studied.

The distinction between verification and validation in steps 7 and 8 of this procedure is important. Step 7 checks that the program statements faithfully execute the model, while step 8 ensures that the model replicates reality to the necessary degree of accuracy. Verification and validation together establish the credibility of the computer model. This credibility is fundamental in engineering modeling, especially with computer models.

2–6 SUMMARY

This chapter has introduced the fundamentals of modeling as it pertains to the study and practice of engineering. We have defined a model, described several means of classifying engineering models, and examined in some detail the iconic, analog, and symbolic characteristics of engineering models. We have given particular attention to graphical models because they are especially important in engineering design and analysis and because they are immediately useful to engineering students. We have also introduced computer modeling, a subject that will occupy much of our attention in subsequent chapters.

In the remaining chapters of this book, we shall frequently examine the themes developed in this chapter. Chapters 3 and 6 will present the basic elements of the computer programming language—FORTRAN—that we shall use to formulate and manipulate computer models in a wide variety of engineering problems. We shall also treat the deterministic–probabilistic structure of engineering models. Finally, we shall examine two very important classes of models in engineering—economic models and feedback models.

PROBLEMS

P2–1 Classify each of the following models as either iconic, analog, or symbolic. State your reasons for this classification.

(a) The top view and left-side view of a supersonic transport aircraft where the scale is 1 inch = 2 feet

(b) The circular recording chart that gives the time–temperature profile of a soaking pit furnace in a steel mill

(c) The relationship for gas-phase transfer in a gas–liquid column,

$$H_g = \frac{1.01G^{0.31}}{L^{0.33}}$$

where H_g = height of the transfer unit, ft
G = rate of gas flow, lb/hr-ft²
L = rate of liquid flow, lb/hr-ft²

(d) A diagram showing the forces acting on a body of weight W situated on an incline at angle ϕ with the horizontal, where the coefficient of static friction between the body and the surface is μ_s

(e) The set of n outcomes of a random experiment, stated as $S = \{a_1, a_2, \ldots, a_n\}$

(f) A flow chart showing the movement of patients in the outpatient clinic of a hospital

(g) A graph of the relationship between the load on a spring (newtons) and the length of the spring (cm)

(h) The equation for the attractive force between two bodies of masses m_1 and m_2, respectively, separated by a distance d, which is stated as

$$F = \frac{km_1m_2}{d^2}$$

where k is a constant of proportionality.

(i) A flow chart describing the steps in an algorithm for computing the geometric mean of a set of n quantities, given by

$$m_g = \left[\prod_{i=1}^{n} x_i\right]^{1/n}$$

where the symbol $\Pi_{i=1}^{n}$ denotes the product of n quantities $(x_1, x_2, \ldots, x_n)$.

P2–2 Select any mechanical or electrical item in your room. Prepare a drawing on standard-sized, unlined paper. Select and state a scale that enables you to show three views (top, front, and side) of the object. Show dimensions. Label the drawing with any pertinent information.

P2–3 Develop a flow chart showing the steps in a computational algorithm for calculating the geometric mean of a set of n quantities, as given by the equation in Problem P2–1(i).

P2–4 Estimate your expenses for this academic year: tuition, fees, room and board, books, clothing, travel, entertainment, and other expense categories. Prepare a circle graph showing each of these expenses as a fraction of your total expenses for the year.

P2–5 Samples of boiler plate undergo strength tests several times during a production process. One measurement is force applied in thousands of pounds force per square inch (x); another is elongation in inches (y). The results for 10 test pieces are as follows. Plot the elongation (y) as a function of the force applied (x).

Test piece	*Force applied x, 1000 psi*	*Elongation y, inches*
1	2.6	0.027
2	5.4	0.050
3	7.2	0.067
4	9.0	0.083
5	10.7	0.101
6	12.5	0.117
7	14.3	0.134
8	19.8	0.154
9	19.6	0.188
10	21.4	0.206

P2–6 From data supplied by your instructor, construct a circle graph showing the freshman enrollment in the several engineering disciplines taught at your school.

P2–7 From the following data, construct a bar graph showing the frequency with which automobile bodies arrive at an engine-mounting station during a particular 168-hour work week. (The frequency represents the number of one-hour periods during the week in which a given number of automobile bodies arrived at the station.)

Hours	*Arrivals*
1	0
15	1
53	2
38	3
30	4
19	5
8	6
4	7

P2–8 A projectile is fired at an initial velocity V_0 at an angle ϕ with the ground. The equations governing the flight of the projectile are

$$y = (V_0 \sin \phi)\, t - \tfrac{1}{2} g t^2, \qquad x = (V_0 \cos \phi)\, t$$

where y is the vertical distance of the projectile above the ground (feet), x is the horizontal distance from the point of firing (feet), t is the time from firing (sec), and g is the gravitational constant (32.2 ft/sec^2). The initial velocity is 900 ft/sec and ϕ is 30°. Construct a graph showing the position of the projectile from the instant of firing until it strikes a target on the ground. [*Hint:* Refer to Figure 2–11.]

P2–9 A civil enginer testing a new concrete mixture obtained the following experimental data relating the compressive strength S (lb$_f$/in^2) to hardening time t (days). Plot strength versus hardening time on a suitable choice of scales.

S (lb$_f$/in^2)	148	1510	3640	5560	6240	6780	6800
t (days)	1	10	30	70	100	300	1000

P2–10 A mechanical engineer who studied the flow rate f of liquid from a cylindrical tank as a function of the liquid height h obtained the following data. Plot flow rate as a function of liquid height.

f (gallons/hour)	380	270	120	85	42
h (feet)	10	5	1	0.5	0.1

P2–11 Plot efficiency (percent) as a function of output load (horsepower) for a certain ¼-hp, 110-volt, direct-current electric motor. Use the following test data.

Output load, hp	*Efficiency, %*
0.02	25.0
0.04	38.0
0.06	44.0
0.08	45.6
0.10	48.0
0.12	49.3
0.14	50.8
0.16	53.0
0.18	55.5
0.20	57.4
0.22	58.0
0.24	58.4
0.26	57.8
0.28	55.9
0.30	53.6

P2–12 In Problem P2–8, suppose you have to prepare a computer program to compute the vertical distance y and the horizontal distance x of the projectile as functions of time t according to the equations given. Prepare a flow chart showing the steps in this program for any given initial velocity V_0 and angle of inclination ϕ.

P2–13 Develop a flow chart for the process of preparing and executing a computer program at your institution.

P2–14 Develop a flow chart for the calculation of the quantity e^x, using the approximation

$$e^x = 1 + x + \frac{x^2}{2!} + \frac{x^3}{3!} + \cdots + \frac{x^7}{7!}$$

where $k!$ is read "k factorial" and means the product $k(k - 1)(k - 2) \cdots (2)(1)$. Use this expression to evaluate e.

CHAPTER THREE
AN INTRODUCTION TO FORTRAN

READ FOR FRIDAY

In Chapters 1 and 2 we developed two major themes. The first centers on the importance of engineering models. The second concerns the value of mathematical analysis of these models when designing systems. Since most engineering tasks involve the use of existing models, we shall focus our attention in this chapter on analyzing models rather than formulating them. In Chapter 7, we shall consider the formulation of models.

Mathematical analysis has two parts, as the following example illustrates. The problem is to find the two roots of the quadratic equation,

$$x^2 - 3x - 4 = 0$$

The first step in solving the equation is to find an appropriate method. For a quadratic equation, one such method is to apply the quadratic formula.

$$\text{If} \qquad ax^2 + bx + c = 0$$

$$\text{then} \qquad \frac{-b \pm \sqrt{b^2 - 4ac}}{2a} \qquad \text{are roots.}$$

We can easily apply this method to our problem by making the following identifications.

$$a = 1, \qquad b = -3, \qquad c = -4$$

The second step is to substitute the numerical values for a, b, and c into the formula for the roots and carry out the indicated mathematical operations. Carrying out this step in the example shows that the two roots are 4 and -1.

All problems of analysis break into these two parts. The first step is to find the appropriate technique to solve the equations. Such problems range from the trivial (find the value of x that satisfies the equation $x - 5 = 0$) to the more challenging (solve 1000 simultaneous equations in 1000 unknowns). Whatever the method of analysis, the engineer frequently needs to carry out mathematical calculations with ease and accuracy. The digital computer is one tool that facilitates this part of an engineer's job. However, the computer cannot perform the first step of the analysis—selecting the method. Despite its great speed and accuracy in carrying out mathematical operations, the computer offers no substitute for sound, experienced judgment by a human being.

In the following sections, we shall first describe the organization of a modern digital computer system and then examine the basic rules of FORTRAN, the language we have selected to use in communicating with digital computers.

3–1 OVERVIEW OF A DIGITAL COMPUTER

We expect any computer to be able to perform three functions: (1) the input function (the method for communicating questions to the machine), (2) the output function (the method for returning answers to the user), and (3) the internal "scratch-pad" function (the method for processing input information to produce a suitable output).

The three major elements for performing these three functions are shown in a block diagram in Figure 3–1. A separate block is used for each of the three major items. For the time being, we can interpret this sketch simply. The user "asks" a question of the computer system, using the input device as the communication channel. The machine works on the question, using the internal scratch pad, and then it returns the answer by means of the output device.

But we can see that a simple scratch pad is not sufficient, especially if a problem is long and requires intermediate answers to be stored and used in later calculations. For example, imagine trying to solve the following problem using only addition, subtraction, multiplication, and division.

$$\text{Evaluate} \qquad y = \frac{x^3}{q^2} + 2 \qquad \text{for} \qquad x = 2,\ q = 1.$$

The following series of mathematical steps gives a possible solution (it is not unique).

1. Start with q, multiply q by q, store result q^2.
2. Start with x, multiply x by x, store result x^2.
3. Multiply x^2 by x, store result x^3.
4. Take x^3, divide by q^2, store result.
5. Add 2 to x^3/q^2.
6. Stop.

With this method, we must store intermediate answers for later use. You have undoubtedly used this technique many times to evaluate complicated formulas simply by writing an answer on a piece of paper for future use. When you use an electronic calculator, you may choose to store an intermediate answer in the calculator's memory. A given instrument has between 1 and 20 memories. If you have many intermediate values to record and save, you have to develop some sort of labeling system so that you can look at a piece of paper and know which number is where.

The computer has an analogous problem. To do complex calculations, it must have a memory. It needs memory to store values needed for later calculations and to store complicated sets of instructions from the user. The ability to store instructions as well as numbers is important; it is this that distinguishes a digital computer from a nonprogrammable electronic calculator.

We can now refine our simple view of a computer system. Figure 3–2 shows four major elements in a block diagram: an input device, an output device, a memory, and a central processing unit (CPU). Let us now consider each in more detail.

FIGURE 3–1 The basic computational elements

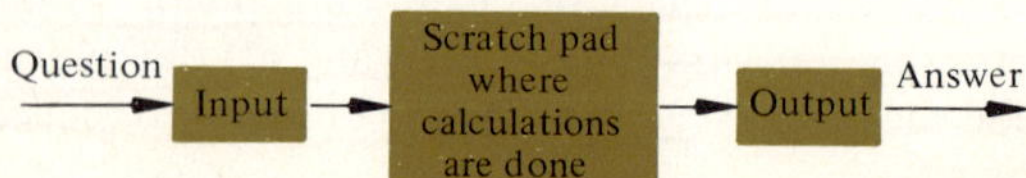

FIGURE 3–2 The four parts of a computer: input, memory, output, and central processing unit (CPU)

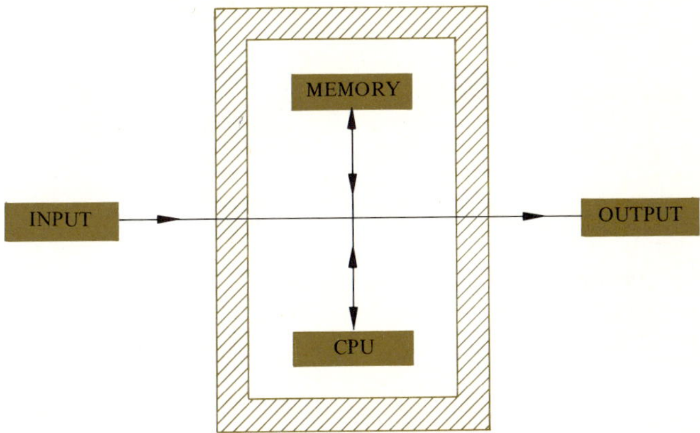

There are several major types of input devices. A commonly used one is the *punched-card reader*. To use a card reader, the user must record instructions as rectangular openings on a set of cards. (The rules for writing such instructions are explained later in this chapter.) The device used to punch cards is similar to a typewriter, but instead of typing letters on a sheet of paper, it punches rectangular openings in cards. Once the user has prepared a suitable deck of cards, the card

FIGURE 3–3 A card reader is one commonly used input device. The stack of cards on the right is read sequentially under control of the computer system.

FIGURE 3–4 An IBM keypunch. Blank cards are stored in the tray above the words IBM and fed sequentially to the punching station. The operator depresses the keys to punch the desired symbols, and the cards move to the left.

reader of the computer system can read them. The card reader is a device in which beams of light "see" the openings in the cards. A set of optical sensors electronically translate these openings into the letters and numbers that correspond to the symbols punched on the cards. Figure 3–5 shows a sample punched card. The openings arranged vertically beneath each character are unique to that character.

The *teletype* is an input device on which statements are typed on a keyboard. The resulting electronic signals are fed directly to the computer. This instrument eliminates the need for punched cards. The teletype can sometimes be connected via a telephone line to a computer. The electrical signals produced by the teletype are converted to audible sounds, picked up by the telephone handset, transmitted across the telephone system, reconverted to electrical signals, and then fed into the computer. Such communication between user and computer can take place over thousands of miles without a significant loss of accuracy in transmission.

Other types of input devices include readers of punched paper and magnetic tape and cathode-ray screens (television-like devices through which the computer can "see" sketches). Technology has, unfortunately, not yet evolved to the point at which computers can routinely understand the spoken word (science and engineering have not yet caught up with science fiction).

In summary, the primary input devices include punched-card readers, paper and magnetic tape readers, and teletypes. We shall consider only the card reader here.

FIGURE 3–5 A sample punched card showing the punch patterns corresponding to the printed symbols

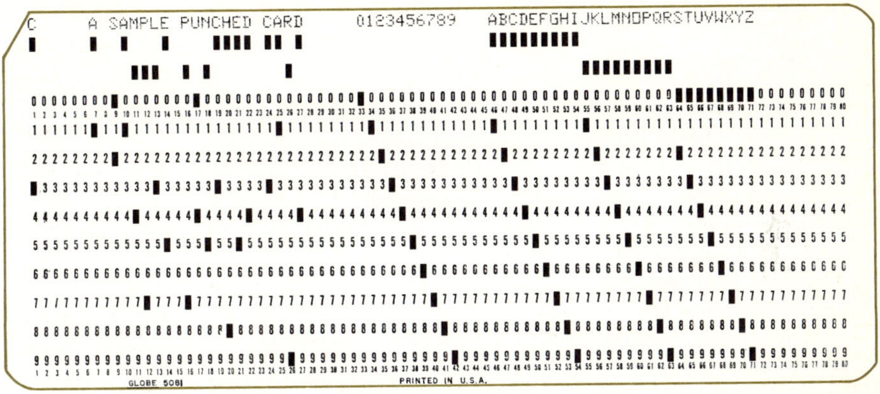

It should be no surprise that output devices are quite similar, in nature and operation, to input devices. After all, both devices are used for communication between user and machine. Output devices include the following: card punch, printer, paper-tape punch, magnetic-tape recorder, and cathode-ray screen. Each of these devices works in exactly the way its name implies. The card punch produces computer output (i.e., answers) in the form of punches on a set of cards. This form of output is most useful when the answers have to be fed into the system at some later time for further processing. The printer is a typewriter-like device that prints answers on sheets of paper. The paper-tape punch produces a series of holes on a strip of paper, and the cathode-ray tube is a television-like screen that displays sketches, graphs, and numerical data. In this introduction to computing systems, we shall focus attention on the printer as the primary output device.

We have little need to consider the details of computer memories at this point. Such memories are based on complex physical phenomena, normally magnetic or electrical, and the precise details of their operation rarely concern the users of a computing system. It is sufficient to recognize that the memory is available for storage of instructions and numerical values. The details of the fourth element of the computer—the central processing unit in which arithmetic calculations are carried out—are quite complex, as Figure 3–8 illustrates.

We can further refine the general view of a computer shown in Figure 3–2 by considering the question of languages. Anyone reading this text is familiar with English. As students progress in their engineering educations, they begin to speak more and more in "engineerese," a language that mixes English and the mathematical and physical concepts engineers use frequently. Since this language is most useful for formulating problems that require a computer for a solution, it would be helpful if computers understood engineerese. Unfortunately, no such computers now exist.

Your next question may be: Exactly what language does a computer understand?

FIGURE 3–6 The printer is a commonly used output device,

FIGURE 3–7 A magnetic disc is one form of computer memory.

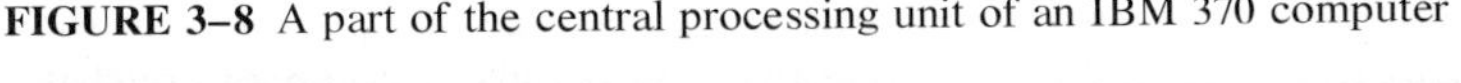

FIGURE 3–8 A part of the central processing unit of an IBM 370 computer

The answer to this question will disappoint you if you are not familiar with computers. The only language that the memory and the central processing unit of any computer understands is a mathematical one based on the binary number system. Any computer can intrinsically understand only binary numbers in special sequences. This limitation on the "comprehension" of a computing system is due to the physical behavior of the electrical or magnetic devices that make up the memory. Essentially the computer's memory consists of a series of switches—much like electric light switches. Each switch can be in one of two possible states—on or off. The *on* state represents a binary "one" and the *off* state represents a binary "zero."

The problem inherent in communicating with a computer should now be clear. The prospective user speaks in engineerese, and the computer comprehends only binary. The only apparent solution is for the user to formulate problems in binary. Needless to say, this process is laborious and very unpleasant. If we had found no other solution to the problem of communicating with computers, we would not use them as extensively as we do today.

A better solution than having all users learn binary is having a translator which can convert an engineerese-like language into a sequence of binary numbers that the computer can understand. Computer engineers call this translator a *compiler.*

Developing a compiler is a significant task in the overall design and use of a computer system. In this text, we need not cover the details of compiler construction because we can use digital computers with little knowledge of compilers.

Figure 3–9 shows yet another view of the total system, with the compiler and its effects added. The language being used at any given point is shown adjacent to the lines with arrows on them. Translation takes place during both the input and the output processes. Without translation during output, answers would be binary numbers, not easily comprehensible. Imagine the problems that would occur if all paychecks prepared by computers had names and amounts printed in the form of strings of binary digits.

Compilers cannot understand just any language. They can translate only statements that are written according to certain rules of grammar. These rules of grammar constitute a computer language. There are many such languages, but we shall use only one—FORTRAN (FORmula TRANslator)—because we believe that FORTRAN is, and will continue to be, the most widely used in engineering.

The rules governing FORTRAN depend to some extent on the particular computer that you are using, because compilers differ from one system to another. Thus you will have to follow some local ground rules. The following sections provide

FIGURE 3–9 Languages associated with various stages involved in solving problems on a digital computer. The sequence is engineerese to binary to engineerese. Translation occurs during input and output.

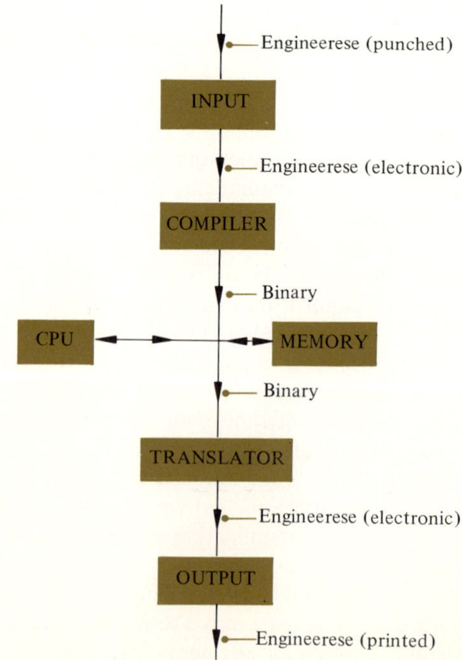

blank spaces at appropriate points to record these specific rules. The two chapters in this book devoted to FORTRAN will contain statements and illustrations that apply to all systems and will leave space for local variations. Thus you can customize this book and adapt it to the particular system you will be using. With these warnings out of the way, we are going to throw you into computer programming as suddenly as possible. A complete program will serve to illustrate some of the major concepts of FORTRAN.

3–2 SAMPLE OF A FORTRAN PROGRAM

Here is a sample program in general form.

```
CARD FOR ACCOUNTING PURPOSES
CARD TO SELECT THE COMPILER
OTHER SYSTEM CARDS (IF ANY)
bbbbbbA = 4.
      B = 3.
      SUM = A + B
      WRITE (   ) SUM
      STOP
      END
CARD TO INITIATE EXECUTION IF NO ERRORS
CARD TO INDICATE FINISH
```

The symbol b indicates a blank space on the punched card containing the statement A = 4. The first six columns of the five subsequent statements must also be blank, although we have not shown this explicitly. Figure 3–10 shows these statements as they appear on punched cards. Now let's discuss the purpose of each statement.

SYSTEM CONTROL CARDS

The first card in any program is normally used for accounting purposes. It tells the system the user's name and the account number to which the cost of the time used is to be charged. Some schools do not charge students for the time they use, and a course number rather than an account number may be given on this first card. In the blank space below, record the specific form of the accounting card used in your local computer system. Your instructor will supply the proper form of this card.

ACCOUNTING CARD

//b JOBbbbbE6-1208JMA

Since a computer system can understand several possible languages, the user has to indicate explicitly the language used to compose the instructions in a given program. The word "program" denotes the complete set of instructions written to solve a specific problem. The second card normally identifies the language used. It is called a compiler selection card. Use the following space to record the exact form of the compiler card you will be using.

COMPILER CARD

//ɃNDFØR

Some computer systems need additional instructions to ensure the proper translation and processing of a program. In the blank space below, enter any additional instructions that your computer system needs.

ADDITIONAL CARDS (IF ANY)

FORTRAN STATEMENTS

The fourth line in the sample program listing is an example of an arithmetic statement.

bbbbbbA = 4.

The columns that the symbols occupy on a punched card are very important. Figure 3–11 shows two computer cards with the same statement punched on them. One has six blank columns before the statement begins and displays the normal proper form. (We shall point out some exceptions to this rule of six blank columns later in the chapter.) The second statement in Figure 3–11 will generate an error in all computer systems that we know of. In our sample program, we showed the need to leave six blank columns by putting six lower-case b's in the first arithmetic statement.

What exactly does this statement (A = 4.) accomplish? It tells the computer to reserve a slot in its memory and to assign the label A to that slot. You should think of these labels as addresses to individual physical locations within the memory. During subsequent processing, this first statement will also cause the number 4. to be stored in the memory location with the label A. Whenever another statement, during either the translation or execution of a program (and it is important to distinguish between these two cases, as you will see later), refers to the symbol A,

FIGURE 3–10 The punched-card versions of the FORTRAN statements in the sample program. Each statement is on a separate card.

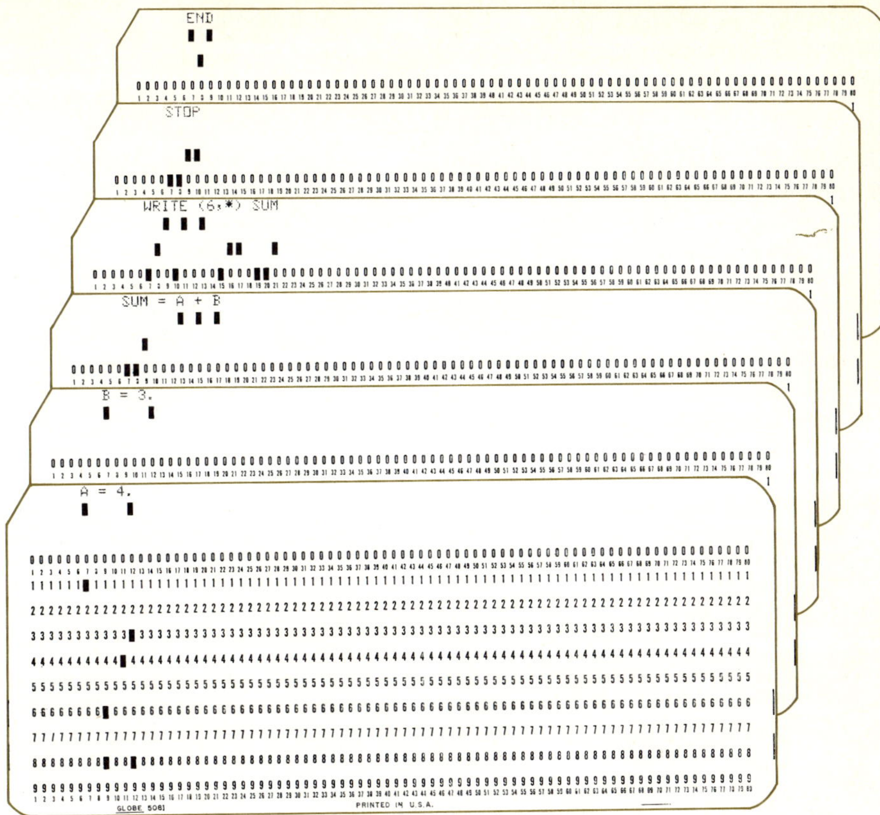

FIGURE 3–11 Properly and improperly punched versions of the FORTRAN statement $A = 4$.

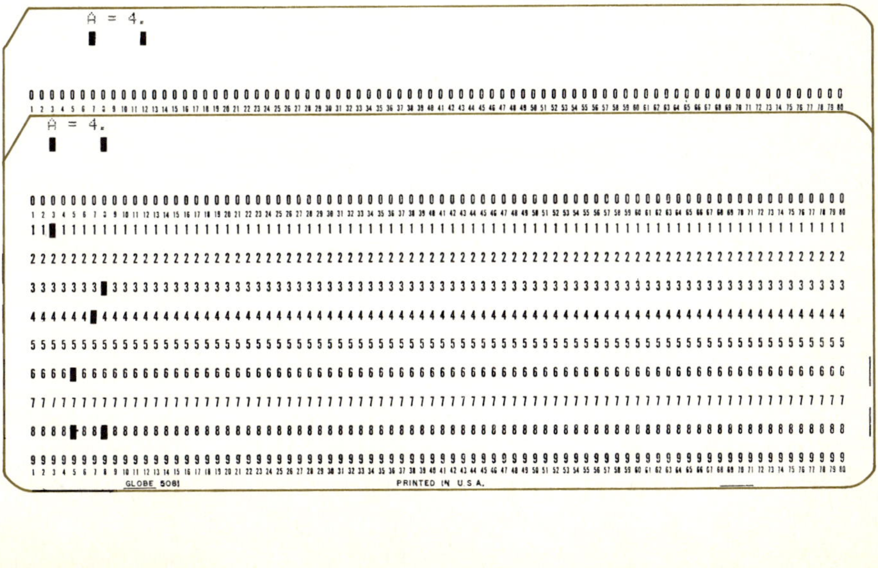

the computer must have a location in memory labeled A in order to continue with the problem.

The general arithmetic specification statement is of the form:

label = number

This statement labels a particular memory slot and eventually stores the number on the right hand side of the statement in that slot. You don't have to worry about choosing a memory slot; the computer system takes care of this internal bookkeeping without any action on your part.

The fifth line in the sample program,

```
B = 3.
```

assigns the label B to another memory location (different from that to which the label A was attached). The computer will subsequently store the number 3. in this location.

The sixth statement,

```
SUM = A + B
```

is different from the first two, since it contains no numbers. With such statements, FORTRAN comes close to engineerese. The meaning is reasonably clear. The label SUM is attached to a new slot in memory. When the program is executed, the numbers found in the locations labeled A and B are added (that is the meaning of the plus sign), and the result is placed in the slot labeled SUM.

The general form of an arithmetic statement is as follows:

Label = an expression involving defined quantities,
numbers, and mathematical operators

Examples of such statements include

```
PROD = A * B
QUOT = A/B
DIFF = A - B
```

As the labels on the left-hand side of these statements imply, the asterisk in FORTRAN means multiply, the slash means divide, and the minus sign retains its algebraic meaning.

The basic algebraic symbols in FORTRAN are as follows.

FORTRAN *Symbol*	*Operation*
+	addition
−	subtraction
*	multiplication
/	division

Exercises

3–1 Write a FORTRAN statement to store the numerical value of your age (rounded off to the nearest year) in a memory location with the label AGE.

3–2 Write a set of three FORTRAN statements that will assign the value 1. to A, 2. to B, and 3. to C.

3–3 Assume that you have written statements that stored numerical values in three different memory locations, A, B, and C. Write a FORTRAN statement that will calculate the product of these three numbers and place the result in a memory location labeled PROD.

3–4 Assume that numbers have been stored in memory locations X and Y. Write a FORTRAN statement that will calculate the ratio of the number in X to the number in Y and store the result in U.

3–5 After the following set of statements are processed (compiled and executed), what number will be stored in the location labeled ANS?

```
A =  .5
B = 3.2
C = 7.0
D = B + C
ANS = D * A
```

3–6 What number will be stored in RESLT after the following five statements are translated and executed?

```
X = 15.
Y =  8.
Q =  3.
DIFF = Y - X
RESLT = DIFF/Q
```

Let's look at our sample FORTRAN program again.

```
CARD FOR ACCOUNTING PURPOSES
CARD TO SELECT THE COMPILER
OTHER SYSTEM CARDS (IF ANY)
bbbbbbA = 4.
      B = 3.
      SUM = A + B
      WRITE (   ) SUM
      STOP
      END
CARD TO INITIATE EXECUTION IF NO ERRORS
CARD TO INDICATE FINISH
```

The seventh statement is

```
WRITE (   ) SUM
```

This statement enables the user of the computer to obtain the answers generated by the program. The WRITE statement directs the computer system to display, on the output device chosen by the operator, the current values stored in the memory slots specified in the WRITE statement. In our example, the WRITE statement literally instructs the computer system to fetch the number stored in the location labeled SUM and display this number on the indicated output device.

The number stored in SUM is, of course, 7. Consider some of the ways in which this number could be written:

$$7{,}000(10^{-3})$$
$$7.0$$
$$0.7(10^{1})$$
$$0.700(10^{1})$$

We assume that you are familiar with scientific (powers-of-ten) notation.

We can dismiss the first way of writing the number stored in SUM, because FORTRAN compilers do not allow commas to be used in numbers. However, the other three forms are perfectly acceptable. The exact form in which the output appears depends on the precise WRITE statement you use. That is, there exists a set of rules that allow you to specify the number of significant figures (constrained, of course, by the ability of the machine to supply them), to choose whether a decimal point is to be printed after the number, and even to decide whether the number will be written with or without a power-of-ten factor. We shall discuss these rules in a later section of this chapter; for the moment, their details are not important. We shall use only a WRITE statement in which each answer is printed as a six-digit number between zero and one multiplied by a suitable power of ten. The exact form of the statement is dictated by local ground rules.

You must also choose the particular output device you want. Do you wish output to be printed on a sheet of paper, punched on a card, or displayed graphically on a cathode-ray screen? You usually choose the output device by placing an integer in the parentheses after the word WRITE in the output statement. Each possible output device is specified by a different integer. The integer code varies from computer system to computer system. In the blank space below, record the particular form of the output WRITE statement that you will use.

WRITE STATEMENT FORM

WRITE (6, *)SUM

PAPER PRINT   FREE FORM

There is one additional aspect of the output statement to be considered:

WRITE ( ) SUM

The label SUM that occurs after the parentheses in the WRITE statement indicates that only the value found in the memory location labeled SUM is to be printed. If the following WRITE statement were used instead,

WRITE ( ) A,B,SUM

the values stored in the slots A, B, and SUM would be printed on one output line in a left-to-right order identical to that in the WRITE statement.

Now let's review the first seven statements in our sample program. The first card identifies the user in the manner appropriate for the given computer system, and the second selects the language desired. The next two cards contain specification statements that assign numerical values to two memory locations, labeled A and B. The third statement is an arithmetic one that labels a memory location SUM. The sum of the numbers found in A and B will be stored in this location while the program is being executed. The next card contains the WRITE statement that will direct the number stored in SUM to be printed during execution.

The eighth line in the sample program contains the command STOP. This means that, during execution, the sequence of steps specified by the user has been completed. This card says: "That's the end of the numerical computations and other instructions."

The ninth line in the sample program also contains a one-word command, END. Although the STOP and END commands may appear to serve the same purpose, their functions are significantly different. The STOP statement signals the end of computations and instructions. But the END statement indicates to the compiler (translator) that all the FORTRAN instructions have now been read and that—if the translator has uncovered no errors and has been able to "understand" all the statements up to the END—the program has been successfully translated and may now be executed. We shall clarify the differences between the STOP and END statements more thoroughly later.

At this stage in running the sample program, all the cards up to and including the END card have been electronically read and translated. *No numerical computations have yet been performed.* The compiler has simply been checking all these statements for proper grammar. If the statements do not violate any of the rules of the FORTRAN language, the stage is now set for executing the program.

The tenth line in the sample program initiates the execution process.

CARD TO INITIATE EXECUTION IF NO ERRORS

This means that the computer is now to actually perform the operations that were specified in the sequence of steps that compose the program. The exact form of the statement necessary to initiate execution differs from system to system. In the blank space that follows, record the precise form of the execution statement.

EXECUTION STATEMENT

Now let us look at the processes that occur in a computer system once the execution statement has been read. Figure 3–12 illustrates the execution process in a series of steps reading from left to right. Although this sketch contains the statements present on the original deck of punched cards, it is crucial that you recognize that these statements represent electronic "images" of the cards. These images reside within the memory of the computer system. We have shown them here only to help you understand the execution process. The arrow shown in each column indicates the point reached during execution. Time is increasing from left to right across the figure. The first statement causes the number 4. to be stored in a memory location labeled A. The second statement causes the number 3. to be placed in slot B. The next line initiates an internal process in which the contents of A and the contents of B are copied into the central processing unit; there they are added and the result is placed in memory slot SUM. This process does not alter the numbers stored in A and B. The fourth statement directs the printer to display the result in the form

0.700000E 01

This result shows the promised six significant figures after the decimal point times the appropriate power of ten. The computer uses the letter E to symbolize the number 10 in output, while the integer appearing after the letter E denotes the exponent of 10. Examine a few computer printouts to become familiar with this convention.

FIGURE 3–12 Step-by-step execution of the sample program along with the associated changes in computer memory

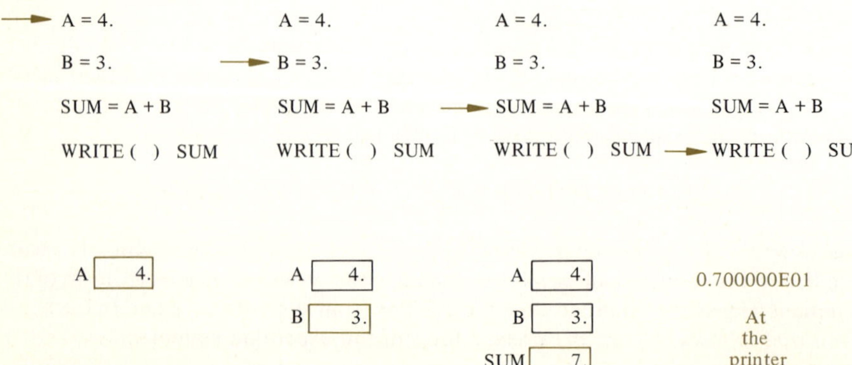

The last statement in the sequence of operations specified by the sample program is STOP; this means that all required steps have been carried out.

Once execution has been completed, the computer reads the next card in the input deck, which normally indicates that the user's job has been finished. You can use the blank space below to record the exact form for the finish card used on your own computer system.

FINISH CARD

We have only one last aspect of the compilation and execution of this program left to discuss. To run the program, the user inserts the complete deck of cards into the appropriate cavity in the card reader. The first card is then read; that is, it is mechanically pulled through the reader and the rectangular openings are sensed optically. Each subsequent card is read until the EXECUTE card is reached. *The preceding cards are no longer in the card reader input cavity when the execution process begins.* Their meanings have been translated into binary form and stored in memory. This is an important point for you to understand. We shall need it for discussions in subsequent sections.

Exercises

3–7 Keypunch and run the sample program. Use the accounting, compiler selection, execution, and finish cards appropriate for your computer system.

3–8 Modify the sample program so that it will print out the numbers stored in the memory slots B, A, and SUM on one line in the order stated. Keypunch and run this modified program.

3–9 Modify the sample program so that it will print the number stored in A on the first line of output, the number stored in B on the second line, and the number stored in SUM on the third line. Keypunch and run this program. [*Hint:* Each separate execution of a WRITE statement results in a new line of printed output.]

3–10 Write, keypunch, and run a program that calculates and prints the difference between two numbers, A = 4., B = 3.

3–11 Write, keypunch, and run a program that calculates and prints the product and ratio of two numbers, X = 23., Y = 15. Arrange the program's output so that the numbers stored in X and Y appear on one line, while the product and ratio appear on another.

3–12 Write, keypunch, and run two programs that calculate the product of two numbers, C = 2., B = 6. In one, use the label PROD to denote the product; in the other, use the label MULT.

The results of Exercise 3–12 will be important in the next section of the text. Before reading any further, you should take some time to work in the computer laboratory and obtain hands-on experience with the keypunch equipment, and, if possible, the computer itself.

3–3 CONSTANTS AND VARIABLES IN FORTRAN

The exercises in the last section should leave you with many questions. In fact, the more precisely and carefully you considered them, the more questions you should have. These questions will pave the way for a deeper understanding of FORTRAN concepts. We shall begin our more thorough examination of FORTRAN by carefully considering the allowed forms of FORTRAN constants (numbers), variables (labels), and the rules governing their definition and use.

CONSTANTS

FORTRAN constants are simply numbers that the FORTRAN compiler can read and understand. We shall give the allowable forms of constants below, but one firm rule is that you may not use commas in the numbers. In addition, you may or may not use a decimal point in constants, but the consequences of omitting a decimal point can sometimes be significant, as we shall see. You may write constants in scientific form when convenient. Here are some examples of acceptable and unacceptable FORTRAN constants.

Acceptable	*Unacceptable*
−4.3E01	$4.3(10^{10})$
1.E-3	3,153
.0003	127523
−127	$4\frac{5}{8}$
15.32	π
127.00	$-1.2(10^{-2})$

The first and last numbers in the right-hand column are unacceptable because it is not possible to keypunch superscripts. In addition, they contain parentheses, which are not allowed in FORTRAN constants. The second number is unacceptable because it contains a comma. The fourth example on the right contains a fraction, and the rules do not allow this. The fifth is a familiar symbol, but it does not exist on a keypunch and the FORTRAN compiler does not recognize it. The third number (127523) in the right column is unacceptable because the absolute magnitude of a number written without a decimal point is limited severely by the storage method used in the computer memory. The details of this process are of no immediate concern to you in an engineering concepts course, but you should recognize that such a limit exists. Use the space below to record this maximum value for your system.

MAXIMUM INTEGER SIZE

When you write FORTRAN constants without an explicit sign, the compiler assumes that the numbers are positive. This is also true for exponents, i.e., the numbers appearing after the E in expressions written in scientific notation. Other than plus signs, minus signs, letter E's, and decimal points, no other punctuation or nonnumerical characters are valid in FORTRAN constants.

VARIABLES

Next we consider the allowable forms of FORTRAN variables or labels. In many cases, we select a FORTRAN variable to remind us of its meaning. For example, it is natural to use the variable (or label) SUM to represent the arithmetic sum of two other quantities. It is a good idea to select such descriptive variables as much as possible, since they will help you recall their meanings. This is particularly important as the complexity of a program increases.

The rules for forming FORTRAN variables are precise and simple. *A valid FORTRAN variable must consist of between one and five alphanumeric characters, and the variable must begin with a letter.* The term "alphanumeric character" means any letter in the English alphabet, as well as any number between 0 and 9. Although punctuation symbols are available on the keypunch, they may not be used in FORTRAN variables. Some computer systems allow FORTRAN variables consisting of more than five characters, but restricting variables to five characters is good practice, since it ensures that programs can run on relatively small computer systems. In this text, restricting variables to five or fewer characters will cause no inconvenience. The following two columns illustrate acceptable and unacceptable FORTRAN variables.

Acceptable	*Unacceptable*
X	S,M
SUM	S*-M
A130	1X
VALID	INVALID
O1	01

Note that VALID is a valid FORTRAN variable, while INVALID is invalid. These examples also raise an important point concerning clarity of notation. The last variable in the acceptable column may appear, on first glance, identical to the last entry in the unacceptable column, but the first symbol in the left variable is the letter O and the first symbol in the right variable is the number 0. You can distinguish O from 0 only by carefully considering the type style used in this book. There is no easy way to distinguish the capital letter O from the number 0. In any handwritten FORTRAN statements, the possibility of confusion and error is great.†

In the remainder of this book, we shall use the following convention for clarity. Whenever the character is written with a slash,

Ø

the letter is intended. Like so many other things, this convention is not universal, and anyone keypunching cards from written statements must recognize this convention. When the symbol is slashed, the letter is intended; when the symbol is written alone, the number is meant.

MODE

Recall that Exercise 3–12 asked you to write two programs to calculate the product of two numbers. In the first, the variable PRØD was used to denote the product, and in the second, the variable MULT was substituted. The program segments necessary to solve this exercise follow, and the outputs produced by each are also shown. The WRITE statements appear without the necessary information in parentheses; you will have to supply this information before running the programs.

Program segments

```
A = 2.                    A = 2.
B = 3.                    B = 3.
PRØD = A * B              MULT = A * B
WRITE (  ) PRØD           WRITE (  ) MULT
```

Printed output

```
0.600000E 01              6
```

The segment on the left produces output in scientific notation, that is, a number between 0 and 1 with six significant figures multiplied by a suitable power of 10. The segment on the right results in output in the form of an integer.

The reason for this difference is the way the system treats numbers with and without decimal points during both the translation and execution stages. This also explains a feature of the rules for naming variables. The FORTRAN compiler assumes that any valid FORTRAN variable beginning with the letters I through and including N represents a quantity without a decimal point. In other words, variables beginning with these letters are used to label locations in memory in which only integer numbers can be stored. The computer memory storage is different for

†Similar difficulties may occur with 1 and I, 5 and S, and 2 and Z.

integers and nonintegers. Some computer systems use 16 switches, or bits, to store an integer and 32 switches for a number with a decimal point.

The term FORTRAN *integer* is used to describe variables beginning with the letters I through N, as well as integers. The term *real variable* or *number* is used to describe variables beginning with other letters of the alphabet (A through H and Ø through Z), as well as numbers with decimal points. Do not confuse this use of the word real with mathematical definitions involving real, imaginary, and complex numbers. Real in FORTRAN simply means a quantity with a decimal point.

This difference between integer and real quantities is termed *mode*. It is important to the programmer because selecting the wrong FORTRAN variables and/or constants will result in numerical errors during some calculations. We shall give an example of this in a moment, but first let's consider the output of our two sample program segments in view of this concept of mode.

For the segment on the left, the compiler reserved a memory slot labeled PRØD. It stored a real number (with a decimal point) in that location after the product of A and B was calculated during execution. The output produced by this program segment is in the familiar form.

For the segment on the right, the compiler reserved a slot labeled MULT, and, according to the rules of FORTRAN, assumed that this location was intended to store only integer numbers. In other words, this location was composed of only 16 binary switches. During execution, the number 6. was calculated when the quantity stored in A was multiplied by the quantity stored in B, but the system was able to store only an integer in the memory slot labeled MULT. The integer stored was 6, and when the WRITE statement was executed, only an integer value was printed. The different modes of the two variables PRØD and MULT account for the differences in the behavior of the two program segments.

In this example, the choice of label made no difference because the numbers were whole numbers even though decimal points were written after them. Then this point may not seem worth worrying about, but the following program segments indicate that much more serious problems can befall the programmer who unwittingly uses FORTRAN integer variables when FORTRAN real variables are necessary.

Program segments

```
A = 2.                    A = 2.
B = 3.6                   B = 3.6
PRØD = A * B              MULT = A * B
WRITE (   ) PRØD          WRITE (   ) MULT
```

Printed output

```
0.720000E 01              7
```

The segment on the left gives the correct value for the product, but the segment on the right gives an incorrect answer. In the program on the right, the correct answer 7.2 resulted when the quantities found in A and B were multiplied, but the memory slot available for MULT could accommodate only an integer. When this situation occurs, the FORTRAN compiler follows very inflexible rules. It simply stores only the digits to the left of the decimal point in the appropriate slot. In our example, the result is that the number 7 appears on the printed output sheet.

You can see from this example that it is important to remember the distinction between integer and real mode. Failure to do so can produce inaccurate answers, as in the right-hand segment of our example. Remember that integer variables begin with I, J, K, L, M, and N.

It is useful to have a general rule to categorize integer and real variables by their first letters, but it is also helpful to be able to deviate from this rule in some programs. Two special statements in FORTRAN allow such deviations. They should appear in a program after the initial accounting and compiler selection cards. Examples of these two statements are

```
bbbbbbREAL I, J1
bbbbbbINTEGER X, XØ
```

The first statement is an example of a REAL specification; it instructs the compiler to treat the variables I and J1 as real quantities in this program. The INTEGER specification tells the compiler to treat the variables X and XØ as integers. The order of the two statements in a given program is unimportant. That is, the INTEGER statement may come before the REAL statement. You may find some minor use for these statements in your early programming endeavors, but you should not feel compelled to use them. The normal rules for FORTRAN variables provide a sufficient number of both integer and real variables.

In the next section, we shall consider arithmetic statements in detail.

Exercises

3–13 Are the following quantities valid FORTRAN constants? Give a reason for each invalid quantity.

−1.005	17.36E + 01	1,705.36
17. * 3.	34681	−3.03E(1326/401)

3–14 Which of the following quantities are invalid FORTRAN variables? Why?

SELF	X12Ø4	L2/4	GØØD
HELP	12BA	LXYRU1	012
Ø12	A * B	C = D	MAP9

3–15 Each of the following is a valid FORTRAN quantity. Identify each as either a constant or a variable and identify the mode of each.

A13	M	SNAP	FUN
15	LULU	1.7E02	291
−154.	HEN	NEH	D13
ØL	6.31E-01	MASS	TEMP

3–16 What output will each of the following program segments produce?

```
A = 6.5
B = 3.
C = A + B
D = A − B
L = C/D
WRITE (   )L
```

```
U = 9.9995
V = 2.
W = 3.
X = V + W
M = U/X
WRITE (   )M
```

```
A = 4.2
B = 2.1
M = A * B
N = 2
J = M/2
WRITE (   )J
```

3–17 What output values will each of the following program segments produce?

```
REAL L
A = 6.5
B = 3.
C = A + B
D = A − B
L = C/D
WRITE (   )L
```

```
REAL M
U = 9.9995
V = 2.
W = 3.
X = V + W
M = U/X
WRITE (   )M
```

3–4 ARITHMETIC STATEMENTS

Despite their appearance, arithmetic statements are not equations. The most vivid example of this is the perfectly valid FORTRAN statement

```
J = J + 1
```

As an algebraic equation, this statement makes no sense. It implies that 1 is equal to 0, an absurdity. But in computer programming, this statement does have a valid role. When a program in which it appears is executed, it causes the quantity stored in the memory slot labeled J to be added to the number 1, and the result is then placed in the memory slot labeled J. Thus the execution of this statement increases the value of the integer stored in J by 1. This illustration shows the need to avoid the logical trap of treating arithmetic statements in precisely the same way as algebraic equations. Some arithmetic statements may be accurately interpreted as equations, but caution is always in order.

The general form of an arithmetic statement is:

```
variable = something
```

The word "something" means any expression involving valid FORTRAN variables, constants, arithmetic operators, and other valid FORTRAN quantities that we shall introduce later. Several examples illustrate the meaning of this definition.

```
DISC = B * B − 4. * A * C
Y = A + B/C
X = X − 1. + Q
AN = A + B * C/D + 5.
M = N + 5 * L
I = J/K − 300
```

These six examples are valid arithmetic statements, since the left-hand side of each expression is a valid FORTRAN variable and the right-hand side is "something," a mixture of constants, variables, and arithmetic operators. Although we have not explicitly said so, we are treating the plus and minus signs as operators.

MIXED-MODE EXPRESSIONS

The expression "something" must be entirely of one mode. In our first four examples, the right-hand sides of each of the expressions were real; in the last two, they were integers. Real and integer quantities cannot be mixed within any one right-hand side. When such mingling occurs, as in the following examples, the expressions are said to be *mixed mode*.

```
DISC = B * B - 4 * A * C
Y = A + B/I
X = X - 1 + N
AN = A + I * J/D + 5
M = N + 5. * Z
I = J/K - 300.
```

While some FORTRAN compilers accept such statements, others reject them and indicate a program error during the translation process. Expressions using mixed modes are always poor programming practice because they can lead to inaccuracies in calculations. You should avoid mixed-mode expressions as a matter of course. The most difficult thing for a novice to remember in writing arithmetic statements is the necessity of adding a decimal point after all whole numbers used in real expressions. The first example in this set is a mixed-mode expression because no decimal point appears after the number 4.

Exercises

3–18 Which of the following arithmetic statements contain mixed-mode expressions? Give reasons for each of your choices.

```
ANS = A/B/C - 15.
BETA = A - 3.E10 + B - 5/C
ME = L + I/A
KY = LA + MØ + TX/5.
ZØØ = LIØN + TIGER + LYNX + BEAR
```

3–19 Each of the following arithmetic statements contains one error that results in mixed modes on the right-hand side. Identify and correct the error in each.

```
DISC = B * B - 4 * A * C
ZØØ = BEAR + ØWL + WØLVES + MØØSE
M = L + N + Z
```

```
Z = R - 3 * S * T
I = J/K/L - 3.E02
```

MODE-CONVERSION STATEMENTS

So far we have used the term mixed mode to mean that both real and integer quantities appear on the right-hand side of an arithmetic statement. There is another way in which the two modes may both be present in the same statement. Examples include the following.

```
J = Y
Z = I + L - 5
I = X * Y - 1.E02
```

These are not mixed-mode expressions in the sense defined earlier; we call them *mode-conversion statements*. In general, these statements are of two distinct types: One type has an integer variable on the left-hand side and a real expression on the right, and the other has a real variable on the left and an integer expression on the right. To illustrate the effects of such expressions, consider a program segment that contains the first type, and the output of this segment.

Program segment	*Output*
Y = 8.3	8 0.830000E 01
J = Y	
WRITE ( ) J,Y	

In the FORTRAN arithmetic statement J = Y, the values stored in the memory slots labeled J and Y are different. This difference again points out that we cannot treat FORTRAN statements simply as algebraic equations. By using this mode-conversion statement, we have lost information. When the statement is executed, the central processing unit fetches the number stored in Y (8.3) and then attempts to store this value in location J. Since this location has room only for an integer quantity, the computer stores only the part of the number to the left of the decimal point (8). That is why the WRITE statement in this program segment produces two different forms of output.

Consider another example of the effects of mode conversion.

Program segment	*Output*
L = 15	0.500000E 01
M = 3	
Z = L/M	
WRITE ( )Z	

In this case, the quantity L/M is evaluated (5), and this number is stored in the memory location labeled Z. Since the number is an integer and the designated storage slot is for a real number, the computer adds a decimal point to the number (in this case, converting 5 to 5.) and stores the result in Z.

The danger of carrying out some arithmetic operations using integer variables becomes clear if we modify this program segment slightly.

Program segment	*Output*
L = 14	0.400000E 01
M = 3	
Z = L/M	
WRITE ( )Z	

When we wrote this segment, we may have intended to divide 14 by 3. But since we wrote these two numbers as integers and defined the ratio as the division of one integer by another, the mathematical result is incorrect. In particular, when the central processing unit evaluates L/M, it is working with integer quantities and preserves only the integer part of the mathematical result. This means that L/M (14/3 = 4.66667) is stored as the integer 4 only.

When this number is placed into the memory slot labeled Z, it is converted to a real quantity, and the output shown is the result of the WRITE statement.

This example indicates that using integer variables in some types of arithmetic statements may result in numerical inaccuracy and loss of information. The natural response to these considerations is to always work with real variables and constants. However, integer constants and variables are required in some very useful FORTRAN statements. That is why we considered them here, although their ultimate value will not become clear until Chapter 6.

Exercises

3–20 In each of the following program segments, what value is stored in the memory slot labeled L?

```
A = 1.E05         M = 10       M = 10.
B = 2.E - 05      N = 11       N = 10.
L = A * B         L = M/N      L = M/N
```

3–21 In each of the following program segments, what number is stored in the memory slot labeled Z?

```
I = 5              Z = 0.
I1 = 4             M = 1
I2 = -3            M = M + 1
Z = I1/I2          M = M + 1
Y = I/I1           M = M + 1
L = Z/Y            Z = M
Z = I - I1 + L     Z = Z/2.
```

OPERATOR PRIORITIES

We shall discuss one last aspect of arithmetic statements before we conclude this section—the order of operations. Consider the following program segment.

```
A = 2.
B = 3.
C = 4.
D = 2.
AN = A + B * C/D + 5.
WRITE (   ) AN
```

When this is executed, what value of AN will be printed? This question does not have a unique answer because the arithmetic statement that produces the number finally stored in AN has several possible interpretations. Three such interpretations are:

$$a + \frac{bc}{d} + 5, \qquad \frac{(a + b)c}{d + 5}, \qquad \frac{a + bc}{d + 5}$$

These expressions exhibit for the first time the convention of using lower-case letters to represent the algebraic equivalent of FORTRAN variables. We shall frequently need to write the mathematical equivalents of FORTRAN statements and shall use lower-case symbols. This convention minimizes the possibility of confusing a mathematical expression and a FORTRAN statement.

For each of these three possibilities, a different numerical value will be stored in the memory location labeled AN. The three possible values stored in AN are:

13.   2.85714   2.

Each possible stored value has a different output:

0.130000E 02   0.285714E 01   0.200000E 01

The quantity actually printed after this program segment is executed is a result of the rules governing operator priorities. These rules, which hold for all FORTRAN compilers, may be stated in two parts.

First, multiplication and division have equal priority. They are executed from left to right across a FORTRAN statement. Second, addition and subtraction have equal priority. They are also executed from left to right. But addition and subtraction have a lower priority than multiplication and division.

These rules mean that, during the execution of an arithmetic statement, the computer system scans the electronic image of the statement in memory and first carries out all multiplications and divisions, beginning at the left. After completing all such operations, the system carries out the additions and subtractions, again scanning from left to right.

Now by applying these rules consistently, we can see which algebraic interpretation is correct.

```
AN = A + B * C/D + 5.
```

A	2.	C	4.	AN	
B	3.	D	2.		

These rectangles with labels attached simply represent the contents of part of the computer's memory. Such diagrams help us understand the sequence in which a given set of FORTRAN statements is executed. We shall now apply the rules governing operator priority to our arithmetic statement.

First, the system scans from left to right and performs all multiplications and divisions in the order in which they appear. The first operation in our expression, denoted by the asterisk, is the multiplication of the numbers found in memory slots B and C. The result is 12. The next priority operation encountered is the division, represented by the slash. In this expression, the quotient to be found is the ratio of the value produced by B * C and the contents of memory location D. The result of this division is 12./2. or 6. The computer encountered no multiplications or divisions in the rest of the statement.

The following set of three equivalent statements illustrates the step-by-step execution of the multiplication and the division.

```
AN = A + B * C/D + 5
AN = A + 12./D + 5.
AN = A + 6. + 5.
```

To finish the evaluation, the system again scans the statement from left to right and carries out all additions and subtractions. The final result is 13. This number is then stored in the memory slot labeled AN. The proper algebraic interpretation of this FORTRAN statement is thus:

$$an = a + \frac{bc}{d} + 5$$

These priority rules are unambiguous; they make possible a unique algebraic interpretation of any arithmetic statement containing only constants, variables, and the basic four operators (+, −, *, /).

Exercises

3–22 Using appropriate lower-case symbols, write the algebraic equivalents of each of the following real arithmetic statements.

```
X = A + B − C * D + E/R − S
Y = A * B − 5./D + 3.
Z = A * B − C * D/E/F * R/Q − 3.
```

3–23 Use appropriate lower-case symbols to write the algebraic equivalents of the following integer arithmetic statements.

```
I = M + 3 * 5 - 7/K
J = I1 * I2/I3/I4
K = M - 8 * 7 + N/LØØN
```

3–24 After the execution of each of the following program segments, what numbers are stored in Z?

```
R = 2.                  A = 2.
PI = 3.1416             B = -1.
A = PI * R * R          C = 2.
C = 2. * PI * R         D = 0.5
Z = A/C                 E = 4.
                        Z = A * B - C * D/E
```

3–25 What numbers are stored in L after each of the following program segments is executed?

```
I = 5                   M = 10/2 - 2
J = 3                   N = 166/80 - 2
K = 2                   I = M * N
L = 3 + I/J * 2         L =I + 1
```

EXPONENTIATION

Besides addition, subtraction, multiplication, and division, FORTRAN has a fifth basic operator—exponentiation. This is used to raise a quantity to a power. The FORTRAN symbol used for this operation is a *double asterisk*. The following statement and its algebraic equivalent illustrate exponentiation:

Y = A ** B $\qquad y = a^b$

FORTRAN does make possible statements in which a real quantity is raised to an integer power. For example,

Algebraic	FORTRAN
$y = x^4$	Y = X ** 4
$y = (u + v)^7$	Y = (U + V) ** 7

Of course, it is possible to carry out the same numerical calculations by raising real quantities to real powers. Statements corresponding to those shown above that do this are:

Y = X ** 4., $\qquad$ Y = (U + V) ** 7.

There are significant differences between these two types of statements. To raise a quantity to an integer power, the FORTRAN compiler performs successive multiplications. To raise a quantity to a real power, it uses an algorithm based on logarithms. These differences mean that an expression in which a real quantity is raised to an integer power is not a mixed-mode expression. Further, the process of successive multiplication is faster than the logarithmic method. Thus, as a matter of good programming practice, it is to be preferred. It also follows from this discussion that an expression such as

Y = (−2.) ** 3

will execute properly, while an expression such as

Y = (−2.) ** 3.

will not execute because the logarithm of a negative number is not defined.

This operation—raising one quantity to another quantity—has the highest priority within an arithmetic statement. In processing an expression, the computer scans from left to right and first carries out all exponentiations, then all multiplications and divisions from left to right, and finally all additions and subtractions. The following illustrates the operator priorities:

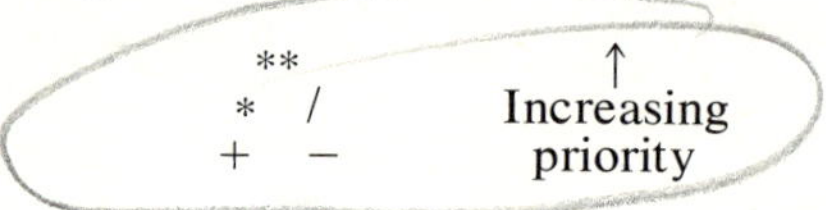

Exercises

3–26 Write expressions for the algebraic equivalents of each of the following FORTRAN statements.

```
Y = 1. + X ** 2./2. + X ** 4./24.
Y = 1. + X * X/2. + X * X * X + X/24.
Y = A + B/U − V * W/X/Y * Z
Y = A + B ** C − D/C
```

3–27 Write FORTRAN statements equivalent to each of the following algebraic equations.

$$y = a + \frac{b^c}{d} - 3, \qquad y = ab^d + \frac{c}{e}$$

$$d = \frac{a}{bc}, \qquad e = ab - ed + \frac{x^y}{z}$$

PARENTHESES

An additional element is used in constructing arithmetic statements. If you try to write a single FORTRAN statement equivalent to the algebraic equation,

$$x = \frac{a + b}{e + d}$$

you will find that there is no way to do so using one arithmetic statement and the FORTRAN concepts we have discussed so far.

If we could not write such statements in FORTRAN, we would certainly question the usefulness of this language. After all, such expressions naturally arise in the formulation and solution of many engineering problems, and FORTRAN must provide a means for easily writing statements equivalent to such expressions. In fact, FORTRAN does—through the use of parentheses. We shall examine several examples, but we first restate the priority rules with the effects of parentheses included.

The priority is:

Operators	
**	↑
* /	Increasing
+ −	priority

Operators are executed in the order of decreasing priority from left to right within parentheses, starting from the innermost parentheses contained in the FORTRAN statement and working outward. Expressions within parentheses are calculated before the operators adjacent to the parentheses are carried out. A few examples will illustrate these priorities.

Y = (A * (B + C))/(E + F)

A	2.	C	−1.	F	2.
B	6.	E	2.	Y	

The rectangles show a segment of the computer's memory before this statement is executed. Following the priority rules, the system first evaluates expressions within parentheses, beginning with the innermost parentheses. In this example, the first expression is B + C, which is 5. Then the quantity within the leftmost of the two remaining parentheses is evaluated A * 5.; this results in the number 10. The expression in the rightmost parentheses is next evaluated and found to be 4. Finally the division operation is carried out, and the result (2.5) is stored in the memory slot labeled Y. In *all* cases, the expressions in parentheses on either side of any operator are evaluated before the operation itself is carried out. It is this property that makes parentheses so useful for grouping and rearranging expressions

Exercises

3–28 Find algebraic equivalents of the following arithmetic statements.

```
Y = 1. + X ** (2./2.) + X ** (4. * 0.5)
Y = (A + B)/(U − V) + W/X/U * Z
Y = C ** ((Z + B)/X) + Z
```

3–29 Write the FORTRAN equivalents of the following algebraic equations.

$$z = \left(\frac{a+b}{c+d}\right)^q, \qquad z = \left(\frac{a}{c} + \frac{b}{d}\right)^{(q+r)/s}$$

A FORTRAN compiler, during translation, always counts the number of left and right parentheses in each statement. When it encounters an unequal number of left and right parentheses in a given statement, it generates an error message and suppresses execution. You should always check to make sure that parentheses are present in pairs, and that they are paired correctly. For example, the following statement has equal numbers of left and right parentheses, but it makes no sense:

(A + B)) * ((C + D)

Some compilers are sufficiently sophisticated to catch such errors. Thus simply counting parentheses is not sufficient; their order is also important.

We have now considered FORTRAN constants, variables, and arithmetic statements in some detail. To complete the discussion of our sample program, we now have to consider the output statement in detail.

3–5 OUTPUT STATEMENTS

One function of the output statement is to indicate which of the available output devices is desired by the user. Throughout this text, we shall use only the printer to provide output from a system.

The general form of a WRITE statement is:

bbbbbbWRITE (Integer, Symbol) list of variables separated by commas

The identifying feature of this output statement is the word WRITE followed by a pair of parentheses which normally contain an integer, then a comma, followed by a symbol. The integer uses the local ground rules to denote the output device desired by the user. Let's consider this statement from the perspective of the computer system. Every system assigns an integer to each of the output devices available. Since several output devices are available, the user must indicate unambiguously which device is to be used. Use the space below to record the integer associated with the printer in your system.

PRINTER INTEGER

6

Once you have specified the output device, you must tell the system the precise *form* in which you want the numerical output. How many significant figures do you want? Should the answer be expressed in scientific notation (powers of ten)? Where on a given line should the answer appear? The symbol following the comma within the parentheses of the WRITE statement provides the computer with the answers to these questions.

If the precise form and arrangement of answers is not important, some systems have a special symbol for producing real output in the form of a number between 0 and 1 with six significant figures after the decimal point multiplied by a power of 10. This form was used in earlier examples. When this symbol is used, numbers stored in memory slots with real labels are printed in this form, while integers appear directly as integers with no change in form.

This symbol specifies the *free* FORMAT *option.* The word FORMAT denotes the form in which answers are desired. The use of this special symbol instructs the compiler to use only the forms discussed above for real and integer quantities. [*Note:* A real number is a number between 0 and 1 multiplied by a power of 10. An integer is just the number; that is, 1478, and so forth.] In the space below, record the symbol for free FORMAT output used on your own computer system.

FREE FORMAT SYMBOL

If this option is not available, more complicated methods must be used to specify forms of output. We explain these in the next section, but you should complete this section before going on to the next.

The last element in the general form of the WRITE statement is the list of variables appearing after the parentheses. To understand the function of this list, consider it from the viewpoint of the computer. We have told the computer which output device to use and what form to use for the answers, but we have not yet specified the numbers desired (in particular, their locations in memory). The variable list that follows the parentheses does this.

We now come to a major problem concerning local ground rules. There is a trend throughout the computer industry toward standardizing the WRITE statement, but it is not yet complete. For example, the integer 6 is *almost* universally used to denote the printer in a WRITE statement, and the symbol * is *widely* used to indicate free FORMAT in computer systems that offer this option. Because of this trend, we have decided to use these two symbols in the WRITE statements in the remainder of the text. But if local ground rules differ, you will have to modify each WRITE statement appropriately on the blank line beneath the statement.

The following program segment contains a typical free FORMAT WRITE statement.

```
A = 4.
B = 2.
PRØD = A * B
WRITE (6,*) PRØD
                    ← Modified WRITE, if needed
```

This elicits the following output:

```
0.800000E 01
```

As indicated, the value printed is the product of a number between 0 and 1, with six significant figures, multiplied by an appropriate power of 10. Again, since local ground rules may vary, you should check the form of answers under the local free FORMAT option (if any).

The following program segment and its associated output show the effects of using more than one WRITE statement and of including more than one variable in the list after the parentheses.

Program segment

```
A = 4.
B = 3.
C = 2.
X = A + B - C
Y = A - B + C
Z = -A + B + C
X1 = A * B * C
Y1 = A * B ** C
Z1 = A * B/C
WRITE (6,*) X,Y,Z
WRITE (6,*) X1, Y1, Z1
```

Output

```
0.500000E 01    0.300000E 01    0.100000E 01
0.240000E 02    0.360000E 02    0.600000E 01
```

Each WRITE statement produces a separate line of output on the paper.

To illustrate the different arrangements of output possible using free FORMAT, let us examine the following two program segments; they differ only in the WRITE statements used and their associated output.

Program segment (first version)

```
A = 4.
B = 3.
C = 2.
SUM = A + B + C
DIFF1 = A - B
```

```
        DIFF2 = A − C
        QUØT = A/C
        WRITE (6,*) A,B,C, SUM
        WRITE (6,*) DIFF1, DIFF2, QUØT
                        Output
0.400000E 01    0.300000E 01    0.200000E 01    0.900000E 01
0.100000E 01    0.200000E 01    0.200000E 01
```

Program segment (second version)

```
        A = 4.
        B = 3.
        C = 2.
        SUM = A + B + C
        DIFF1 = A − B
        DIFF2 = A − C
        QUØT = A/C
        WRITE (6,*) A,B,C
        WRITE (6,*) SUM, DIFF1
        WRITE (6,*) DIFF2, QUØT
                        Output
    0.400000E 01    0.300000E 01    0.200000E 01
    0.900000E 01    0.100000E 01
    0.200000E 01    0.200000E 01
```

Free FORMAT is convenient, but it has severe limitations. For one, it is inflexible; its use in preparing payroll checks, for example, could cause difficulty for both employees and their banks. People prefer to see amounts expressed in dollars and cents rather than in scientific (powers of ten) notation. Another difficulty is that free FORMAT cannot be used to label output with appropriate headings and to arrange results in convenient patterns. When you select free FORMAT, you largely surrender control of output arrangement to the computer. You may well wonder, then, why we ever introduced this form of output. The reason is that free FORMAT is the simplest method to use when you are first gaining experience with a computer.

The free FORMAT option produces six significant figures in each numerical answer. A reasonable question is: How many significant digits can I request from the computer? There are two parts to the answer. The first is that after mastering the rules of output specification, you may request as many digits as you want. But the second is that any digits beyond six will normally be physically meaningless. Some computers can provide more than six significant digits if you specifically request such precision (normally termed *double precision*). In this text and indeed for just about all undergraduate engineering courses, six-figure accuracy will suffice.

As a rule of thumb in using a computer, there is little or no value in requesting more than six digits of accuracy. The computer will supply more digits if requested, but these digits have no significance and are random numbers created by the computer to provide the requested digits. It is not unusual to find a beginning programmer demanding ten to twelve places of accuracy. In most computer systems, four to six of these digits have no physical or mathematical meaning. We repeat: The computer will normally produce six significant figures, and the free FORMAT option is designed to exploit this accuracy limit.

Exercises

3–30 Write the necessary output statement(s) so that the following program will produce the values of X1, X2, X3, and X4 on four different lines. Show the output produced.

```
A = 7.
B = 6.
X1 = A + B
X2 = A - B
X3 = A * B * (X1 - 10.)
X4 = X3/X2
```

3–31 Repeat exercise 3–30 so that the output occurs on only one line.

3–32 Write the necessary output statement(s) so that the following program will produce the values of L1, L2 on one line and L3, L4 on a second line. Show the output.

```
I = 4
J = 3
K = -11
L1 = I + J/K
L2 = L1 + I/J
L3 = L1 * L2
L4 = L1 ** (I - J)
```

3–6 FORMAL RULES FOR SPECIFYING OUTPUT

The computer user who is limited to a free FORMAT output technique will sometimes run into difficulties in arranging output in specific ways. Fortunately, the FORTRAN grammar provides a formal set of rules for specifying the precise form of the output, its arrangement on the printed page, and any column headings or other literal phrases desired.

GENERAL FØRMAT STATEMENTS

The following are general forms of the two types of FORTRAN statements required to do this.

bbbbbbWRITE (Output device integer, SFI) List of variables separated by commas
bbSFIbFØRMAT (output arrangement in symbolic form)

(The abbreviation SFI stands for specification form integer. We shall discuss its use in a moment.)

The first of these two FORTRAN statements is the familiar WRITE. The first integer within the parentheses specifies the output device, just as it did in the previous free FØRMAT statements that we have discussed. We shall continue to use the integer 6 to denote the printer. The second integer, however, replaces the free FORMAT symbol that we used before. This integer tells the system to look for another statement, a FORMAT statement, which specifies the exact form and arrangement of the output the user requires. This second integer is like a post-office box number. The system reads this second integer and then looks for this label on a suitable FORMAT statement. The list of variables in the WRITE statement indicates the values to be printed.

The second statement, the FORMAT statement, specifies the desired form and arrangement of the quantities to be printed. It also must have a label so that the computer system will be able to recognize it. This label may be any integer between 1 and 99999. Again, local ground rules vary, and in some small systems the maximum value for this label is 32767. This label, or statement number, must be punched in columns 1 through 5. When the statement number is less than five integers, it can be punched in any columns within the field defined by column 1 through column 5. The following are examples of valid statement numbers. The first line indicates the column numbers:

```
1  2  3  4  5  6    (Column numbers)
1  b  b  b  b  b
b  1  b  b  b  b
b  9  7  6  2  b
b  b  b  1  5  b
1  5  b  b  b  b
```

After the statement number, the specification card has a blank in column 6, and the word FØRMAT begins in column 7. The parentheses that follow the word FØRMAT contain the detailed rules specifying the form and arrangement desired for the output, that is, the numbers stored in the memory locations labeled by the variables in the WRITE statement. The general forms for the symbols appearing within the parentheses following the word FØRMAT are:

(Valid letter integer . integer)
(Integer valid letter integer . integer)

To understand what these general forms mean, let's look at some examples. The following program segment contains two WRITE statements, each with a different FØRMAT statement, and the output of these two WRITE statements.

Program segment

```
       X = 3.
bbbbbb WRITE (6,100) X
bb100b FØRMAT (F8.2)
bbbbbb WRITE (6,101) X
bb101b FØRMAT (E10.2)
```

Output

```
bbbb3.00
bb0.30E 01
```

Observe the different ways in which the number 3 appears on the two output lines. In the first WRITE, the number 100 is the specification form integer. This instructs the system to use the output specifications found in the FØRMAT statement labeled 100. These instructions were:

F8.2

The letter F indicates that the user wishes the output to appear as a real number (i.e., one with a decimal point) *without* a power of ten. The 8 signifies that eight columns on the line of output are to be used, and the 2 specifies that two numbers are to appear after the decimal point. This produces the following output:

```
1 2 3 4 5 6 7 8
b b b b 3 . 0 0
```

The system uses all eight columns allowed, even though it only needs four and has to fill the others with blanks. Chapter 6 will show that the first blank is not printed, but for now we can safely ignore this complication.

The second WRITE statement contains 101 as the specification integer, and the parentheses in the corresponding FØRMAT statement hold the following symbols:

E10.2

The 10 indicates that ten output columns are to be used, and the 2 specifies two numbers after the decimal point. These meanings are unchanged from the first example. The E specifies a number in exponential form (scientific notation), i.e., a number between 0 and 1 multiplied by a power of 10. The output produced is:

```
                  1
1 2 3 4 5 6 7 8 9 0
b b 0 . 3 0 E b 0 1
```

The E symbolizes 10, and the trailing b01 is the exponent. The blank following the E is reserved for a negative sign when necessary. Plus signs are not printed in exponents.

As with many FORTRAN topics, the only way to understand FØRMAT specification is to practice and work examples. We encourage you to experiment with simple programs using various FØRMAT specifications. This is the most rapid route to understanding this method of arranging output.

COMBINING OUTPUT SPECIFICATIONS

Other kinds of specifications are also available for use in FØRMAT statements; the following program segment illustrates some of these.

Program segment

```
bbbbbbJ = 2
      X = 3.
      WRITE (6,15) X, J
bbb15bFØRMAT (F12.2, I6, 'bAREbANSWERS')
```

Output

```
bbbbbbbb3.00bbbbb2bAREbANSWERS
```

This example shows that only one FØRMAT statement is needed to specify the output for two numerical values. In the example, F12.2 indicates how the number in memory location X is to be printed, and I6 does likewise for the value in slot J. Note that the two specifications are separated by a comma.

I6 specifies six columns and an *integer* form for the associated numerical output. As a result, the number 2 is printed without any decimal point.

This FØRMAT statement also contains the first *literal* output, that is, nonnumerical output. This literal phrase was obtained by using apostrophes. Whatever is keypunched within apostrophes inside the parentheses in a FØRMAT statement appears as printed output.

This example also shows that several output specifications can be combined within the parentheses of one FØRMAT statement. When this is done, a comma must separate each field or specification from the next.

Now we shall introduce yet another FØRMAT concept: Consider a simple variant of the preceding example, in which the literal output AREbTHEbANSWERS is to appear 21 spaces to the right of the numbers. Modifying FØRMAT statement 102 in an obvious way would, of course, accomplish this.

```
bb102bFØRMAT (F12.2,I6, 'bbbbbbbbbbbbbbbbbbbbbAREbTHEbANSWERS')
```

This is a lot of typing on the keypunch; fortunately, there is a simpler way:

```
bb102bFØRMAT(F12.2,I6,21X, 'AREbTHEbANSWERS')
```

The specification 21X tells the printer to skip 21 spaces on the line. In general, the phrase nX within the parentheses of a FØRMAT statement tells the printer to skip n spaces.

It is common for beginning programmers (and, indeed, for one of us, even after 12 years of programming experience!) to try to whittle specification fields as small as possible and simultaneously use the X phrase to space numerical values. An example of such poor programming practice is:

```
      WRITE (6,103) X,I
bb103bFØRMAT (13X, F8.7, 10X, I2)
```

Consider the absurdity of the phrase F8.7. It tells the computer to print seven numbers after the decimal point, but it allows the system only eight columns to print the entire number. It needs to print the decimal point itself (which requires one column) and the values to the left of the point (if any), and it must allow a space for a sign. Imagine that

X [33.] I [121]

indicates the state of memory when this WRITE statement is executed. The system tries to satisfy the FØRMAT specifications by filling columns 1 through 13 and 22 through 31 with blanks.

```
         1         2         3
123456789012345678901234567890123
bbbbbbbbbbbbb        bbbbbbbbbb
```

Next the system inserts the seven significant figures after the decimal point of the number in slot X, along with the decimal point itself.

```
         1         2         3
123456789012345678901234567890123
bbbbbbbbbbbbb.0000000bbbbbbbbbb
```

There is no room left for 33 or 121 at this point. All this "fitting" is carried out electronically; nothing has yet appeared at the printer. The computer concludes it cannot do what was asked, so it prints the following line:

```
bbbbbbbbbbbbb********bbbbbbbbbb**
```

The asterisks symbolize the system's inability to cope.

This problem results from careless use of the X specification. The following segment would work to produce the desired output.

```
       WRITE (6, 104) X,I
bb104b FØRMAT (F21.7,I12)
```

Exercise

3–33 Keypunch and run a simple program that defines values of X and I and uses FØRMAT statements 103 and 104 from the preceding examples to print their values. Compare the results produced by the two different WRITE statements.

As a general rule, do not use the X phrase to separate numerical fields (E, I, or F). The above example illustrates the unnecessary difficulties that can arise from this bad habit.

Other kinds of specifications also exist, but they need not concern you in your first experiences with FORTRAN. This section is rather sketchy, and you will need to supplement it by writing simple computer programs with varied forms of output specifications. But this practice is the only effective way to learn how to use these specifications. There is no shortcut, textbook method of communicating the intricacies of FØRMAT specifications.

3-7 INPUT; THE READ STATEMENT

The sample problems and programs segments that we have considered so far have been somewhat useless. You may be wondering if FØRTRAN and computers are good for solving problems more complex than adding, subtracting, multiplying, dividing, and raising numbers to powers. We assure you that the usefulness of FORTRAN will appear as we learn more about the language. In the remainder of this chapter, we shall introduce higher-level statements that will extend the power of the language.

First we shall introduce some input techniques—techniques for feeding numerical values or variables into a computer and processing them. Consider the simple program that we used before.

```
CARD FOR ACCOUNTING PURPOSES
CARD TO SELECT THE COMPILER
bbbbbbA = 4.
      B = 3.
      SUM = A + B
      WRITE (6,*) SUM

      STØP
      END
CARD TO INDICATE EXECUTION IF NO ERRORS
CARD TO INDICATE FINISH
```

We have left the blank line in the program listing so that you can enter the WRITE statement appropriate for your own system if it is different from the WRITE expression in the sample. This program takes two numbers, adds them, and prints their sum. Consider a problem in which several hundred pairs of numbers have to be added and several hundred sums printed. How can this be done?

One way to do it would be to prepare several hundred programs like this one. The only difference between any program and any other would be the numerical values assigned to A and B. Besides being a monumental waste of your time, this procedure would waste computer time and punched cards. If the computer is to be at all useful to the practicing engineer, there must be a simpler way. After all, engineers often have to use formulas repetitively to calculate one physical quantity as a function of some other variable. For example, a civil engineer would use a formula to calculate the load-carrying ability of a given girder in terms of its cross-sectional area and material composition. To analyze a given design, the engineer might have to evaluate this formula for many different cross-sectional areas. Hence, he or she needs to carry out similar calculations over and over again for different values of one or more variables. This design problem is roughly equivalent to the simplified case we are examining.

What additional FORTRAN statements would make it possible for us to solve this problem simply? We would need statements to read numerical values for variables from punched cards and use the same program, again and again, to calculate and print the sum. All computer systems can read numbers and assign them to variables using a FORTRAN statement that is an input analog of the output WRITE. The general form of this input statement is:

READ(   ) list of variables to be read, separated by commas

The similarity of the READ and WRITE statements is clear. READ statements can use both free and specified FØRMAT, but there is rarely any advantage in reading numerical values of variables with specified FØRMAT statements. For this reason, we shall discuss only the free FØRMAT READS. This statement has the form:

READ(5,*) variable list

The integer 5 denotes the punched-card reader and the * instructs the computer to read numbers, separated from each other by at least one blank space, from left to right. The variable list in READ statements serves the same function as in WRITE statements.

Our sample READ statement is widely used, but some systems may require other forms. Use the space below to record the exact form of the free FØRMAT READ that you will use.

READ STATEMENT

To illustrate the value of the READ statement, examine the following modification of our sample program.

```
CARD FOR ACCOUNTING PURPOSES
CARD TO SELECT THE COMPILER
bbbbbbREAD (5,*) A,B

      SUM = A + B
      WRITE(6,*) SUM

      STØP
      END
CARD TO INDICATE EXECUTION IF NO ERRORS
CARD TO INDICATE FINISH
```

We shall now go through each card to describe the translation and execution of this program.

The accounting card identifies the user, and the compiler card selects the language in which the program was written. The READ statement causes the following actions during translation. The computer "observes" that two variables, A and B, will be read during execution and reserves two memory slots for these quantities; it labels one A and the other B. The next card creates a slot labeled SUM. The output card instructs the computer (using the proper sequence of binary numbers) to write the value in SUM during execution. The STØP card indicates that the calculations are complete, and the END statement denotes the end of the FORTRAN statements.

If the compiler finds no errors (in this case, there are none), the card used to initiate execution is pulled into the card reader and begins the sequence of calculations. When execution begins, the computer goes to the beginning of the electronic "image" of the set of FORTRAN statements stored in its memory. The first instruction is a command to READ numerical values for A and B. The system satisfies this instruction by ordering the card reader to pull another card in and assign the first number on the card to memory slot A and the second number to B. These numbers must be punched on a card, but where should the cards be placed in the original deck? The card or cards containing the numerical values to be read must be placed immediately after the EXECUTE card and before the FINISH card. Two possible sets of cards to supply the necessary data are:

EXECUTE CARD		EXECUTE CARD
4.b3.6	or	4.
FINISH CARD		3.6
		FINISH CARD

Either of these will work. The READ statement initiates a command that tells the computer to use the card reader to look for two numbers. They may occur on the same card or on separate cards, but the system must be able to find two numbers at the card reader. If it finds only one value or only a FINISH card, it will generate an error message and suppress execution of the program.

After the READ statement is executed, the computer's memory "looks like" this:

A [4.] B [3.6] SUM [empty]

Since A appeared first in the variable list of the READ statement, the leftmost number (the first encountered by the system at the card reader) was assigned to that memory location.

In this example, the two values on one card must be separated by at least one blank space. If they are not, the computer will not be able to tell which number was intended for which variable. For example, suppose that the card carrying the numbers was punched as

4.3.6

Is 4.3 stored in A and .6 in B, or is 4. in A and 3.6 in B? There is no way to tell. The computer cannot unambiguously assign values to the two memory slots in this case. If the computer comes across such a mistake, it will generate an error message and suppress execution.

We could have typed more than two numbers on the card, but since we instructed the system to look for just two numbers, it will ignore any additional values when it reads the card containing them. Further, these extra values will be lost for any later use, since the card is pulled through the reader when the READ command is executed.

Exercises

3–34 Calculate by hand the numerical values of A and B produced by the following program. Then keypunch and run the program to check your hand-generated answers.

```
ACCOUNTING CARD
COMPILER CARD
REAL INT
READ(5,*) X,Y,Z

INT = (X - Y)/Z
A = INT ** X + Y
B = A ** 2 - Z
WRITE(6,*) A,B

STØP
END
EXECUTION CARD
3.0b - 1.0
2.0
FINISH CARD
```

3–35 Calculate by hand the numerical values of R1 and S1 produced by the following program. Keypunch and run the program to check your hand-generated answers.

```
ACCOUNTING CARD
COMPILER CARD
READ (5,*) X

READ (5,*) Y

READ (5,*) Z

R1 = X * Y * Z
S1 = X - Y ** Z
```

```
WRITE (6,*)R1,S1
STØP
END
EXECUTION CARD
1.0b2.0b3.0
1.0b127.3
2.0
FINISH
```

We began by recognizing that FORTRAN programs would be more useful if numerical values of variables could be read into the system while a program was being executed. To enter numerical data, we place cards containing numbers between the execution and finish cards. These cards are known as *data cards*.

Figure 3–13 is a diagram of the program considered at the beginning of this section. As this sketch indicates, the program executes the desired steps essentially in only one direction. The computer begins execution with the READ statement and proceeds through the electronic "image" of the FORTRAN statements until it reaches the STØP. Thus the computer reads only one pair of values for A and B, calculates only one value for the FORTRAN variable SUM, and prints only one line of output.

In our original problem, we wanted to add hundreds of pairs of numbers. The READ statement enables us to enter values from "outside" the system while execution is proceeding, but we have not yet described a way to do the desired calculations repetitively. In the next section, we shall describe techniques to break the straight-line pattern of Figure 3–13 and find a simple, compact solution to the problem of adding hundreds of pairs of numbers.

FIGURE 3–13 A unidirectional flow chart. The processing flows from top to bottom only.

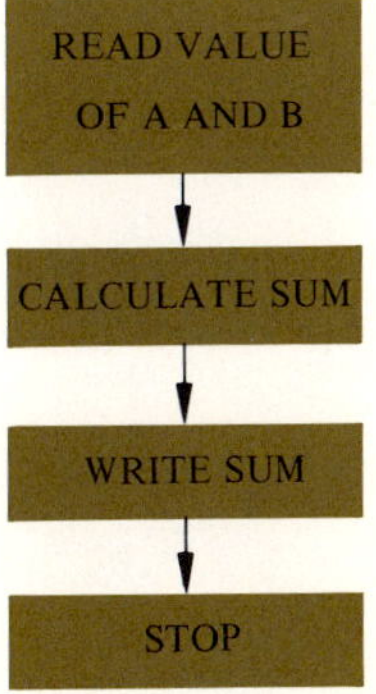

3–8 THE UNCONDITIONAL GØ TØ

To see how we can efficiently add hundreds of pairs of numbers, let's begin by considering a slight modification of the program flow of Figure 3–13 shown in Figure 3–14. We have modified the one-way flow from top to bottom by adding another line from the WRITE box back to the READ box. For the moment, we have omitted the STØP. We have done this for clarity, and we will replace it later. Adding the new line forms a *loop*, a complete cycle through a set of instructions. When a program segment includes a loop, the computer encounters the instructions in the loop over and over again. This happens *during execution*. Note that no looping, or cycling, occurs during the compilation stage.

A loop is most simply established by means of a FØRTRAN statement known as a GØ TØ. The general form of this statement is:

```
GØbTØbinteger
```

The integer referred to in this general form is a statement number as discussed in Section 3–6. Statement numbers may take values between 1 and 99999 (on some systems 32767). The statement to which the GØ TØ is directed must be labeled with the integer referred to in the GØ TØ statement. This is done by keypunching the appropriate integer in columns 1 through 5 of the card containing the FORTRAN statement to which the flow should branch during execution. The following set of instructions will realize the program sketched in Figure 3–14.

```
bbbb5bREAD (5,*) A,B
      SUM = A + B
      WRITE (6,*) SUM
      GØbTØb5
```

During the execution sequence, the READ statement will read two numerical values from cards and assign them to A and B, respectively. The next statement will calculate and store the sum of these numbers, and the WRITE statement will print the current value of SUM. Then the GØ TØ statement will complete the loop. The computer will return to the instruction labeled with the integer 5, in this case, the

FIGURE 3–14 A flow chart creating an endless loop. The GØ TØ statement causes a branch from bottom to top.

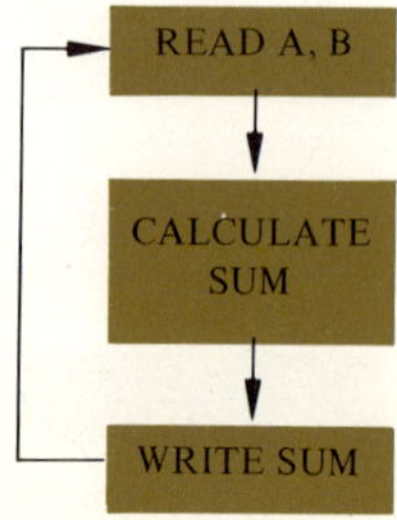

READ statement. If additional cards containing numbers are still in the card reader, the computer will read another pair of values, calculate their sum, and print it. The computer will then loop back again.

The next section will introduce FORTRAN techniques to terminate such loops after they have been executed the desired number of times. Most computers terminate an endless loop when no more data cards remain at the card reader. Some systems also limit the total time that any one program can take to be translated and executed.

We have, in this section, added two important new tools to our inventory. The first is that a statement number may be used as an integer label for any statement to which there is a need to branch. In our example, the computer system could not return to the READ statement unless it had a statement number. Note: There is no need to label statements unless other statements refer to them; in fact, it is a waste of time.

The second new concept is the FORTRAN GØ TØ statement. This instruction is often referred to as an unconditional GØ TØ. When the computer system encounters this instruction during execution, it immediately branches to the statement number given.

We have also mentioned the need to terminate such loops after they have been executed the required number of times. However, we have solved the crux of the problem of adding hundreds of pairs of numbers. The loop set up by the GØ TØ statement obviates any need to keypunch the basic program hundreds of times, although hundreds of data cards are still needed. We have intentionally restricted this discussion to a simple calculation, but the principles of a loop can be usefully applied to more realistic problems. In the next section, we shall develop a FORTRAN technique for terminating loops.

3–9 THE LOGICAL IF

Figure 3–15 shows yet a second modification of the basic program that appears in Figures 3–13 and 3–14. Consider the differences between these three figures carefully. This third and final version of the program contains both a loop and a STØP statement. It also contains a totally new element represented by the box labeled TEST. We shall explain how FORTRAN realizes such a test statement momentarily, but for now, consider how such a statement affects the overall flow of the program's execution.

This box has two possible paths leading away from it. The first is labeled No and proceeds to STØP, while the second is labeled Yes and completes the programming loop back to the READ statement. We can explain the effect of such a test in words, although the "translation" will be imprecise. Roughly speaking, it asks: "Is a specified condition satisfied?" If the answer to this question is no, the execution proceeds to STØP. If the answer is yes, the loop is completed, and the computer returns to the READ statement.

FIGURE 3–15 A flow chart using a test to terminate a loop. The test statements provide two options: a branch from bottom to top or a continuation down.

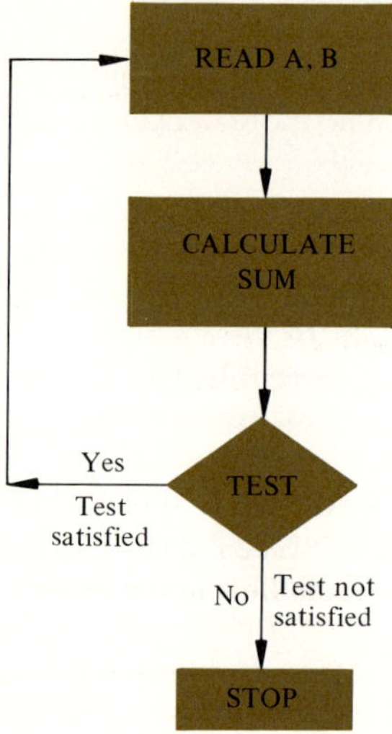

Several FORTRAN statements can carry out such tests. One is the logical IF statement. To see how it works, consider yet another variation of our sample program.

```
  ACCOUNTING CARD
  COMPILER CARD
         J = 0
  bbbb5b READ (5,*) A,B
         J = J + 1
         SUM = A + B
         WRITE (6,*) SUM
         IF (J.LT.10) GØ TØ 5
         STØP
         END
EXECUTION CARD
DATA CARDS
FINISH CARD
```

This program introduces three new concepts; each is necessary to satisfactorily terminate a loop after a finite number of repetitions. These three concepts appear in three lines:

```
J = 0
J = J + 1
IF (J.LT.10) GØ TØ 5
```

The first two are simple arithmetic statements in the integer mode. While these are not new, their purposes in this program are. But before we find out what they are, we shall discuss the third statement. This statement is a logical IF; it serves the purpose of the test box in Figure 3–15. In words, this FORTRAN statement says: "If the integer value stored in the memory labeled J is less than 10, go to statement number 5; if J is greater than or equal to 10, continue." The two branches leaving the test box in Figure 3–15 represent the two possible outcomes produced by the logical IF. When the condition within the parentheses of this statement is satisfied, the program branches to statement number 5. When this condition is not satisfied, the program simply continues to the next statement, a STØP.

To understand this program, we must now examine the behavior of the integer variable J. Imagine that the compilation has been completed and that the EXECUTE card has been read. The next cards in the reader will be the data cards. Assume that each data card contains one pair of numbers. Now execution begins, and the integer constant 0 is stored in the memory slot labeled J. The READ statement causes the first pair of values to be read from the first data card (which is pulled from the stack of cards remaining in the reader) and stored in the slots labeled A and B, respectively. The next statement causes the value stored in J to increase by 1. At this point, then, the integer 1 is stored in that memory location. Next the sum of the numerical values in A and B is calculated, and the result is printed.

Now the logical IF statement appears. The value stored in J (1) is compared to the integer quantity 10. Since the test is satisfied (1 is indeed less than 10), execution branches to the READ statement, thereby completing the loop within the program for the first time. It is not accidental that the value of the variable J is 1 after the *first* trip around the loop has been completed.

Next another pair of numbers is read and stored in the slots labeled A and B. The value of J is again increased by 1, so it becomes 2. Another sum is calculated and written, and the test is carried out for the second time. Again, 2 is less than 10, and control is transferred through the loop, back to the READ statement. The value stored in J is 2, and the loop has been repeated twice.

We could rewrite this last paragraph seven more times to describe the next seven cycles through the loop. The sentences would be exactly the same, but the number stored in J would increase by 1 each time. Instead of doing this, we shall assume that the ninth pass through the loop has been completed. The READ statement will be executed for the tenth time as the tenth pair of numbers is read and stored in A and B. Then J will be increased by 1 to the value 10. A tenth sum will be calculated, and a tenth answer will be printed. Now, for the first time, the logical test specified within the parentheses of the IF statement will not be satisfied, because 10 is definitely not less than 10. Control will not branch back to the READ statement; instead it will pass to the statement below the logical IF, the STØP. This marks the end of execution. If the user was careful to ensure that only ten data cards were between the execution and finish cards, the card remaining in the reader indicates finish, the end of the user's job. A normal termination, without error, will result, and the program will have successfully added ten pairs of numbers and printed the ten answers desired. Generalizing this program to the case of hundreds of pairs of numbers is not difficult.

Three statements were required to establish this loop and terminate the number of cycles at the desired point. The quantity stored in the memory location labeled J serves as a counter. The sole purpose of this FORTRAN variable is to keep track of the number of times the program, during execution, goes around the loop. For this reason, the initial value stored in J is 0 and the statement, J = J + 1, is embedded within the loop to ensure that J is increased by 1 each time the loop is traversed.

Now let us examine the general form of logical IF statements.

IF(expression .operator. another expression) statement

In this form, the two periods set off an allowed logical operator; the following list gives valid forms of these operators.

.Operators.	*Meaning*
.LT.	less than
.GT.	greater than
.LE.	less than or equal
.GE.	greater than or equal
.EQ.	equal
.NE.	not equal

The operator expressions serve as mnemonics for their mathematical content; you do not need to memorize them, since using them a few times will firmly implant them in your inventory of FORTRAN grammar rules. The statement after the parentheses in the logical IF does not have to be a GØ TØ statement. It can be any executable FORTRAN statement except a DØ statement (to be introduced in Chapter 6) or another logical IF. We have not discussed executable and nonexecutable statements; in this text, only FØRMAT statements will be categorized as nonexecutable. With these three exceptions, any other FORTRAN statement may be written after the parentheses of a logical IF.

Exercise

3–36 Write and execute a program that will produce three columns of output, J, J^2, J^3 for J = 1,2,3, . . . ,10. Use literal output to label the columns INTEGER, SQUARE, CUBE, respectively.

3–10 MISCELLANEOUS

In some FORTRAN statements in the preceding sections, blanks were carefully indicated by the use of a b. Fortunately, blanks are critical in only a few situations in FORTRAN. As a general rule, the compiler ignores blanks when it is translating FORTRAN statements into the strings of binary bits that the computer will use. There is little point in spelling out the exceptions, since experience will show you when blanks are critical.

The sixth column of a punched card should be left blank nearly all the time. If you review the FORTRAN programs in this chapter, you will see that the sixth column never contained a symbol. The only exception to this rule that will ever concern you occurs when a single FORTRAN statement is too long to fit on one card. You may continue such a statement onto a second card by typing any symbol in the sixth column of the second card. Thus a symbol in this column says that a statement continues. On some computer systems, not every symbol may be used in column 6 for this purpose. Again check the applicable local ground rules.

When the letter C appears in the first column, the compiler ignores any other symbols on the card. This feature may at first seem useless, but you may want to use such a card to include comments for later reference. For this reason, cards with a letter C punched in the first column are called *comment cards*. We shall use such cards later in this text. It is convenient to use a comment card with the user's name on it as the first card in each deck of FORTRAN statements. Note that this card must come after the accounting and compiler selection cards.

We must also mention error messages. Every compiler that translates a FORTRAN deck into binary has a set of rules against which it checks each of the statements. If one of these rules is violated (for example, by using illegal FORTRAN variables), the computer stops execution of the program and prints an error message. These error messages differ from system to system, and we shall not attempt to cover them in this book. You should be aware—and will undoubtedly be made aware by the unforgiving compiler—of the existence of such error messages. Beginners spend much of their initial computer operating time trying to understand the meaning of such error messages. Just about anyone who has ever tried to learn computer programming has done so.

This winds up our initial discussion of FORTRAN. In the next two chapters, we shall apply these concepts to engineering problems. In Chapter 6, we shall return and expand the range of FORTRAN tools.

PROBLEMS

P3–1 Write a FORTRAN statement to cause a computer to store the numerical value of your height (expressed in meters, to three significant figures) in a memory location labeled HGT.

P3–2 Prepare a 15-digit number by writing on one horizontal line your area code, local telephone number, and zip code. Write a set of FORTRAN statements to cause a computer to store the first three digits of this number in a memory location labeled I, the second three digits in J, the third three digits in K, the next three digits in L, and the last three digits in M.

P3–3 Assume that a decimal point appears after the last digit of the number found in P3–2. Write this number in the form of six significant figures, each with a magnitude between 0 and 1, times an appropriate power of 10. Use the letter E to symbolize the number 10.

P3–4 Numbers are stored in memory locations I, J, and K. Write a FORTRAN statement that will calculate the sum of these three numbers and store the result in the memory location M.

P3–5 Four numbers have been stored in memory locations A, B, C, and D. Write a FORTRAN statement that will calculate the ratio of the sum of the first two to the sum of the second two and store the result in a memory location labeled Z.

P3–6 After the following set of FORTRAN statements are processed (i.e., compiled and executed), what number will be stored in the location ANS?

```
A = 12.
B = 3.
C = 4.
D = 2.
Q = A + B
R = C - D
ANS = Q/R
```

P3–7 After the following set of statements are compiled and executed, what number will be stored in the memory location labeled MULT?

```
I = 3
J = 10
K = -4
L = 2
M = I + J
N = K + L
MULT = M * N
```

P3–8 What number will be stored in HYP after the following four statements are translated and executed?

```
A = 3.
B = 4.
SUM = A * A + B * B
HYP = (SUM) ** .5
```

What geometric figure does this program segment call to mind?

P3–9 Write a two-statement FORTRAN program segment that will store your zip code in a memory location labeled ZIP and store the square root of your zip code in A.

P3–10 What number will be stored in UAVE after the following five statements have been translated and executed?

```
U1 = 10.
U2 = 20.
U3 = 30.
```

```
U4 = 40.
UAVE = U1 + U2 + U3 + U4/4.
```

P3–11 Keypunch and run the program segment given in P3–10, using the accounting, compiler selection, execution, and finish cards appropriate for your computer system. What output does this produce?

P3–12 The answer to the last question in P3–11 should have been "No output at all." Now write a suitable output statement; keypunch and run the program of P3–10 with the output statement added. Your program should print only the value stored in UAVE.

P3–13 Modify the program of P3–12 so that it will cause the numbers stored in the memory slots U4, U3, U2, U1, UAVE to be printed out on one line in the order stated. Keypunch and run this modified program.

P3–14 Modify the program of P3–13 so that the values of U4 and U3 appear on the first line of output, the values of U2 and U1 appear on the second line, and UAVE on a third line. Keypunch and run this modified program.

P3–15 An engineer works for a manufacturer of pizza plates. As part of a quality control effort, the engineer must take a representative sample of plates, measure their thicknesses, and calculate the average thickness. During one week, the engineer selects 10 plates with the following thicknesses (in inches).

.049	.045	.048	.049	.048
.039	.038	.039	.035	.035

Write, keypunch, and execute a program that will store the measurements in memory locations labeled T1, T2, . . . , T10, calculate the average of these thicknesses, and print it.

P3–16 Identify each of the following as either a valid or an invalid FORTRAN integer constant. Give a reason for each invalid quantity.

−18	18.	1265E − 02	−32800
18	18.01	32676E02	15 * 4
+18	15,326	32676	16/4
0	0.0	0.E01	−1764
2 + I	1976	−200	1776

P3–17 Identify each of the following as either a valid or an invalid FORTRAN real constant. Give a reason for each invalid quantity.

−18	5.3E02	−18.	$2.7(10)^4$
2. + A	2. + 18./3.	0	2753.E + 12
1,121.2	0.	682E − 4	682.E − 127
10.5	74.	682.E − 04	1.E(15./3.)
−1.53 − 3	π	126. * 4.	1.5E17.

P3–18 Identify each of the following as either a valid or an invalid FORTRAN constant. Give a reason for each invalid quantity, and identify the mode of each valid quantity.

\$2.	0	42.E12.	1,521,806.E − 06
−15.2	15E01	$2.^{2.}$	146782.
7,357	15. + 11.	−1326	3.1416
162.261	38.E − 02	401.	2.718
1200	38.E − 2	8. * E02	32600

P3–19 Express each of the following numbers in scientific notation—the product of a number with magnitude between 0 and 1 times an appropriate power of 10. Use E to represent 10 in your expressions.

−3.2188	89.938	−5526.43
72.742	.0000893	30969.4
0.26993	0.26993	.00107

P3–20 The left column is a set of algebraic equations, attempts at equivalent FORTRAN statements are shown on the right. Indicate which FORTRAN statements contain errors, and write corrected versions.

Algebraic equations	FORTRAN *statements*
$x = 15.2$	X = 15.2
$y = 108721.$	Y = 1.08721E 04
$z = -674.2$	Z = −6.742E 02
$a = .0000002$	A = .2E − 04
$b = 9.12(10^4)$	B = .912E 05
$c = a + 9123.$	C = A + 9123
$d = 2b(a - c)$	D = 2. * B * A − C

P3–21 Are the following valid FORTRAN integer variables? Give a reason for each invalid quantity.

I123	N1J2L	1I4	LAMPOON
IJKLM5	VALID	IMP	CAMP
EGØ	MAYBE	5.87	Ø(1.)
\$M	NØ	I − J	MØØ
M,3	YES	IDØ	END

P3–22 Are the following valid FORTRAN real variables? Give a reason for each invalid quantity.

A11	A11 + 2	ISN'T	A = B
B(12	A11P2	END.	C * D
TWØ	TAX	ABC	A) +
ELEVEN	PAIN	ANS	* ANS
YOU2	I1A	PRØD	RSLT

P3–23 Each of the following is a valid FORTRAN quantity. Identify each as either a constant or a variable, and identify the mode of each.

1562	ØRIF	LEVEL	0
UV21	0.271	STACK	0.
−7.E02	ABC	−.1E − 3	Ø
IA3	L12E	T1	−775

P3–24 What output will the execution of each of the following program segments produce?

```
VEL0 = 10.
ACC = 15.
TIME = 5.53
VEL = VEL0 + ACC * TIME
WRITE (6,*) VEL
```

```
REAL MASS
MASS = 127.
ACC = 15.
FØRCE = MASS * ACC
WRITE (6,*) FØRCE
```

P3–25 What values of C will be printed when the following two program segments are executed?

```
A = 4.
B1 = A ** 2. − 2.
C = B1 + 2. * A
WRITE (6,*) C
```

```
U = .125E02
V = U ** 2. − 100.
C = (V) ** .5
WRITE (6,*) C
```

P3–26 Which of the following arithmetic statements contain mixed-mode errors? Give reasons for your choices.

```
Y = A + B − 3. * C/3.25E − 02
Z = A * B/C − 5 * Q
Z0 = A ** (C − D/6) + 14.
Y1 = 3. * S − T ** 4.
I = J/K + 7. * L
J = (M1 + M2 + M3)/3.
```

P3–27 Which of the following arithmetic statements are correct as written? Identify the error or errors in the incorrect expressions.

```
      Y = A + B * /C
     1Y =A − B/C
 Y + 8. =Z + 1S.
      Q =(S + T)U − V + 18.
      R =T1/S4 * 132 − U
```

P3–28 What value of Z will be printed when each of the following program segments is executed?

```
I = 4
J = 10
K = J * (I − 2)
L = K − 98
X = −L
Z = X ** .5
WRITE (6,*) Z
```

```
I = 2
J = (5 + I)/(5 − I)
K = 8 ** J
L = K − 52
Z = 2 * L
WRITE (6,*) Z
```

P3–29 Use appropriate lower-case symbols to write the algebraic equivalents of each of the following FORTRAN arithmetic statements.

```
Q = R + S/T * U - 3. * (A + B)
Q1 = A - B/(C + C * A)
Q2 = C ** (D - E/F * (S - T))
Q3 = A - (B - (C - (D * A1) + B2) - C2)
M = I/(J - 2) + K * (L + 8)
N = I ** ((N + 14)/(M - 12))
```

P3–30 Use suitable upper-case symbols to write FORTRAN expressions equivalent to the following algebraic equations.

$$x = \frac{3a}{4b + \frac{c}{d} - 3} \qquad i = j^k - m^{n+5}$$

$$y = \frac{(a + b)(7c + d)}{(e - f)} \qquad m = (6 + n^{2i})(k + m^{3n})^{5-i}$$

$$z = \frac{a^{(c+d/2)}}{st - \frac{uv}{2}} \qquad a = \frac{(7b + c)(e - 3f)}{q*(s - 5t)}$$

$$\text{Area} = \pi r^2 \qquad \text{Perim} = 2\pi r$$

$$\text{Vol} = \frac{4}{3}\pi r^3 \qquad \text{Area} = 4\pi r^2$$

P3–31 Part of a computer's memory is in the following state.

A	14.	D	9.8
B	3.	E	-5.
C	7.2	F	-20.

What value will be stored in the slot labeled X after each of the following arithmetic statements is executed?

(a) X = A - B ** 3./2.
(b) X = A - B ** (3./2.) + C/D - E * (F + 10.)
(c) X = (C + 7. * D) ** (3. - B) - 1.
(d) X = A + B/C - D + (7. * E)/F

P3–32 Each of the following arithmetic statements contains one pair of parentheses that may be removed without changing the meaning of the expression. Rewrite each without the extra parentheses.

```
Y = (A + B) * (C + D) + E - (F * G)
Y = Q ** Y * (A)/(A - B)
Y = (((A + B)/(C + D)) ** E) - F
```

P3–33 Write the necessary output statement(s) so that the following program will produce the values Q1, Q2, and Q3 on three different lines. Show the output produced.

```
A = 15.
C = 11.
X = A - C
Y = A ** (C - 9.)
Z = Y ** .55
Q1 = Y * (Z - 15.)
Q2 = X * (Z - 15.)
Q3 = Q1 - Q2
```

P3–34 Repeat P3–33 so that the three ouput values occur on only one line.

P3–35 Repeat P3–33 so that the values of Q1 and Q2 appear on the first printed output line and the value of Q3 on the second line.

P3–36 The state of a part of computer's memory is

X [762.964]

I [203]

Write a FØRMAT statement that will produce each of the output lines shown below when the following statement is executed.

WRITE (6, 100) X,I

(a)	7	6	2	·	9	6			2	0	3							
(b)				0	·	7	6	2	E		0	3		2	0	3		
(c)			7	6	2	.		I	=	2	0	3						
(d)			*	*	*	*	*		2	0	3							
(e)						7	6	2	.	9	6	4				2	0	3

↑ (First column on sheet)

P3–37 For the numbers stored in memory locations X and Y, indicate the output produced by each pair of statements. In your answers, show blanks explicitly as b's.

X [3.335] Y [6.167]

(a)
```
      WRITE (6, 100) X,Y
100bFØRMAT (1X, 2F8.2)
```
(b)
```
      WRITE (6, 101) Y,X
101bFØRMAT (1X, 2F8.2)
```
(c)
```
      WRITE (6, 102) X,Y
102bFØRMAT (1X, F12.2, E10.3)
```

(d)
```
      WRITE (6, 103) X, Y
103bFØRMAT (1X, 2E10.2, 'bbARE ANSWERS')
```

P3–38 The following indicates the state of memory desired by a computer user.

A	B	C
5.5	4.4	3.3

For each of the following READ statements, show the data card or cards necessary to produce the desired storage.

(a) READ (5,*) A,B,C
(b) READ (5,*) C,A,B
(c) READ (5,*) A,B
    READ (5,*) C
(d) READ (5,*) A
    READ (5,*) B
    READ (5,*) C

P3–39 Using an electronic calculator, determine the numerical values of C and D produced by the following program. Keypunch and run the program to check your hand-generated answers.

```
READ (5,*) X,Y

READ (5,*) U,V

A = X − Y
B = U − V
C = A * B
D = A/(B + 2.)
WRITE (6,*) C,D

STØP
END
EXECUTION CARD
3.3b7.9b9.2b17.41
10.7b4.16b8.3
FINISH CARD
```

P3–40 Determine the value or values of X printed out by the execution of each of the following program segments.

```
   I = 0
2bI = I + 2
   Y = I
   X = Y/2.
   IF (I .LT. 6) GØ TØ 2
   WRITE (6,*) X
   STØP
```

```
   I = 1
   K = 4 − I
   J = 2
1bX = I + J
   I = I + 1
   J = J + 1
   WRITE (6,*) X
   IF (J .LT. K) GØ TØ 1
   STØP
```

P3–41 Indicate whether the following statements are valid. For each incorrect statement, give a reason.

```
IF (X = 10.) GØ TØ 4
GØ TØ STØP
IF (I .LT. 4.) GØ TØ 7
IF (A .GT. B) STØP
GØ TØ 1
```

P3–42 For the memory storage shown below, calculate the value stored in X after each of the following four statements is executed.

A	B	N	I	Y
3.	0.5	2	6	4.2

```
(a) X = (A ** N - 4.)/5. - B
(b) X = N/6 + 1
(c) X = A - B * 4. ** N + 7.
(d) X = Y * (6. - B ** 3.)/A + 7.9
```

P3–43 Give the value that will be stored in Z or L (whichever is pertinent) after each of the following program segments has been executed.

```
     X = 3.
     C = 0.5E01
     Y = X ** 2/2. - C
     IF (Y .LT. 0.) GØ TØ 2
     Z = Y/(C * Y + 2.)
     GØ TØ 2
 2bZ = 0.
 1bWRITE (6,*) Z
```

```
     A = 2.0
     B = 30.E-01
     C = 4.0
     D = B * B - 4. * A * C
     IF (D .LT. 0.) GØ TØ 30
     A = 2. * B
30bA = B.
     Z = B - C/A
```

```
     N = 3
     K = 0
     L = 1
 1bK = K + 1
     L = L * K
     IF (N .LT. K) GØ TØ 1
     WRITE (6,*) L
```

```
     M = -5
     N = 3
     L = 5
 1bL = L - 1
     IF (M .LT. 0) GØ TØ 11
     IF (M .EQ. 0) GØ TØ 12
     N = N + 1
12bN = N + 2
11bM = N - L
     IF (M .LE. 3) GØ TØ 1
     STØP
```

P3–44 Write a program that will accept any number of data cards, each containing the base and hypotenuse of a right triangle, and will compute the height of the triangles. The program should print out the base, hypotenuse, and height of each triangle, and it should stop after reading the last data card. Show all control cards and run the complete program for the following data. Draw a flow chart.

Base	*Hypotenuse*
76.2	114.37
3.0	5.0
4.9	5.3
82.1	253.6

P3–45 Because of increasing national concern over air pollution, engineers and scientists have been investigating new methods of measuring pollutant content in the atmosphere. One such method uses thin films of semiconductors as simple air-pollution monitors. The electrical resistance of a film R (ohms) is a function of the temperature T (°C) and the pollutant concentration C (parts per million, ppm). The relationship is given by the symbolic model

$$R = A(T + 273)(2.72)^{aC}$$

where A and a are constants with the following numerical values.

$$A = 2 \text{ ohms/°C}, \qquad a = 1$$

If you placed such a thin film in a region in which you wanted to know the pollutant concentration (C) and then measured the film resistance (R) and temperature (T), you could use the equation relating R, T, and C to determine the level of pollution. Such devices would make possible simple and inexpensive measurements of pollutants because instruments already exist to measure R (ohmmeters) and T (thermometers and thermocouples).

(a) For a temperature of 20°C, write and execute a FØRTRAN program that will print out corresponding values of R and C for the following values of C.

$$C = 0.01, 0.02, 0.04, 0.1, 0.2, 0.4, 1.0, 2.0, 4.0 \text{ ppm}$$

(b) Repeat part (a) for a temperature of 125°C.

(c) Repeat part (a) for a temperature −20°C.

(d) Plot your results and show R as a function of C for each of the three different temperatures, using only one sheet of graph paper.

P3–46 The Poisson probability distribution is defined by the following model:

$$p(n) = \frac{\lambda^n (2.71828)^{-\lambda}}{n!}, \qquad n = 0, 1, 2, \ldots$$

where n denotes the number of occurrences of a phenomenon in a unit of time, λ is the average number of occurrences in a unit of time, and $p(n)$ is the probability of n occurrences in a unit of time.

This distribution is very useful in engineering design, especially for problems involving automated and semiautomated production lines. For example, in an automobile assembly plant, the rate at which car bodies arrive at the engine-mounting station is Poisson-distributed with an average rate of λ bodies/hour. The number of bodies arriving at the station in any given hour is the random variable n. The probability of n arrivals in any given hour is $p(n)$.

Develop formulas and write and execute a computer program that will read any number of different values of the average arrival rate (λ) and then compute and print out $p(n)$ for $n = 0, 1, 2, \ldots, 20$ for each value of λ. Use the following three values of λ:

$$2.4, 5.8, 7.6$$

The average number of bodies on the assembly line is 5.8 per hour. What percentage of the time will 10 bodies arrive at the engine-mounting station in one hour?

P3–47 The surface area and volume of a circular cylinder are given by the following formulas:

$$\text{Area} = 2\pi RH + 2\pi R^2$$
$$\text{Volume} = \pi R^2 H$$

where R is the radius and H is the height. Write and execute a FORTRAN program that will calculate the area and volume of a circular cylinder for the following four cases.

R (meters)	1	2	3.5	4
H (meters)	1	1	2	3

Your program should read the values of R and H from data cards, and it should arrange the output so that R, H, area, and volume are printed on the same line for each of the four cases.

P3–48 Write a FORTRAN computer program that will find the average of a set of examination grades. The grades should be read from data cards (one grade per card), and the program should count the number of grades read in each set. Execute the program for the following sets of grades. Draw a flow chart before writing the program.

Set	61	73	95	91	75	89	67	92	80	86	87	90
1	65	97	71	88	81	43	93	72	54	90	88	92

Set	84	95	82	64	92	34	85	98	48
2	99	62	64	78	72	84	58	96	78

P3–49 In testing quality control, a manufacturer measures thicknesses of a plate at four places and makes a quality-control decision based on the average of these readings. For a given buyer, the following specifications must be met.

Maximum average thickness = 5.5 mm
Minimum average thickness = 4.5 mm

In addition, a plate must be rejected if any one or more of the four thickness readings are zero, since this indicates the presence of a hole.

(a) Construct a flow chart showing the algorithm of the solution.

(b) Write and run a FORTRAN program that will test the following sets of data in order and print one of the following messages for each plate.

THE PLATE IS GOOD
THE PLATE IS TOO THIN
THE PLATE IS TOO THICK
THE PLATE HAS A HOLE

Plate number	*Thicknesses measured (in mm)*			
1	5.1	5.2	5.3	5.6
2	5.0	0.0	3.1	6.4
3	4.9	4.8	4.8	4.6
4	4.6	4.4	4.4	4.4
5	4.5	5.0	5.1	5.6
6	4.5	4.6	4.7	4.8
7	4.5	3.9	4.1	4.3
8	4.5	4.0	4.1	0.0
9	4.6	4.2	4.7	4.8
10	5.5	4.5	4.1	4.4

P3–50 Modify the program in P3–49 so that it calculates and prints the yield of the manufacturing process where

$$\text{Yield} = \frac{\text{Number of good plates}}{\text{Total manufactured}}$$

Execute your modified program with the data given in P3–49.

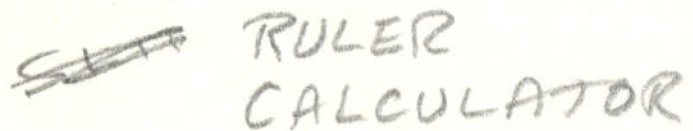

CHAPTER FOUR
MEASUREMENTS, DIMENSIONS, AND UNITS

One of the most fundamental engineering activities is measuring the physical quantities that characterize an engineering system or design. Through measurements, engineers can better understand the basic mechanisms of physical systems, and develop more accurate descriptive and prescriptive models to represent these systems. The ability to obtain, analyze, interpret, and communicate measurements is essential to successful performance in engineering, both to an engineering student and to a practicing engineer.

This chapter discusses the purpose of engineering measurements, describes several types of measurements that engineers encounter, and examines the components of measurements, including dimensions and units. Two different systems of units, the FPS (foot–pound–second) system and the International System of Units (SI) are described. After completing this chapter, you should have a better understanding of measurements and their relationship to engineering models.

4–1 MEASUREMENTS

The purpose of *measurement* is to obtain quantitative data and information about the physical quantities that characterize an engineering system or design. Such data and information may then be incorporated into a descriptive or prescriptive model of the system or design, or they may be used to effect control of an ongoing process or project. Engineering measurements are used for both the entities and the phenomena that comprise an engineering system or design. That is, engineers need to measure the physical quantities associated with both the objects and the activities that define a system. For example, measuring the length of an electrical cable characterizes an attribute of an entity, the cable. Measuring the electrical current in the same cable characterizes a phenomenon, the flow of current.

The utility of measurements depends on their accuracy and precision. *Accuracy* reflects how closely the measurement comes to the true value; *precision* is the smallest fraction (or decimal) that an instrument or device can provide in determining the magnitude of a physical quantity. A high degree of precision does not assure one of a high degree of accuracy, but high accuracy requires high precision. The engineer has great confidence in measurements if they are accurate and precise, and is more willing to utilize them in the design and control of physical systems. The accuracy and precision of measurements depend on the capability and limitations of the instruments used to obtain them. The engineer must be careful not to record, report, or manipulate measurements with any greater accuracy or precision than the process of measuring permits. This chapter will examine the principles of recording, reporting, and manipulating measurements. By *manipulation,* we mean carrying out mathematical operations with measured quantities.

EXPRESSING MEASUREMENTS

It is important to immediately establish a standard practice for stating measurements. Incompletely or poorly expressed measurements can be as misleading and as useless to the engineer as erroneous or inaccurate measurements. Properly recording, manipulating, and communicating measurements is just as important to the engineer as care in obtaining them. The practice described in this section might, at first, appear overly cautious, but early in your training, you should adopt the practice of recording and communicating measurements as clearly and definitively as possible. The practice described here offers a standard framework for doing so.

A measurement relates or compares the extent of a physical quantity to that of an accepted standard or reference. The physical quantity itself is a *dimension;* an observation of the dimension consists of a *magnitude* and *units*. The magnitude numerically conveys the relative extent of the dimension, while the units define the basis on which the comparison is made. Units are basically labels for identifying a measured quantity. For example, the measurements 6 inches and ½ foot represent the same observation of the dimension length. The magnitudes (6 and ½) are different simply because the selected units (inches and feet) are different.

The recorded statement of an engineering measurement should (1) clearly and concisely describe the entity or phenomenon from which a measurement is taken, (2) identify the physical characteristic measured, (3) record the magnitude of this observed physical characteristic, and (4) state the units used. For example, suppose that the inside diameter of a nominally 2½-inch 40S steel pipe actually measures 2.469 inches. The following pertinent information should be recorded:

Entity:	2½-inch 40S steel pipe
Characteristic:	inside diameter
Magnitude:	2.469
Units:	inches

This recorded information describes a particular *type* of pipe; additional information about it might be obtained from an appropriate engineering handbook or catalog. The physical characteristic measured is the *inside diameter,* which is obviously different from the outside diameter, wall thickness, or length of the pipe. The *magnitude* indicates that the inside diameter is 2.469 units, conveying the number of such units measured. The *unit* inch† is universally accepted to be one-twelfth of a foot, which has historically been the basic unit of length in the FPS and American engineering systems of units. As we shall see later in this chapter, the International System of Units of Measurement, abbreviated SI, is replacing the FPS and American engineering systems. (SI comes from the French *Système Internationale.*)

Implicit in the measurement 2.469 inches is the idea that the measuring device is precise to thousandths of an inch, because three significant digits are carried *beyond* the decimal. Usually we retain one doubtful or uncertain digit as a significant digit, and we must not interpret the recorded measurement as being absolute.

ERROR IN MEASUREMENT

Error in engineering work expresses the uncertainty in measured data. The word error is meant to convey the notion of uncertainty; it does not mean "mistake" or "blunder." We must account for error in both absolute terms, called *numerical error,* and relative terms, called *percent error.* For example, suppose that the

†The *inch* was originally defined in A.D. 1324 by King Edward II of England to be equal to three barleycorns taken from the middle of the ear and placed end to end.

measurement of the inside diameter of 40S steel pipe is accurate to ±0.003 inch. Thus the error could be as large as 0.003 inch. The numerical error and percent error in this case are as follows:

$$\text{Numerical error} = 0.003 \text{ inch}$$

$$\text{Percent error} = \frac{(\text{Numerical error})(100\%)}{\text{Measured value}} \tag{4–1}$$

$$= \frac{(0.003)(100)}{2.469} = 0.12\%$$

If we know the percent error in a measured process and want to find the numerical error, we simply rearrange Equation (4–1) to yield

$$\text{Numerical error} = \frac{(\text{Percent error})\ (\text{Measured value})}{100\%} \tag{4–2}$$

For example, suppose that the process of measurement contains 0.5% error. If the measured value is 11.450 kilograms, how large can the error be?

$$\text{Numerical error} = \frac{(0.5)(11.450)}{100\%} = 0.057 \text{ kilogram}$$

Therefore the measured value should be properly stated as 11.450 ± 0.057 kilogram. We must be careful not to show more digits in stating the numerical error than the uncertain digit in the measured value.

Exercises

4–1 Compute the percent error for: (a) 182.5 ± 0.2, (b) 0.06 ± 0.01, (c) 0.00054 ± 0.00005, (d) 2.54 ± 0.02, (e) 1024 ± 3.

4–2 Compute the numerical error for: (a) 0.0424 ± 0.6%, (b) 525.6 ± 2%, (c) 15.65 ± 0.5%, (d) 13.5 ± 1%, (e) 0.000857 + 10%.

4–3 A steel shaft 4.47 cm in diameter is designed to fit into a sleeve with an inside diameter of 4.50 cm. If both these dimensions have the same percent error, how large can this error be and still have the largest shaft fit the smallest sleeve? What is the greatest mismatch of shaft and sleeve that can result under this percent error?

4–4 Draw a flow chart for a FORTRAN program that will read a measured value and a numerical error from one data card and then calculate and print the associated percent error. The program should be able to branch back and repeat the process as many times as the available data permit. Write, keypunch, and execute this program, using the data in Exercise 4–1.

4–2 DIMENSIONS AND UNITS

The quantities that engineers measure are *dimensions,* the basic building blocks on which the laws of science are based. Among these quantities are length, mass, time, temperature, electric current, light intensity, angles, energy, force, and pressure. Many of these terms are part of our everyday vocabulary, and we may have formed vague notions of their scientific meanings through their everyday uses. The following sections will attempt to establish a more precise concept of these dimensions.

We may divide physical quantities into fundamental dimensions and derived dimensions. *Derived dimensions* are those which are defined in terms of two or more fundamental dimensions, whereas *fundamental dimensions* are the most basic physical quantities. Examples of fundamental dimensions are length (L), mass (M), and time (T). Examples of derived dimensions are volume (L^3), velocity (L/T), and density (M/L^3). Engineering measurements involve both fundamental and derived dimensions. As we stated before, measured dimensions are expressed in terms of both magnitude and units. Table 4–1 lists several fundamental dimensions, two so-called supplementary dimensions, which are basically fundamental quantities, and a number of derived dimensions. Table 4–1 states the *name* of the dimension and an *abbreviation* in terms of its fundamental dimensions. These abbreviations apply regardless of the unit system applied. The next sections describe the more important of these dimensions as well as procedures for measuring them.

LENGTH

The process of measuring length, area, volume, and angle is called *mensuration,* which is a branch of geometry. The fundamental dimension involved in this measurement process is *length,* denoted L. Figure 4–1 is a graphical representation of length. *Angles* are considered supplementary dimensions. Angles are of two types, *plane* angles ϕ and *solid* angles Ω. Figure 4–2 shows these. *Area* (L^2) and *volume* (L^3) are derived dimensions based on the fundamental dimension length (L). Certain areas and volumes involve the supplementary dimension of angle. Lengths are stated in such units as inches, feet, miles, meters, and so forth. Angles are expressed in degrees or radians.

Figure 4–3 illustrates some areas and volumes. The rectangle shown in Figure 4–3(a) is a second-order dimension involving the fundamental dimension length (L); it is denoted by the dimension L^2. Similarly, the cube shown in Figure 4–3(b) has the third-order dimension (L^3) when we measure its volume. But this three-dimensional entity also possesses a two-dimensional attribute, its surface area, which is the sum of the areas of its faces. The circle in Figure 4–3(c) is clearly a two-dimensional entity; its area A depends on the fundamental dimension length, represented by the radius R of the circle ($A = \pi R^2$). The area of a segment of the circle depends on the angle ϕ included in the segment. The volume of the sphere shown in Figure 4–3(d)

FIGURE 4–1 The fundamental dimension *length:* (a) plane angle, (b) solid angle

a *b*

TABLE 4–1. SELECTED FUNDAMENTAL, SUPPLEMENTARY, AND DERIVED DIMENSIONS IN THE INTERNATIONAL SYSTEM OF UNITS OF MEASUREMENT (SI)

Quantity	*Dimensions*
Fundamental dimensions	
Length	L
Mass	M
Time	T
Temperature	Θ
Electric current	I
Supplementary dimensions	
Plane angle	ϕ
Solid angle	Ω
Derived dimensions	
Acceleration	LT^{-2}
Angular velocity	T^{-1}
Angular acceleration	T^{-2}
Area	L^2
Energy (work, heat)	ML^2T^{-2}
Force	MLT^{-2}
Frequency	T^{-1}
Mass density	ML^{-3}
Period	T
Power	ML^2T^{-3}
Pressure	$ML^{-1}T^{-2}$
Velocity	LT^{-1}
Volume	L^3
Capacitance	$M^{-1}L^{-2}T^4I^4$
Charge	TI
Current density	IL^{-2}
Electrical potential (voltage)	$ML^2T^{-2}I^{-1}$
Inductance	$ML^2T^{-2}I^{-2}$
Resistance	$ML^2T^{-3}I^{-2}$

involves the third-order dimension L^3, while its surface area involves the second-order dimension L^2.

Various instruments and techniques are available for measuring the fundamental dimension length. Still others are available for measuring area and volume. To measure length, we compare the distance between two specified points on the entity being measured to that on some widely accepted standard. For instance, to measure the length of a steel beam on a construction site, we might use a tape measure, a thin, semiflexible strip of steel or plastic material marked in meters, centimeters,

FIGURE 4–2 The supplementary dimensions *plane angle* and *solid angle*

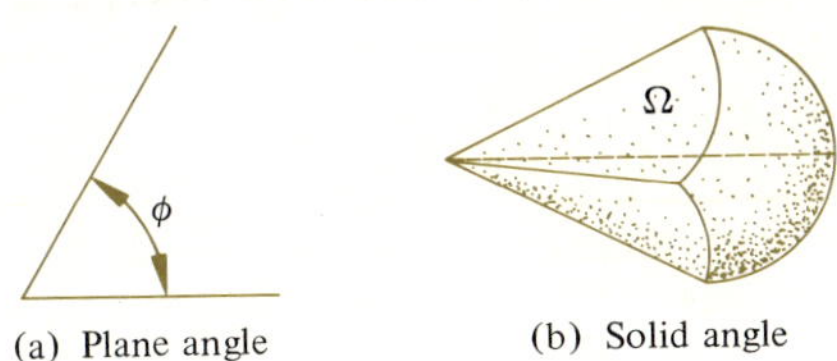

(a) Plane angle (b) Solid angle

and millimeters. We measure the length of the beam by aligning one end of the tape with one end of the beam and then recording the mark on the tape that coincides with the other end of the beam. Such a procedure for measuring length involves direct comparison of the dimensions with a reference; an unknown length is compared to a known length. Other direct-comparison devices, including micrometers and calipers, are available for measuring lengths within specific ranges.

One objection to direct comparison is that, even with the most precise instrument, accuracy is limited to one part in 10^7. By using an indirect-measurement procedure such as a light interferometer, however, we can obtain measurements accurate to one part in 10^9. The light interferometer is an optical device in which light of precisely known wavelength is used as to compare lengths. This instrument can be used only to measure objects no more than a few centimeters in length, however.

MASS

Mass is a measure of the quantity of matter in an entity. We do not usually measure mass directly; instead we measure its relationship to a property called *weight*. The

FIGURE 4–3 The derived dimensions *area* and *volume*

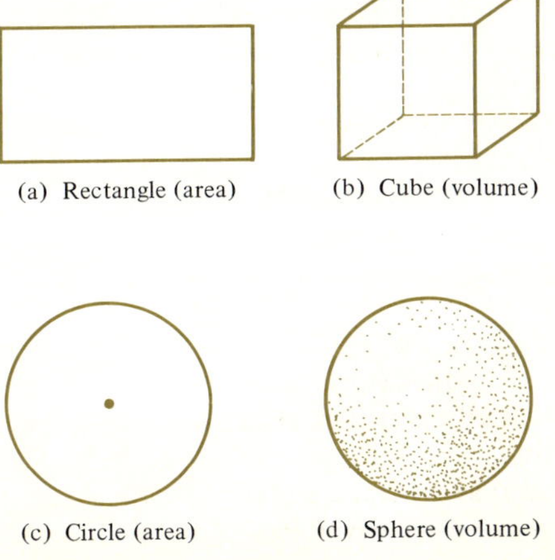

(a) Rectangle (area) (b) Cube (volume)

(c) Circle (area) (d) Sphere (volume)

weight of a body is the gravitational force that the Earth exerts on the body. Thus weight (W) is related to mass (M) by the equation

$$W = Mg \tag{4–3}$$

where g is the gravitational constant equal to 32.2 ft/sec^2, or 9.807 m/sec^2. This constant g varies with the latitude on the Earth's surface as well as the distance above the Earth's surface. We can avoid these terrestial variations in the weight of a body by comparing the weight of the body with a matching known weight on a beam type of balance. This technique involves *direct comparison* of the unknown quantity with a known quantity. Another technique for measuring weight is to place the object on a scale and observe the linear deflection caused by the compression of a carefully calibrated spring. This is an *indirect comparison,* because we are measuring the predictable effect of the *force W* on the length of the spring.

The units of mass are pounds mass (lb_m), kilograms, grams, and so forth, depending on the system of units employed.

TIME

We measure *time* in two different ways. On the one hand, we may want to know the time of day or the day in the year, so that we can properly sequence events. On the other hand, we may need to measure the *duration* of an activity or process. In the first case, we are viewing time as an *entity;* in the second, as a *phenomenon.* We may be interested in an *instant* in time (an entity) or the *lapse* of time (phenomenon).

The passage of time has historically been measured by observing the movements of celestial bodies, including the Earth itself. A solar day is equivalent to one complete rotation of the earth on its axis. For centuries, a solar day was crudely measured by the angle of the sun with the Earth's surface. (A sidereal day is one revolution of the Earth relative to the stars.) For both civil and scientific purposes, time has traditionally been measured by noting the number of rotations of the Earth relative to the sun during the time being measured. A scale consisting of 24 hours in a day (one rotation of the Earth), 60 minutes in an hour, and 60 seconds in a minute has been adopted world-wide for measuring time. Further, the time required for the Earth to orbit the sun, 365.242 days, establishes the sidereal year. Because this quantity is rather inconvenient, we have three 365-day years followed by one 366-day year (leap year). The phases of the moon are the basis for yet another convention for noting the passage of time, the month. We are all aware of the cumbersome structure associated with this unit of time.

The fundamental invariable unit of time is the ephemeris second, which is 1/86,400 of the mean solar day. Other familiar units of time are the minute and the hour.

ELECTRIC CURRENT

Electric current is the rate of transfer of charged particles, usually electrons, in a conductor. Thus current is a phenomenon involving the movement of charged particles (entities) in a conductor (entity). The unit describing the number of charged particles is the *coulomb,* which is about 6.24×10^{18} electrons. The practical unit of current is the *ampere*. This unit is defined as the current that, if maintained in each of two parallel wires separated by one meter in free space, would produce a force between the two wires (due to their magnetic fields) of 2×10^{-7} newton for each meter of length.

Electric current is measured using a device called an ammeter. To determine the current in a wire using the ammeter, we cut the wire, attach the newly formed ends to the ammeter, and observe the deflection of a needle on a carefully calibrated scale as the current passes through the ammeter. This is an indirect comparison.

TEMPERATURE

Temperature is the condition of an entity that determines its potential to transfer heat to or from another entity. Heat flows from a hot body to a cold body. In terms of temperature, heat flows from a body at a high temperature to one at a low temperature. These terms, hot and cold, high and low, are used simply to convey a relative condition—heat flows from an entity at one temperature to one at a lower temperature.

The basic unit of temperature is the *Kelvin,* defined as the fraction 1/273.16 of the thermodynamic temperature of the triple point of water. The temperature 0 K is called "absolute zero." We discuss temperature scales later in this chapter.

We cannot measure temperature directly. Fortunately, there are measurable physical properties that vary directly as temperature varies. Among these are the volume of a fluid, the length of a rod, the electrical resistance of a wire, and the color of a lamp filament. Any of these properties can be used to construct a thermometer—a device used to measure temperature.

We can measure the temperature of a fluid by placing a mercury thermometer in the fluid and observing the volumetric expansion or contraction of the mercury as it is heated or cooled in the fluid. The height of a column of mercury in a glass tube increases or decreases as the mass of mercury in the base of the thermometer expands or contracts. After some brief period, the height of the mercury column stabilizes at a particular level, and the marking on the thermometer corresponding to that level is recorded. Thus measuring temperature involves an indirect comparison.

VELOCITY

Velocity is the time rate of motion of an entity. It has the dimensions L/T (also expressed as LT^{-1}). Actually, velocity is defined in physics as having a magnitude and a direction. Thus an airplane might be said to have a velocity of 700 miles/hour (magnitude) due east (direction). We conventionally call the magnitude of the velocity of an entity its *speed*. When motion is in a straight line, the word velocity then essentially refers to the speed of the entity.

The direction component of velocity becomes very important when we study motion in a plane. For example, the motion of a projectile can be viewed as motion in the x-y plane, where x represents the horizontal and y the vertical direction. Thus a projectile fired straight upward has a positive vertical velocity as long as the projectile moves upward. At some point, the force of gravity will overcome the upward motion of the projectile, and the projectile then moves downward. Hence its vertical velocity becomes negative, connoting its negative direction. You will come across this concept again when you study physics and engineering mechanics, specifically in those segments of mechanics called kinematics and dynamics. *Kinematics* is the study of an entity's motion without regard to the forces that may be imposed on it. *Dynamics* is the study of the relationships between the motion of an entity and the forces applied to it.

We measure velocity by measuring its fundamental dimensions, length and time. Any device used to measure velocity directly must measure both these quantities. We often measure the velocity of a fluid (liquid or gas) indirectly by observing the behavior of parallel columns of liquid in a manometer, or in some similar device that measures the difference in pressure between two points in the flow stream. The units used to measure velocity are those of its fundamental dimensions, length and time. Thus a velocity might be expressed in miles/hour, feet/second, or meters/second.

ACCELERATION

Acceleration is the rate at which the velocity of an entity changes with time. It is a derived dimension, L/T^2 or LT^{-2}. A positive acceleration means that velocity is increasing with time; the entity moves faster and faster. An important quantity is the acceleration due to gravity, 32.2 ft/sec² or 9.807 m/sec².

To measure acceleration, we take successive measurements of the entity's velocity. Hence we directly measure the fundamental dimensions, length and time. Typical units for acceleration are feet/second² and meters/second².

FORCE

The concept of *force* derives from Isaac Newton's second law of motion

$$F = Ma \tag{4–4}$$

where F is the dimension force, M is the mass of the entity, and a is the acceleration to which the entity is subjected. Force has the dimensions MLT^{-2}. Force has the units pounds force, abbreviated lb_f, or newtons, abbreviated N.

WORK AND ENERGY

Work is performed when an entity is displaced a distance d as a result of having a force F applied to it. Work is the product Fd, and has the dimensions ML^2T^{-2}. An entity in motion has *energy* equal to the work it can do as it is brought to rest. Hence energy is equivalent to work and has the same dimensions, ML^2T^{-2}. Work and energy have the units (ft)(lb_f). In the SI system, the unit of work and energy is the *joule* (newton-meter).

POWER

Power is the time rate at which work is done, and therefore has the dimensions ML^2T^{-3}. Units of power include horsepower and the watt. One horsepower equals 746 watts. One *watt* is one joule per second.

PRESSURE

Pressure is the force applied to a unit of area; it has the dimensions $ML^{-1}T^{-2}$. Units for pressure include lb_f/in^2 and pascals. One *pascal* is one newton/m^2.

DENSITY

Density is the quantity of matter in a unit volume of a gas, liquid, or solid. Hence density is mass per unit volume and has the dimensions M/L^3. Units for density include lb_m/ft^3 and kg/m^3.

4–3 MANIPULATING AND RECORDING UNITS

CONSISTENCY OF DIMENSIONS

As engineers formulate and manipulate models of systems, entities, and phenomena, they must be sure that the units used for the physical quantities in the model are *consistent*. By this we mean that the operation itself is valid, and that the proper units are assigned to the constants and variables. For instance, the following operations are valid:

$$120 \text{ miles} + 36 \text{ miles}$$
$$12 \text{ feet}^2 - 5 \text{ feet}^2$$
$$8.0 \text{ meters} + 3.2 \text{ meters}$$

The following are invalid:

$$2 \text{ feet} + 12 \text{ seconds}$$
$$40 \text{ feet}^2 - 129.7 \text{ feet}^3$$

The units in each of the first three expressions are identical. The units in the last two are different, and the expressions are meaningless. For addition (and subtraction, since it is simply the addition of a negative quantity), it is necessary to add quantities that have the same dimensions. On the other hand, the division operations

$$\frac{2 \text{ feet}}{12 \text{ seconds}}, \qquad \frac{129.7 \text{ ft}^3}{40 \text{ ft}^2}$$

would be perfectly valid if particular models required them.

The following operations are valid, but we must restate them in consistent units before carrying out the indicated operation.

$$1 \text{ foot} + 2\tfrac{3}{4} \text{ inches}$$
$$4 \text{ horsepower} - 1240 \text{ watts}$$

In the first operation, both terms are lengths, and it remains only to express both quantities in the same units. Either

$$12 \text{ inches} + 2\tfrac{3}{4} \text{ inches} \qquad \text{or} \qquad 1 \text{ foot} + 0.229 \text{ foot}$$

would be proper operations. In the second operation, each term is a measure of the physical quantity power, that is, energy per unit time. To carry out the intended operation, we have to use a *conversion factor* to convert horsepower to watts or vice versa. The conversion factor for converting horsepower to watts is

$$1 \text{ hp} = 746 \text{ watts}$$

The operation then becomes

$$(4)(746) \text{ watts} - 1240 \text{ watts} = 1744 \text{ watts}$$

A more cautious approach would have been to write the operation as

$$(4 \text{ hp}) \left(746 \frac{\text{watts}}{\text{hp}}\right) - 1240 \text{ watts} = 1744 \text{ watts}$$

In this expression we state the original terms, apply the necessary conversion factor, and cancel the hp units in the numerator of the first term that are the same as the units in the denominator of the conversion factor.

MANIPULATING UNITS

The example with horsepower and watts shows that engineers must adopt a standard procedure for manipulating units and dimensions in models. The following example illustrates two such procedures. Either will serve you well throughout your remaining courses, as well as in engineering practice later on. We recommend either of these procedures for homework assignments, laboratory notebooks, and technical reports.

Suppose an aircraft travels at twice the speed of sound (assumed to be 1100 ft/sec). How fast is the plane flying in miles per hour? We can show this calculation either as

$$(2) \left(1100 \frac{\text{ft}}{\text{sec}}\right) \left(\frac{1 \text{ mile}}{5280 \text{ ft}}\right) \left(\frac{3600 \text{ sec}}{\text{hr}}\right) = 1500 \frac{\text{miles}}{\text{hr}}$$

or as

$$\begin{array}{c|c|c|c} 2 & 1100 \text{ ft} & 1 \text{ mile} & 3600 \text{ sec} \\ \hline & \text{sec} & 5280 \text{ ft} & \text{hr} \end{array} = 1500 \frac{\text{miles}}{\text{hr}}$$

The first format uses parentheses to separate the quantities and conversion factors, while the second employs a series of boxlike cells. The first is better for typed material, while the second is more applicable to the carefully documented calculations recorded in laboratory notebooks. In either format, the segregation of terms and explicit recording of units is essential in recording, manipulating, and reporting engineering models. In this example, the first term is the number 2, representing the concept "twice" in "twice the speed of sound." The second quantity is the given value for the speed of sound in the units feet/second. Since we finally need the dimension *length* in miles, we have to employ a conversion factor as the third term in this calculation. Since one mile is equal to 5280 feet, we place this conversion factor in the third cell, taking care to ensure that the unit ft appears in the denominator so that it cancels the same unit in the numerator of the preceding term.

To convert time from seconds to hours, we need yet another conversion factor, which we place in the fourth cell. To perform the calculation, we then follow this procedure.

1. Cancel the appropriate units and collect the remaining units on the right side of the equality.
2. Perform the necessary arithmetic operations, in this case multiplications and divisions, and record the result on the right side of the equality.
3. Record only the appropriate significant figures, so that you do not imply greater accuracy in the result than is in the measurements themselves.

Although this procedure may at first seem overly cautious, using it ensures that the engineer always records both the *magnitude* and the *units*. This procedure affords better communication between student and teacher, and between one engineer and another. In addition, anyone who needs to examine an entry some time after it was recorded can see exactly how the calculation was performed.

Exercises

4–5 The position of an entity moving in a straight line along the x axis depends on the time t according to the equation

$$x = at^2 - bt^3$$

where x is in feet and t in seconds.

(a) What dimensions and units must a and b have?

(b) If $a = 3$ and $b = 1$, at what time does the entity achieve its maximum positive value of x?

(c) What total distance is covered in the first four seconds?

4–6 The moon is approximately 240,000 miles from the Earth. Express this distance in (a) feet, (b) inches, (c) meters, (d) kilometers.

4–7 In the study of heat and mass transfer, we encounter the Grashof number, given by the equation

$$\mathrm{Gr} = \frac{D^3 g \rho^2 \beta\, \Delta T}{\mu^2}$$

Given that $D = 2.5$ ft, $g = 32.2$ ft/sec^2, $\rho = 0.0035$ $\mathrm{lb}_m/\mathrm{ft}^3$, $\beta = 1.2$°F, $\Delta T = 192$°F, and $\mu = 2.1 \times 10^{-5}$ $(\mathrm{lb}_f)(\mathrm{sec})/\mathrm{ft}^2$, find the value of Gr. What are its units?

4–4 SYSTEMS OF UNITS

Engineers have long confronted the perplexing problem of working with several systems of units. A *system of units* is an established convention for associating labels with measured quantities. The FPS (foot–pound–second) system and the American engineering system, also based on foot-pound-second units, have long been used in engineering disciplines. Science disciplines have favored the MKS (meter–kilogram–second) and CGS (centimeter–gram–second) systems, which are the so-called *metric* systems of units. The International System of Units (SI) is a modernized version of the metric system. Another factor also distinguishes these different unit systems: whether *gravitational* or *absolute* units are employed for the quantity *mass*. Table 4–2 summarizes these unit systems according to the gravitational-versus-absolute scheme.

Among the absolute systems of units, the MKS system has been more widely adopted for engineering-scale applications, while the CGS system has been most often used in laboratory-scale scientific work. Each of these unit systems suffers the handicap that the time units—days, hours, minutes, and seconds—are not amenable to decimal changes in magnitude. However, in the FPS (foot–pound–second) system, the traditional system in English-speaking countries, and the basis for the American engineering system of units, none of the units used for the same quantity is derivable from another simply by shifting a decimal point. For example, the standard FPS units for length involve such conversion factors as the following.

$$12 \text{ inches} = 1 \text{ foot}$$
$$3 \text{ feet} = 1 \text{ yard}$$
$$5280 \text{ feet} = 1 \text{ mile}$$

Although we defined *force* as a derived dimension in Table 4–1, Table 4–2 shows force as a fundamental dimension in the gravitational systems and a derived dimension in the absolute systems. Conversely, *mass* is defined as a fundamental dimension in the absolute systems of units, a derived dimension in the FPS system, and a fundamental dimension in the American engineering system. The reason for these differences lies in the relationship between force and mass in the gravitational system of units, based on Newton's law. The next section gives attention to this relationship. We shall see later that we can alleviate this confusion by adopting the SI system of units, which is essentially the same as the absolute MKS system. But for the present, U.S. engineers are caught in a gray zone; they must know and be able to apply any of the systems of units summarized in Table 4–2.

GRAVITATIONAL SYSTEMS

Engineering analysis has traditionally applied a system of units that defined both the units of mass and the units of force. Unfortunately, the same term (pounds) was chosen to represent both quantities, although they are physically different. To distinguish one from another, engineers have used the notation lb_m or lb_f for pounds mass or pounds force, respectively. Analyzing the expression for Newton's law reveals the distinction between lb_m and lb_f:

$$F = Ma$$

where F = force, M = mass, and a = acceleration. In the FPS system of units, the fundamental unit of force is lb_f, and the derived units of mass are

TABLE 4–2. UNIT SYSTEMS

	GRAVITATIONAL		*ABSOLUTE (METRIC)*		
Fundamental dimensions	*FPS*	*American engineering*	*MKS*	*CGS*	*SI*
Force (F)	lb_f	lb_f	—	—	—
Length (L)	ft	ft or in.	meter	cm	meter
Time (T)	sec†	sec	sec	sec	sec
Mass (M)	—	lb_m	kg	gram	kg
Derived dimensions					
Force (F)	—	—	kg-meter sec^{-2} (newton‡)	gram-cm-sec^{-2} (dyne)	newton‡
Mass (M)	lb_f-sec^2-ft^{-1}	—	—	—	—
Energy (LF)	ft-lb_f	ft-lb_f	meter-newton	cm-dyne (erg)	joule
Power $\left(\frac{LF}{T}\right)$	$\frac{\text{ft-lb}_f}{\text{sec}}$	$\frac{\text{ft-lb}_f}{\text{sec}}$	$\frac{\text{meter-newton}}{\text{sec}}$	$\frac{\text{erg}}{\text{sec}}$	watt
Velocity $\left(\frac{L}{T}\right)$	$\frac{\text{ft}}{\text{sec}}$	$\frac{\text{ft}}{\text{sec}}$	$\frac{\text{meter}}{\text{sec}}$	$\frac{\text{cm}}{\text{sec}}$	$\frac{\text{meter}}{\text{sec}}$
Acceleration $\left(\frac{L}{T^2}\right)$	$\frac{\text{ft}}{\text{sec}^2}$	$\frac{\text{ft}}{\text{sec}^2}$	$\frac{\text{meter}}{\text{sec}^2}$	$\frac{\text{cm}}{\text{sec}^2}$	$\frac{\text{meter}}{\text{sec}^2}$
Area (L^2)	ft^2	ft^2	$meter^2$	cm^2	$meter^2$
Volume (L^3)	ft^3	ft^3	$meter^3$	cm^3	$meter^3$

†The accepted SI abbreviation for second is s; but since we are here dealing with so many single-letter variables, we have elected, in this text, to abbreviate second as sec.
‡A newton is the force required to accelerate a 1-kg mass at 1 meter/sec^2. The acceleration of gravity at sea level and 45° latitude has the measured value of 9.807 meters/sec^2.

$$M = \frac{F}{a} \frac{\text{lb}_f}{\text{ft/sec}^2} \qquad \left(\text{or } \frac{\text{lb}_f \text{ sec}^2}{\text{ft}}\right)$$

This derived unit of mass is called a *slug* in the FPS system of units. Thus a force of 1 lb_f will cause a mass of 1 slug to have an acceleration of 1 ft/sec^2. The same force will accelerate 1 lb_m at $g = 32.2 \text{ ft/sec}^2$, where g is the acceleration due to gravity at sea level and 45° latitude.

The derived unit of force is

$$F = Ma \frac{\text{lb}_m\text{ft}}{\text{sec}^2}$$

which is called the *poundal*. A force of 1 poundal will cause a mass of 1 lb_m to have an acceleration of 1 ft/sec^2.

Neither slugs nor poundals are common in engineering calculations. Rather, the weight–mass relationship stated in Equation (4–3) is combined with Newton's law stated in Equation (4–4) to yield the equation

$$F = \frac{W}{g} a \tag{4–5}$$

where g is the acceleration due to gravity. Since both g and a have the units ft/sec^2, both F and W have the same units, lb_f.

Exercises

4–8 The value of the acceleration due to gravity g at any latitude on the Earth may be approximated by the relationship

$$g = 32.09(1 + 0.0053 \sin^2 \phi) \text{ ft/sec}^2$$

Write, keypunch, and execute a FORTRAN program to compute and print the gravitational constant g for every 5 degrees of latitude ϕ from the North Pole ($\phi = +90°$) to the South Pole ($\phi = -90°$). The equator has a latitude of 0°.

4–9 Solve for the lb_m accelerated at 4.25 ft/sec^2 by a force of 288 lb_f.

4–10 Solve for the mass in slugs that is being accelerated at 7.52 ft/sec^2 by a force of 1380 lb_f.

4–11 Solve for the force (lb_f) required to accelerate a body with mass 2.4 slugs at 12 ft/sec^2.

INTERNATIONAL SYSTEM OF UNITS OF MEASUREMENT (SI)

The English-speaking countries are embarking on the difficult task of adopting the *International System of Units of Measurement*, which is essentially the MKS system in Table 4–2. These changes particularly affect engineering, as engineers have long employed the gravitational systems of units. The meter replaces the foot as the basic unit for length. The kilogram becomes the standard unit for mass, replacing the lb_m. The watt is to be used instead of horsepower for measuring power. Envision the millions of changes which must be made in machines, instruments, hardware, building materials, recording devices, and measuring equipment to cope with this transformation!

Table 4–3 gives the name and abbreviation for the SI units used to measure some familiar quantities. Table 4–4 defines the most basic of these quantities in the manner in which they have been adopted in the SI system of units. Table 4–5 presents several of the important prefixes to which we must all become accustomed in dealing with multiples and submultiples of powers of the base 10, on which the SI system is structured.

Table 4–6 gives some factors for converting units in gravitational systems to the SI units and vice versa, as well as for converting dimensions in a given system to similar dimensions in the same system. The following sections describe the SI system as it applies to some of the more important dimensions encountered in engineering.

UNITS OF LENGTH

The *meter* (m) is the standard unit of length in the SI system of units. It is defined as 1,650,763.73 wavelengths of a particular monochromatic light in a vacuum.

UNITS OF TIME

The *second*† is the standard unit of time in all systems of units. It is defined as the time duration of 9,192,631,770 periods of the transition between two hyperfine levels of the ground state of the atom of cesium-133. Useful conversion factors are

1 minute = 60 seconds
1 hour = 60 minutes = 3600 seconds
1 day = 24 hours = 86,400 seconds

UNITS OF MASS

The *kilogram* (kg) is the SI unit of mass. It is defined as the mass of a standard block of platinum maintained by the French Bureau of Standards at Sèvres, France.

$$1 \text{ lb}_m = 0.4535924277 \text{ kg}, \qquad 1 \text{ slug} = 32.2 \text{ lb}_m = 14.594 \text{ kg}$$

UNITS OF FORCE

The *newton* (N) is the worldwide standard unit of force. It is defined as the force required to accelerate a 1-kg mass at an acceleration of 1 m/sec^2. Thus the newton is expressed in the units kg-m/sec^2. To convert from pounds force to newtons, we use the transformation 1 lb_f = 4.448 N.

†Although second is abbreviated s in the SI system of units, we shall use the more familiar abbreviation sec in this text.

TABLE 4–3. UNITS FOR THE INTERNATIONAL SYSTEMS OF UNITS OF MEASUREMENT (SI)

Quantity	*Unit*	*Abbreviation*
	Fundamental units	
Length	meter	m
Mass	kilogram	kg
Time	second	s
Electric current	ampere	A
Temperature	Kelvin	K
Luminous intensity	candela	cd
Plane angle	radian	rad
Solid angle	steradian	sr
	Derived units	
Area	square meter	m^2
Volume	cubic meter	m^3
Frequency	hertz	Hz (s^{-1})
Density	kilogram per cubic meter	kg/m^3
Velocity	meter per second	m/s
Angular velocity	radian per second	rad/s
Acceleration	meter per second squared	m/s^2
Angular acceleration	radian per second squared	rad/s^2
Force	newton	N ($kg \cdot m/s^2$)
Pressure	newton per sq meter (pascal)	N/m^2 (Pa)
Kinematic viscosity	sq meter per second	m^2/s
Work, energy, quantity of heat	joule	J (N·m)
Power	watt	W (J/s)
Electric charge	coulomb	C (A·s)
Voltage, potential difference, electromotive force	volt	V (W/A)
Electric field strength	volt per meter	V/m
Electric resistance	ohm	Ω (V/A)
Electric capacitance	farad	F (A·s/V)
Magnetic flux	weber	Wb (V·s)
Inductance	henry	H (V·s/A)
Magnetic flux density	tesla	T (Wb/m^2)
Magnetic field strength	ampere per meter	A/m
Magnetomotive force	ampere (turns)	A
Luminous flux	lumen	lm (cd·sr)
Luminance	candela per sq meter	cd/m^2
Illumination	lux	lx (lm/m^2)

TABLE 4–4. DEFINED UNITS IN THE INTERNATIONAL SYSTEM OF UNITS OF MEASUREMENT (SI)

Unit (dimension)	*Abbreviation*	*Definition*
Meter (length)	m	1,650,763.73 wavelengths in vacuum of the orange-red radiation in the spectrum of krypton-86
Kilogram (mass)	kg	Mass of the standard platinum-iridium bar kept by the International Bureau of Weights and Measures in Paris
Second (time)	s	Duration of 9,192,631,770 periods of the transition between two hyperfine levels of the ground state of the cesium-133 atom
Kelvin (temperature)	K	The fraction 1/273.16 of the thermodynamic temperature of the triple point of water
Ampere (electric current)	A	The electric current that, if maintained in each of two parallel wires separated by one meter in free space, would produce a force between the two wires (due to their magnetic fields) of 2×10^{-7} newton for each meter of length
Mole (amount of substance)	mol	The amount of substance of a system that contains as many elementary entities as there are atoms in 0.012 kilogram of carbon-12
Candela (luminous intensity)	cd	The luminous intensity of 1/600,000th of a square meter of a blackbody at the temperature of freezing platinum (2045 K)
Radian (plane angle)	rad	The plane angle with its vertex at the center of a circle that is subtended by an arc equal in length to its radius
Steradian (solid angle)	sr	The solid angle with its vertex at the center of a sphere that is subtended by an area of the spherical surface equal to that of a square with sides equal in length to the radius

TABLE 4–5. UNIT PREFIXES IN THE SI SYSTEM

Multiples and submultiples	*Prefixes*	*Symbols*
10^{12}	tera	T
10^{9}	giga	G
10^{6}	mega	M
10^{3}	kilo	k
10^{-3}	milli	m
10^{-6}	micro	μ (Greek mu)
10^{-9}	nano	n
10^{-12}	pico	p
10^{-15}	femto	f
10^{-18}	atto	a

UNITS OF AREA AND VOLUME

Area and volume are derived dimensions based on the fundamental dimension length. Thus m^2, m^3, in^2, in^3, ft^2, and ft^3 are common units for expressing area and volume; those involving meters (m) are standard in the SI system.

UNITS OF VELOCITY AND ACCELERATION

The worldwide standard units of velocity and acceleration are m/sec and m/sec^2, respectively. These units are derived from the SI units for length and time.

UNITS FOR WORK AND ENERGY

Work measures the phenomenon that occurs when a force acts on a body. Work is defined as the product of the force acting on the body and the distance moved against the resistance. The standard international unit for measuring work is the *joule,* which is equivalent to a force of 1 N moving through a distance of 1 m. Thus the joule has the units

$$1 \text{ joule} = 1\,(\text{N}) \cdot (\text{m}) = 1\,\frac{(\text{kg})(\text{m}^2)}{\text{sec}^2}$$

The terms work and energy are synonymous, at least in the units they carry. The joule as a unit of electrical energy is the work expended per second by a current of 1 ampere flowing through a resistor of 1 ohm.

Engineers have traditionally used ft-lb_f and horsepower-hours as units for mechanical work or energy, joules and kilowatt-hours for electrical energy, and calories and British thermal units (BTU) for heat and thermal energy. In the SI system of units, all these forms of energy are expressed in joules.

TABLE 4–6. SELECTED CONVERSION FACTORS

Multiply	*by*	*to obtain*	*for dimension*
Angstrom units	10^{-10}	meters	length
Inches	2.54	centimeters	length
Kilometers	3.28×10^3	feet	length
Meters	3.28	feet	length
Meters	39.4	inches	length
Miles	5280	feet	length
Miles	1.61	kilometers	length
Square feet	9.29×10^{-2}	square meters	area
Square inches	6.45×10^{-4}	square meters	area
Square meters	10.8	square feet	area
Cubic feet	2.83×10^{-2}	cubic meters	volume
Cubic inches	1.64×10^{-5}	cubic meters	volume
Cubic meters	35.3	cubic feet	volume
Gallons (U.S. liquid)	3.79	liters	volume
Gallons (U.S. liquid	231	cubic inches	volume
Liters	61.0	cubic inches	volume
Liters	3.53×10^{-2}	cubic feet	volume
Kilograms	2.21	pounds	mass
Pounds	453.6	grams	mass
Pounds	0.4536	kilograms	mass
Dynes	10^{-5}	newtons	force
Newtons	0.225	pounds (force)	force
Pounds (force)	4.45	newtons	force
British thermal units (BTU)	1.05×10^3	joules	energy
BTU	2.93×10^{-4}	kilowatt-hours	energy
Foot-pounds (force)	1.36	joules	energy
Foot-pounds (force)	1.36	newton-meters	energy
Joules	9.49×10^{-4}	BTU	energy
Kilowatt-hours	1.34	horsepower-hours	energy
BTU/hour	2.93×10^{-4}	kilowatts	power
ft-lb_f/hr	5.05×10^{-7}	horsepower	power
ft-lb_f/hr	3.77×10^{-7}	kilowatts	power
Horsepower	0.746	kilowatts	power
Horsepower	550	ft-lb_f/sec	power
Kilowatts	1.34	horsepower	power
Watts	1.34×10^{-3}	horsepower	power
Watts	1	joules/sec	power
Degrees	1.75×10^{-2}	radians	plane angle
Radians	57.3	degrees	plane angle
Steradian	7.96×10^{-2}	spheres	solid angle
Atmospheres	1.03×10^7	grams/m^2	pressure
Atmospheres	760	millimeters of mercury	pressure
Atmospheres	14.7	lb_f/in^2	pressure
Atmospheres	0.101	pascals	pressure
Pascal	1	newtons/m^2	pressure
Amperes	1	coulombs/sec	current
Coulombs	6.24×10^{18}	electrons	charge

UNITS OF POWER

Power is the time rate at which work is done. The SI unit for power is the *watt,* which is equivalent to 1 kg-m²/sec³. Thus 1 watt is 1 joule/sec, 1 N-m/sec, or 1 kg-m²/sec².

Engineers have often used *horsepower* as units of power. Useful conversions in dealing with power are:

1 horsepower = 550 ft-lb_f/sec (by definition)
1 horsepower = 746 watts
1 watt = 1 joule/sec

UNITS OF PRESSURE

Pressure is force per unit area. The SI unit for pressure is the *pascal.* One pascal is 1 N/m². Engineers have traditionally used the unit pounds-force per square inch, abbreviated either as lb_f/in^2 or psi. Another traditional unit of pressure is the *atmosphere* (atm), which is equivalent to 14.7 lb_f/in^2.

UNITS OF DENSITY

Density is the concentration of matter, measured in mass per unit volume. The SI units for density are kg/m³. Engineers have used the units lb_m/ft^3, while laboratory scientists have employed g/cm³ as standard units.

UNITS OF TEMPERATURE

Temperature may be defined as the condition of a body that determines the transfer of heat to or from other bodies. The SI unit for temperature is the *Kelvin,* although we more often employ the Celsius scale. In this temperature scale, there are 100 degrees between the *freezing point* of pure water, established as 0°C, and the boiling point of pure water, 100°C. On the Celsius scale, *absolute zero* is established at −273.15°C. This temperature corresponds to the zero point on the *Kelvin* scale. Thus

$$°K = °C + 273.15 \qquad (4\text{–}6)$$

Engineers have traditionally used the *Fahrenheit* scale to measure temperature. On the Fahrenheit scale, there are 180 degrees between the freezing point of water (32°F) and its boiling point (212°F), and absolute zero is −459.67°F. This temperature corresponds to the zero point on the *Rankine* scale. Thus

$$°R = °F + 459.67 \qquad (4\text{–}7)$$

To convert back and forth between the Celsius scale and the Fahrenheit scale, you can use the following relations:

$$°F = \tfrac{9}{5}°C + 32 \qquad (4\text{–}8)$$
$$°C = \tfrac{5}{9}(°F - 32) \qquad (4\text{–}9)$$

Exercises

4–12 A body has a mass of 253 lb_m on the Earth. What is its weight in pounds force on the surface of the moon, where the acceleration due to gravity is 5.3 ft/sec^2?

4–13 Develop a flow chart for a FORTRAN program that gives the temperature for every 5 Kelvin between the freezing point and the boiling point of water for each of the following scales: Celsius, Fahrenheit, and Rankine:

$$\text{Kelvin} = \text{Celsius} + 273.15, \qquad \text{Rankine} = \text{Fahrenheit} + 459.67$$

Write, keypunch, and execute the program.

4–14 Given that ϕ is the latitude and H the elevation in centimeters, the acceleration due to gravity in the CGS system is given by Helmert's equation:

$$g = 980.616 - 2.5928 \cos 2\phi + 0.0069 \cos^2 2\phi - 3.068 \times 10^{-6} \text{ H}$$

Write, keypunch, and execute a FORTRAN program that gives g for $0 \leq \text{H} \leq$ 100,000 meters at 45° North latitude.

4–5 SUMMARY

The engineer formulates and manipulates models of physical systems. These models represent the relationship among a number of physical quantities, many of which are obtained through measurements. Much of an engineer's work is acquiring, recording, communicating, and manipulating measured quantities. This chapter discussed some of the basic concepts concerning measurements and how the engineer should approach them.

The measured quantity is a *dimension,* which consists of a magnitude and units. The accuracy and precision of a measurement are both properties of its magnitude. *Accuracy* is the closeness of the result to the true value of the quantity being measured. *Precision* is the smallest increment of a unit that the instrument chosen to measure the quantity can provide. Two basic principles in measuring physical quantities are the following: (1) It is first necessary to establish the required accuracy and precision and to employ the proper instrument to obtain that precision, repeating the measurements and averaging them if necessary to obtain the desired accuracy. (2) A measurement must be expressed to no more than one or two uncertain digits.

In performing calculations with measured quantities, one must be sure that the units associated with each dimension are stated and properly manipulated. The units associated with the various terms in a mathematical model must be consistent, and the appropriate conversion factors must be applied when necessary. A model is a useful tool in engineering design and analysis, but it is no more valuable than a disciplined approach to manipulating it.

PROBLEMS

P4–1 Write the following expressions in terms of their SI defined units.

(a) 12,427 miles
(b) 6,743 pounds mass
(c) 25.4 centimeters
(d) 0.535 liter
(e) 11,100 grams
(f) 1.26 kilometers
(g) 4.2 minutes
(h) 0.6 hour
(i) 5 feet, 10 inches
(j) 210 lb_f/in^2

P4–2 Ohm's law states that the current flow in amperes is directly proportional to the applied electromotive force (emf) in volts and inversely proportional to the resistance in ohms. What is the current flow when an emf of 50 volts is applied to a resistance of 16 kilo-ohms?

P4–3 A mechanical engineer heats an experimental apparatus from 260 K to 410 K. Express the temperature change in (a) degrees Kelvin, (b) degrees Celsius, (c) degrees Fahrenheit, (d) degrees Rankine.

P4–4 The quantity of heat flowing through a material of thickness L, with one side at a temperature T_2 and the other side at a lower temperature T_1, is given by the equation

$$Q = \frac{kA(T_2 - T_1)}{L} \text{ watts}$$

where
k = thermal conductivity of the material, watts/(m)(°C)
A = cross-sectional area of the material, m^2
T = temperature, °C
L = thickness of the material, m

How much heat is lost through a square meter of a wall 10 cm thick if the inside temperature is 21°C, the outside temperature is 6°C, and the thermal conductivity of the wall is 1.41 watts/(m)(°C)?

P4–5 (a) A force F causes a mass of 24 kg to be accelerated at 8 m/sec². What is the value of F?
(b) A force of 160 N is applied to a mass of 40 kg. What is the resulting acceleration of this body?
(c) The Earth attracts a mass of 1 kg with an acceleration of 9.81 m/sec². What is the force of gravity on this mass?
(d) A force of 1200 N acts for 15 seconds to change the velocity of a rocket from 318 m/sec to 1848 m/sec. What is the mass of the rocket?
(e) A satellite moves through space with an acceleration of 48 m/sec² under the influence of a force of 9.6 kN. What is the mass of the satellite?
(f) A mass of 7.4 slugs as measured on the Earth is taken to the moon, where the acceleration due to gravity is 5.33 ft/sec². What is its mass on the moon?

P4–6 The force of attraction between two bodies is directly proportional to the product of the masses and inversely proportional to the square of the distance separating them. Suppose that this relationship is stated as $F = km_1m_2/d^2$; what are the units associated with the constant k in the SI system? (The magnitude of k is $6.67(10^{-11})$ in these units.)

P4–7 Referring to Problem P4–6, calculate the mass of the sun, given that the Earth (whose mass is 6×10^{24} kg) has an orbital diameter of 1.49×10^{10} meters and that the force of attraction between sun and Earth is 1.44×10^{25} N.

P4–8 The expression for the power P in watts in an electrical circuit is $P = EI = I^2R$, where I is the current in amperes, E is the voltage in volts, and R is the resistance in ohms. The volt is defined as the electrical potential between two points when 1 joule of work is required to carry 1 coulomb of charge between these points. The ampere is defined as the flow of 1 coulomb per second in a circuit. Derive an expression for the power in the circuit.

P4–9 What is your mass in slugs? your weight in newtons? your height in meters?

P4–10 (a) The radius of the proton of an atom is approximately 10^{-12} mm. The radius of the observable universe is about 10^{23} km. Identify a meaningful distance that is approximately halfway between these two extremes on a logarithmic scale.
(b) The mean life of a neutral pion is about 2×10^{-16} second. The age of the universe is about 4×10^9 years. Identify a physically meaningful time interval that is approximately halfway between these two extremes on a logarithmic scale.

P4–11 A body dropped from rest falls freely. Its position after time t seconds is

$$y = -\tfrac{1}{2}gt^2$$

where g is the acceleration due to gravity (32.2 ft/sec^2). Suppose that a photographer taking pictures from a helicopter hovering 60 feet above the ground drops a camera. How long will it take for the camera to hit the ground?

P4–12 The ideal-gas equation states that the pressure–volume–temperature relationship for n moles of an ideal gas is given by

$$PV = nRT$$

where P is the pressure, V the volume, and T the temperature of n moles of the gas. R is called the *universal gas constant* and is evaluated at standard conditions of temperature and pressure.
(a) A mole of an ideal gas at 273.16°K and 1 atmosphere pressure is known to occupy 22.4 liters. What are the value and the unit for R? [*Note:* A mole of an ideal gas is 6.023×10^{23} molecules of the substance.]
(b) Given that 1 atmosphere = 1.013×10^5 N/m^2, what are the value and the unit of R in SI units?
(c) Given that 1 atmosphere = 14.7 lb$_f$/in^2, what are the value and the unit of R in American engineering units?

P4–13 The angular velocity ω of a body about its axis of rotation is given by

$$\omega = \omega_0 + \alpha t$$

where ω_0 is its initial angular velocity (radians/sec) and α is its angular acceleration (radians/sec^2). Suppose that a grindstone has a constant angular acceleration α of 3.0 radians/sec^2, and that its initial angular velocity is zero. What is its angular velocity after 4 seconds?

P4–14 The angular displacement ϕ of a rotating body is given by

$$\phi = \omega_0 t + \tfrac{1}{2}\alpha t^2$$

where ω_0, α, and t are as defined in Problem P4–13. What is ϕ after 4 seconds?

P4–15 Curtain rods are manufactured in two tubular sections, one of which slides into the other for mounting on the window. The design specifies that the inside diameter of the outer rod should be 2.00 cm and the outside diameter of the inner rod, 1.95 cm. How large can the percent error be and still allow a minimum clearance of 0.01 cm between rods? With this error, what is the greatest clearance one could obtain between the inner and outer rods? Depict these two extremes with a sketch of a cross section of the fitted rods.

CHAPTER FIVE
OPTIMIZATION AND ENGINEERING MODELS

The purpose of modeling a physical system is to be able to better understand the fundamental mechanisms of that system and to establish the *optimum* conditions for the construction and operation of the system. *Optimum* conditions are those that produce the best, most favorable, or most beneficial results from a system. Results and conditions are key words. The word results suggests that there is some dependent variable by which we compare one solution with another; this dependent variable is called a *criterion.* The word conditions refers to the values we assign to one or more independent variables to achieve some result for the criterion. Thus we manipulate independent variables to achieve the optimum value for a criterion. For example, we might experiment with the amount of curing agent needed in an adhesive to minimize hardening time. Hardening time is the criterion, while the amount of curing agent is the independent variable.

The concept of an optimum solution is important in engineering. Whether faced with determining the optimum temperature for operating a chemical reactor, the optimum mix of ingredients in structural concrete, or the optimum sequence of machining operations on a fabricated part, the engineer wants to achieve the best results. Thus *optimization,* the process of seeking the best solution, is a fundamental part of the engineer's work. This chapter examines the basic concepts of optimization and treats several procedures for determining optimum solutions. The procedures are presented in an elementary way and applied to small-scale problems so that you will be able to understand them with your present mathematical background. The techniques presented in the earlier part of the chapter emphasize numerical solutions, while those encountered later in the chapter employ the concept of the derivative from calculus. Before we study optimization methods, we shall examine several ways of classifying these methods.

5–1 CLASSIFYING OPTIMIZATION METHODS

We have said that *optimization* is the process of seeking the best value, condition, or solution to a problem. In optimization, we first describe a given system in terms of a mathematical model of the form

$$y = g(x_1, \ldots, x_n) \qquad (5\text{–}1)$$

where y is the criterion, which is some measure of system effectiveness, x_i $(i = 1, \ldots, n)$ are n independent variables, and $g(\cdot)$ is some relationship between y and x_i $(i = 1, \ldots, n)$. We then seek the values of the n independent variables that produce the optimum value for the criterion.

One means of categorizing methods of optimization is by the number of independent variables. *Single-variable methods* are used with a function of a single variable $g(x)$, and *multivariable methods* are used for optimizing a function of several variables $g(x_1, \ldots, x_n)$.

A second scheme for classifying methods of optimization depends on whether the method employs derivatives. This is, there are *derivative methods* and *numerical methods.* Derivative methods seek to determine those values of $x_1, \ldots, x_n$ for which the first derivative vanishes; that is, for a single variable, the value of x for which

$$y' = \frac{d[g(x)]}{dx} = g'(x) = 0 \tag{5–2}$$

We shall illustrate this concept later in this chapter for functions of a single variable. We use numerical methods when the derivative either does not exist or is too complicated to manipulate. Numerical methods typically use algorithms, and they are therefore well suited for implementation on a digital computer.

A third way to classify optimization methods is to distinguish between *unconstrained* and *constrained* problems. In constrained problems, the solution must satisfy special conditions. Equation (5–1) gives the unconstrained case. The general form of the constrained problem is as follows:

$$\text{Optimize } y_0 = g_0(x_1, \ldots, x_n) \tag{5–3}$$

subject to

$$a_i \leq x_i \leq b_i, \qquad i = 1, \ldots, n \tag{5–4}$$

$$y_j = g_j(x_1, \ldots, x_n) \begin{Bmatrix} \leq \\ = \\ \geq \end{Bmatrix} d_j, \qquad j = 1, \ldots, m \tag{5–5}$$

Equation (5–3) is exactly like (5–1); it states that we want to find conditions $x_1^*, \ldots, x_n^*$ that optimize a primary criterion to yield y_0^*. However, Equation (5–4) limits the range of values that the ith independent variable can assume; that is, x_i can range between a lower bound a_i and an upper bound b_i. These bounds might be due to limitations in the equipment's capability, to costs in a competitive marketplace, to specifications of the product's performance, or to legal restrictions established by federal, state, and local codes. The expression given in (5–5) states that there are other criteria that must be controlled. These criteria also depend on the independent variables $x_1, \ldots, x_n$. Thus changes in the variables $x_1, \ldots, x_n$ that are favorable to the primary criterion y_0 might be unfavorable to one or more of the other criteria $y_1, \ldots, y_m$. As with Equation (5–4), these restrictions might be due to any number of technical, economic, social, or legal factors.

We shall examine optimization procedures for each of these three classifications of methods. We shall first treat the simplest techniques—numerical methods for optimizing a function of a single independent variable. Later in the chapter, we shall examine linear programming, a technique by which we can solve the constrained problem. Finally, we shall study a method based on derivatives.

5–2 SEARCH METHODS

Nonderivative methods of optimization are generally called *search methods*. With these methods, we numerically or graphically evaluate a function of the form expressed by Equation (5–1). We start at one or more values for the independent variables X and search for better solutions until we find a point $\hat{X}$ that promises to be the best point that can be found. This point $\hat{X}$ is usually close to, but not exactly equal to, the true optimum X^*. (The notation $\hat{X}$ refers to an approximately optimum point, while X^* refers to an exactly optimum point.) The quantity X can be either the single variable x or the set of variables x_i, $i = 1, \ldots, n$. Procedures applied to functions of a single variable are called *unidimensional* search techniques, while those used for functions of several variables are referred to as *multidimensional* search methods.

UNIDIMENSIONAL SEARCH

Unidimensional search is a numerical or graphical procedure for determining the optimum solution to a problem of the form

$$y = g(x) \tag{5–6}$$

We assume that the function $g(x)$ is complicated enough to render derivative methods impractical, if not impossible. As a practical matter, we are usually content in unidimensional search to find a solution $\hat{y}$ at a value $\hat{x}$ that lies sufficiently close to, but is not exactly equal to, the true optimum $x*$. We are willing to isolate some *final interval of uncertainty,* given by L in the relation

$$L = x_u - x_l \tag{5–7}$$

where x_u is the upper end of the final interval of uncertainty and x_l is the lower end. This interval contains both the true optimum $x*$ and the estimated optimum $\hat{x}$. Therefore

$$x_l \leqslant \hat{x} \leqslant x_u \tag{5–8}$$

and

$$x_l \leqslant x* \leqslant x_u \tag{5–9}$$

The purpose of unidimensional search is to identify $\hat{x}$ in the final interval of uncertainty expressed by Equation (5–8).

The simplest unidimensional search technique is *exhaustive search*. Suppose we have a function $g(x)$ and an initial interval $[a,b]$ that we believe contains the optimum solution $x*$. We want to find an estimated solution $\hat{x}$ that lies within a final interval of uncertainty given by Equation (5–8). The procedure for exhaustive search is as follows.

Step 1. Establish the magnitude L of the desired final interval of uncertainty. Let the spacing between the observations of the independent variable x be Δx, where

$$\Delta x = \frac{L}{2} \tag{5–10}$$

Thus $\hat{x}$ and $x*$ both lie in the final interval of uncertainty, as stated in Equations (5–8) and (5–9). Furthermore, we are guaranteed that the worst error we can make is Δx. That is, Δx is greater than or equal to $|x* - \hat{x}|$.

Step 2. Divide the initial interval $[a,b]$ into k intervals of width Δx, where

$$k = \frac{b - a}{\Delta x} \tag{5–11}$$

Step 3. Place $k + 1$ observations at points

$$x^i = a + i\,(\Delta x), \qquad i = 0, \ldots, k. \tag{5–12}$$

Measure or compute the value of $g(x^i)$ at each of these $k + 1$ points. The superscript i here denotes the ith point in the search; it is not to be confused with an exponent or power.

Step 4. Select as a solution the point $\hat{x}$ such that $\hat{y}$ is the maximum (or minimum) computed value among the $k + 1$ values $g(x^i)$, $i = 0, \ldots, k$. The final interval of uncertainty is therefore

$$(\hat{x} - \Delta x) \leqslant x* \leqslant (\hat{x} + \Delta x) \tag{5–13}$$

To illustrate exhaustive search, let's consider the problem of finding the minimum of the function

$$g(x) = x^2 - 6x + 3$$

in the initial interval [0,7], such that the final interval of uncertainty is 1.0. Therefore the spacing between observations is

$$\Delta x = \frac{L}{2} = 0.5, \qquad k = \frac{b - a}{\Delta x} = \frac{7 - 0}{0.5} = 14$$

Hence $k + 1 = 15$ observations must be placed in this initial interval. Table 5–1 gives the values of $g(x^i)$ at each of the points x^i. Figure 5–1 graphically illustrates this exhaustive search. Note that we make all 15 observations in this search simultaneously. We obtain the solution simply by inspecting the values of $g(x^i)$ and selecting the minimum such value. Inspecting the results in Table 5–1, we see that the *estimated* optimum solution is $\hat{x} = 3.0$, yielding the minimum $g(\hat{x}) = -6.00$. This solution also happens to be the *true* optimum. Only occasionally does a search technique produce the exact solution, however.

Exhaustive search is very simple, but it requires many observations to achieve accurate results. It is inefficient because all $k + 1$ observations are taken independently. Hence one foregoes the opportunity to gain information from a few observations and use that information to place subsequent observations more strategically. Sequential search procedures are more efficient because they observe one or two points at a time.

The improved computational efficiency of sequential search procedures over exhaustive search depends on a crucial assumption—that the function $g(x)$ is unimodal† in the interval $[a,b]$. Figure 5–2 illustrates unimodality. Sequential search procedures eliminate an unfavorable interval after each block of one or two search points. Thus we have to be sure that the eliminated interval does not contain

†Unimodal means having a single mode. A mode is a peak or a trough, depending on the nature of the function and the objective of the optimization.

TABLE 5–1. EXHAUSTIVE SEARCH APPLIED TO ILLUSTRATIVE PROBLEM

Point i	x^i	$g(x^i)$
0	0	3.00
1	0.5	0.25
2	1.0	−2.00
3	1.5	−3.75
4	2.0	−5.00
5	2.5	−5.75
6	3.0	−6.00
7	3.5	−5.75
8	4.0	−5.00
9	4.5	−3.75
10	5.0	−2.00
11	5.5	0.25
12	6.0	3.00
13	6.5	6.25
14	7.0	10.00

the optimum. Unimodality assures us that this is the case. We shall examine two procedures for sequential search in later sections. In the first procedure, *dichotomous* search, we observe pairs of points. In the second, called *golden-section* search, we start with a pair of observations, but proceed one point at a time thereafter.

Dichotomous search, like exhaustive search, seeks to determine a final interval of uncertainty L containing both the measured optimum $\hat{x}$ and true optimum $x*$. The procedure for maximizing a criterion consists of the following steps, as Figure 5–3 illustrates.

Step 1. Evaluate the function $g(x)$ at two points separated by $\Delta x = L/2$ and located about the midpoint of the interval $[a,b]$. That is, compute values of the function $g(x)$ at points

$$x^1 = \tfrac{1}{2}(a + b - \Delta x), \qquad x^2 = \tfrac{1}{2}(a + b + \Delta x) \tag{5–14}$$

Thus x^1 and x^2 are $\Delta x/2$ to the left and to the right, respectively, of the midpoint $(a + b)/2$.

Step 2. If $g(x^1) < g(x^2)$, eliminate the interval to the left of x^1. If $g(x^1) > g(x^2)$, eliminate the interval to the right of x^2. Let a be the left end of this reduced interval and b be the right end. (If $g(x^1) = g(x^2)$, eliminate neither interval, since the optimum could lie in either the left interval or right interval. In that case, continue the search in both intervals.)

FIGURE 5–1 Exhaustive search applied to illustrative problem

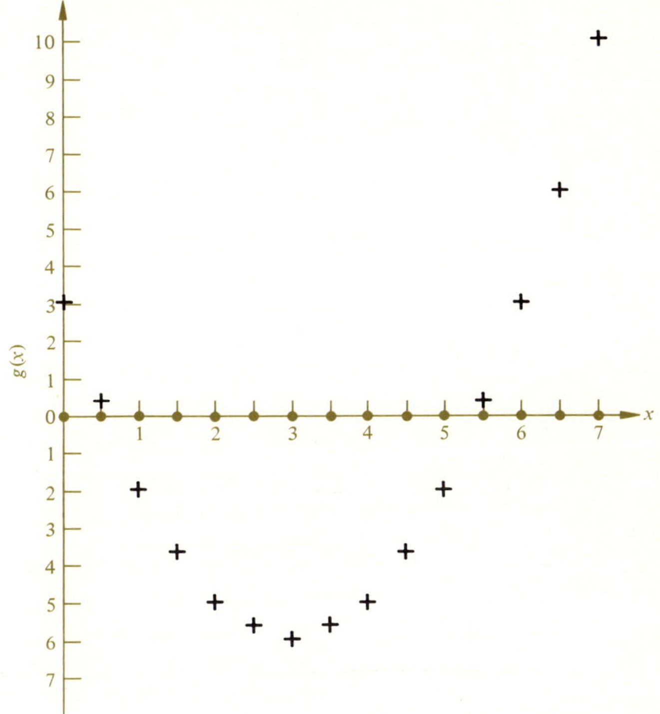

FIGURE 5–2 The concept of unimodality: (a) multimodal, (b) unimodal and concave, (c) unimodal and convex

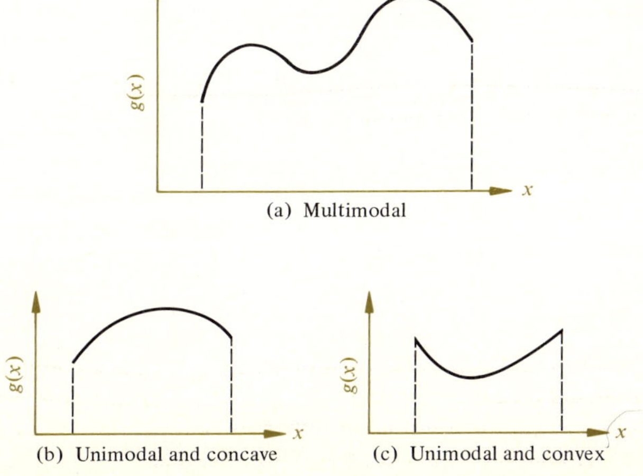

FIGURE 5–3 Two sequential steps in dichotomous search: (a) Eliminate $a \leq x \leq x^1$. (b) Eliminate $x^4 \leq x \leq b$.

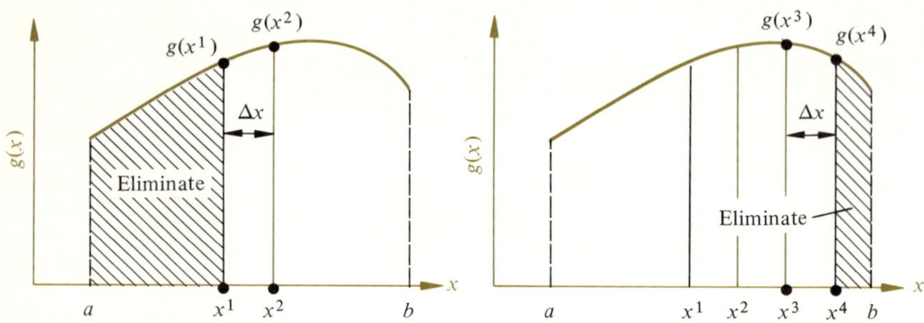

Step 3. Repeat Steps 1 and 2 in the remaining interval of uncertainty $[a,b]$. Terminate the search when $(b - a) \leq L$, and take the value of x that gave the maximum $g(x)$ as the estimated solution.

To *minimize* $g(x)$, reverse the direction of each of the inequalities in Step 2. The estimated solution $\hat{x}$ is then the search point that yields the minimum value of the function $g(x)$.

To illustrate dichotomous search to locate a minimum value, we shall consider the example used to illustrate exhaustive search. That is, we want to find the minimum of the function

$$g(x) = x^2 - 6x + 3$$

in the interval [0,7]. Again suppose that the final interval of uncertainty L is to be no greater than 1.0. Therefore the spacing between the pair of points x^1 and x^2 is

$$\Delta x = \frac{L}{2} = 0.5$$

Figure 5–4 illustrates the dichotomous search for this problem, and Table 5-2 shows the numerical results.

The first two search points are $x^1 = 3.25$ and $x^2 = 3.75$. They bracket the midpoint, 3.50, of the initial interval. Since $g(x^1) < g(x^2)$, as seen in Table 5–2, eliminate the interval $x > 3.75$. The midpoint of the remaining interval ($a = 0$, $b = 3.75$) is $x = 1.88$. The second pair of search points is $x^3 = 1.63$ and $x^4 = 2.13$, so eliminate the interval $x < 1.63$. The third pair of observations is $x^5 = 2.44$ and $x^6 = 2.94$; eliminate the interval $x < 2.44$. Finally, taking the fourth pair of search points, $x^7 = 2.85$ and $x^8 = 3.35$, eliminates $x > 3.35$, and terminates the search, since $(b - a) = (3.35 - 2.44) < 1.0$. The estimated optimum in this interval is $\hat{x} = 2.94$, which yields the value $g(\hat{x}) = -6.00$.

This dichotomous search required 8 search points; exhaustive search required 15. The estimated optimum was only very slightly removed from the true optimum at $\hat{x} = 3.0$. In fact, the value of the function at $g(\hat{x})$ was essentially the same as that at $g(x*)$. This comparison reveals the improved efficiency of placing search points two

FIGURE 5–4 Dichotomous search applied to illustrative problem

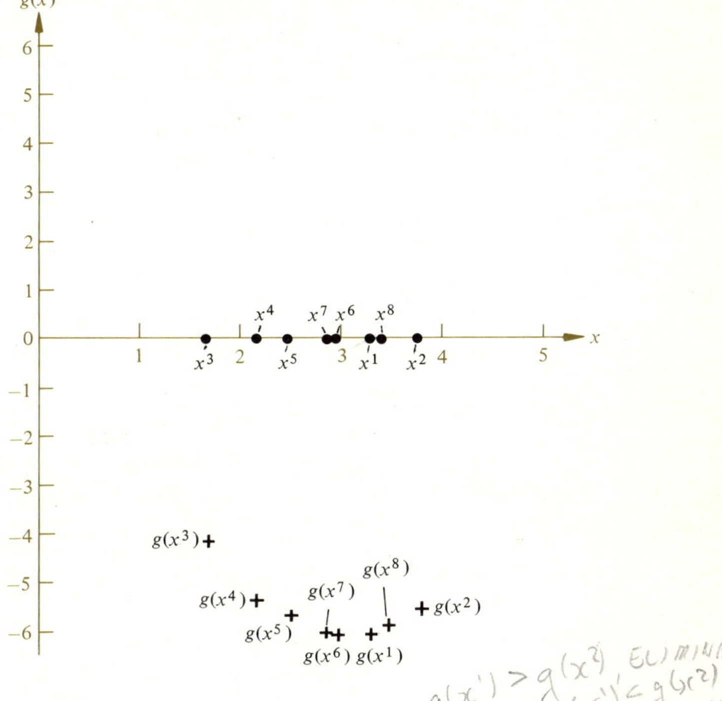

TABLE 5–2. DICHOTOMOUS SEARCH APPLIED TO ILLUSTRATIVE PROBLEM

Point pair	*Midpoint*	*Search point*	*Value of x*	*Value of g(x)*	*Interval eliminated*	*(b − a) remaining*
		x^1	3.25	−5.94		
1	3.50				$x > 3.75$	3.75
		x^2	3.75	−5.44		
		x^3	1.63	−4.12		
2	1.88				$x < 1.63$	2.12
		x^4	2.13	−5.24		
		x^5	2.44	−5.69		
3	2.69				$x < 2.44$	1.31
		x^6	2.94	−6.00		
		x^7	2.85	−5.98		
4	3.10				$x > 3.35$	$0.91 < L$
		x^8	3.35	−5.88		

at a time instead of all at once. One might wonder whether placing search points one at a time would be even more efficient.

Several sequential search methods employ one search point at a time. Perhaps the simplest of these is *golden-section search*. This technique derives its name from the geometric relationship between the dimensions r and s in the rectangles in Figure 5–5. Let us choose r and s so that, if we remove an area s^2 from the original rectangle, we are left with a rectangle having exactly the same length-to-width ratio as the original rectangle. That is,

$$\frac{r}{s} = \frac{s}{r - s} \qquad \text{or} \qquad r^2 - sr - s^2 = 0$$

Solving for r in terms of s, we obtain

$$r = \frac{s}{2}(1 \pm \sqrt{5})$$

Selecting the positive root and forming the ratio r/s gives us

$$\frac{r}{s} = \tfrac{1}{2}(1 + \sqrt{5}) = 1.618 \tag{5–15}$$

This ratio is known as the *golden section,* which we will denote by ρ. This quantity is such that if we place a point 1/1.618 of the distance from one end of an interval, then the distance from that end of the interval to the point is 0.618 or ρ-1 of the entire interval.

The significance of the golden section in unidimensional search becomes clear if we examine the following procedure for maximizing a criterion $g(x)$ in an interval $[a,b]$.

Step 1. Place two observations in the interval $[a,b]$ so that

$$x^1 = b - (\rho - 1)(b - a) \tag{5–16}$$
$$x^2 = a + (\rho - 1)(b - a) \tag{5–17}$$

FIGURE 5–5 The golden section—a geometric rationale

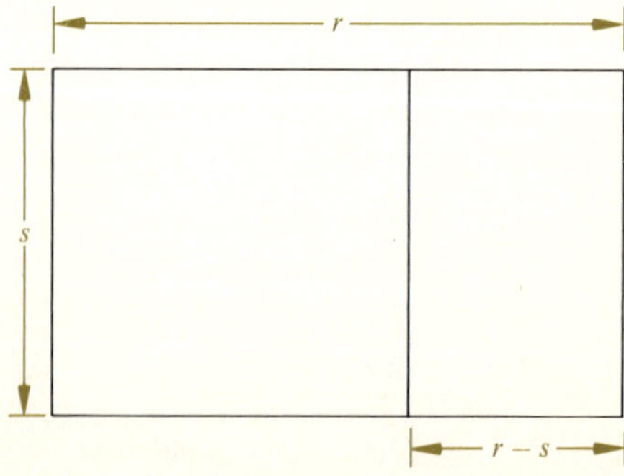

FIGURE 5–6 Symmetry of search points in golden-section search

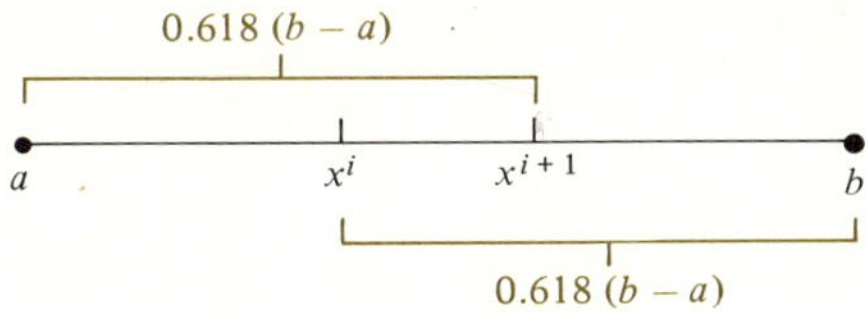

Step 2. If $g(x^1) < g(x^2)$, eliminate the interval to the left of x^1. If $g(x^1) > g(x^2)$, eliminate the interval to the right of x^2. Let a be the left end of this new interval and b be the right end. This new interval is $(\rho - 1)(b - a)$, or 61.8% of the preceding interval. (If $g(x^1) = g(x^2)$, neither interval can be eliminated and the search must be continued in both intervals.)

Step 3. The new interval $[a,b]$ contains a single point at a distance $(\rho - 1)(b - a)$ from one end. Place a new point symmetrically in this interval a distance $(\rho - 1)(b - a)$ from the other end. Figure 5–6 illustrates this process.

Step 4. Repeat Steps 2 and 3 until $(b - a) \leq L$. Take the best point $\hat{x}$ as the estimated optimum solution.

To illustrate golden-section search with our sample problem, we place the first two points at

$$x^1 = b - (\rho - 1)(b - a) = 7 - (0.618)(7) = 2.67$$
$$x^2 = a + (\rho - 1)(b - a) = 0 + (0.618)(7) = 4.33$$

The values of y at these points are

$$g(x^1) = (2.67)^2 - 6(2.67) + 3 = -5.89$$
$$g(x^2) = (4.33)^2 - 6(4.33) + 3 = -4.23$$

Hence we eliminate the interval to the right of x^2, leaving x^1 located $\rho - 1$ from the *left* end of the remaining interval. Placing x^3 a distance $\rho - 1$ from the *right* end of this interval, we have $x^3 = 1.65$ and $g(x^3) = -4.18$, eliminating the interval to the left of x^3. Continuing in this manner, we obtain the results shown in Table 5–3.

TABLE 5–3. GOLDEN-SECTION SEARCH APPLIED TO ILLUSTRATIVE PROBLEM

Point i	*Value of x^i*	*Value of $g(x^i)$*	*Interval eliminated*	*$(b - a)$ remaining*
1	2.67	−5.89	—	—
2	4.33	−4.23	$x > 4.33$	4.33
3	1.65	−4.18	$x < 1.65$	2.68
4	3.30	−5.91	$x < 2.67$	1.66
5	3.70	−5.51	$x > 3.70$	1.03
6	3.06	−6.00	$x > 3.00$	0.63

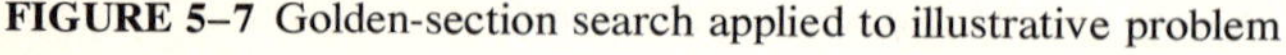

FIGURE 5–7 Golden-section search applied to illustrative problem

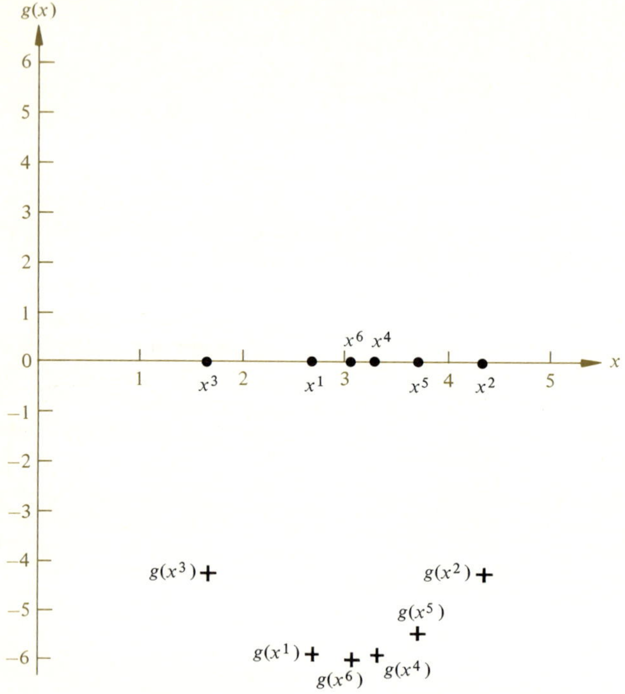

Figure 5–7 illustrates the progress of this golden-section search. We see that only 6 observations are needed to produce the estimated optimum $\hat{x} = 3.06$, $g(\hat{x}) = -6.00$, compared to 8 observations in dichotomous search, and 15 in exhaustive search. Hence we achieve the most efficient search by placing one search point at a time.

We have examined three different techniques for optimizing a function of a single variable x. These techniques are both graphical and numerical. Moreover, each of these techniques is algorithmic; it consists of a definite set of rules that could be applied to any given problem. It is not difficult to write a FORTRAN computer program to carry out each of these procedures.

Exercises

5–1 Develop a flow chart for the exhaustive search procedure. Write, keypunch, and execute a FORTRAN program to carry out exhaustive search. Test your program by maximizing the function $y = x \cos x$ in the interval $0 \leqslant x \leqslant \pi$. Let $L = 0.1$.

5–2 Repeat Exercise 5–1 with a dichotomous search.

5–3 Repeat Exercise 5–1 with a golden-section search.

MULTIDIMENSIONAL SEARCH

When optimizing a function $g(x)$ of a *single* variable, one can be reasonably certain of obtaining a good solution in a manageable number of calculations. However, the problem of locating the optimum of a function $g(x_1, \ldots, x_n)$ of *several* variables is considerably more difficult. This difficulty arises due to the so-called *curse of dimensionality,* that is, the presence of several independent variables $x_1, \ldots, x_n$. In unidimensional search, if we want to locate an estimated optimum $\hat{x}$ in a final interval of uncertainty of 1.0 unit and we start with an initial interval of 10.0 units, the reduction is 1/10. This is demonstrated in Figure 5–8(a). Now suppose that we want to locate the optimum values of *two* independent variables x_1 and x_2. If the final interval of uncertainty is 1.0 unit and the initial interval is 10.0 units, the required reduction is $(1/10)^2$ or 1/100, as shown in Figure 5–8(b). For the three-variable problem, we would need a cube only $(1/10)^3$ or 1/1000 of the original region. In general, if we start with an interval $[a_i, b_i]$ for each of the n independent variables x_i, and we want a final interval of uncertainty L_i for the ith independent variable, the required reduction is

$$R = \left(\frac{L_1}{b_1 - a_1}\right)\left(\frac{L_2}{b_2 - a_2}\right) \cdots \left(\frac{L_n}{b_n - a_n}\right) \tag{5–18}$$

Despite this "curse of dimensionality," there are several multidimensional search techniques that are conceptually simple as well as computationally efficient. The following sections describe two of the simpler multidimensional search techniques—sectional search and simplicial search.

FIGURE 5–8 The "curse of dimensionality" in optimization: (a) single variable, (b) two variables

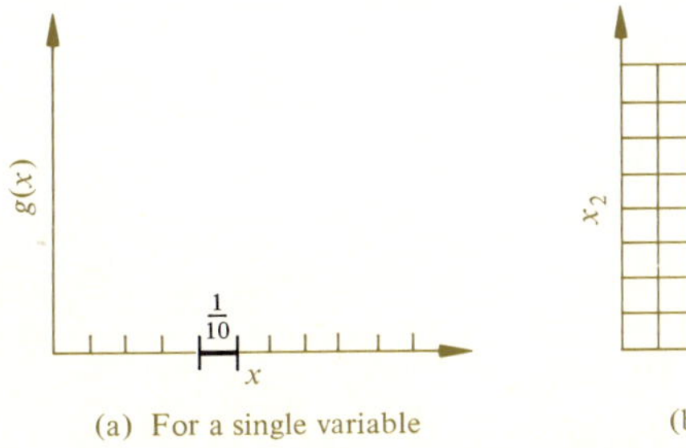

(a) For a single variable

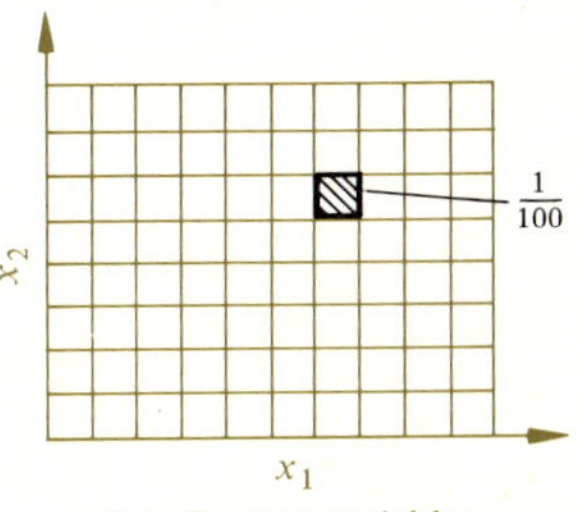

(b) For two variables

SECTIONAL SEARCH

Sectional search is simply unidimensional search applied one variable at a time to a function of several variables $g(x_1, \ldots, x_n)$. Hence it is often called simply *one-variable-at-a-time search*. It is called *sectional* search because it is unidimensional search applied successively to sections through a surface in $n + 1$ dimensions; that is, n dimensions for the n independent variables x_i, plus one dimension for the criterion y. Figure 5–9 illustrates this concept. The procedure for sectional search is as follows.

Step 1. Let $x_2, \ldots, x_n$ remain constant at some starting values $(x_2^0, \ldots, x_n^0)$. Vary x_1 over its permissible range $[a_1, b_1]$, using one of the unidimensional search techniques described in the previous section. Locate the best value of x_1 along this *section;* that is, find the value $\bar{x}_1$ that maximizes the simplified criterion $g(x_1, \ldots)$.

Step 2. Repeat Step 1 in turn for each of the remaining independent variables $x_2, \ldots, x_n$, retaining previous estimates of $\bar{x}_1$, and so forth.

Step 3. After completing a *cycle* of unidimensional searches along the n independent variables $x_1, \ldots, x_n$, repeat Steps 1 and 2 until the value of the criterion function $g(x_1, \ldots, x_n)$ does not improve any further. Terminate the search at the point $\hat{X} = (\hat{x}_1, \hat{x}_2, \ldots, \hat{x}_n)$ at which no further improvement is gained. This point $\hat{X}$ is considered an *estimated optimum*. As with unidimensional search procedures, there is no guarantee that $\hat{X}$ is the same as the *true optimum* X^*.

Figure 5–9 illustrates a sectional search in two dimensions.

As an example of sectional search, consider the problem of minimizing the criterion function

$$g(x_1, x_2) = 6x_1^2 + 3x_2^2$$

Let us arbitrarily start at $x_2 = 2$. Substituting this value into the above equation yields the simplified function

$$g(x_1, \ldots) = 6x_1^2 + 12$$

We see immediately that this function has a minimum value at $\bar{x}_1 = 0$. Substituting this value into the original function produces

$$g(\ldots, x_2) = 3x_2^2$$

We can easily see that this criterion has a minimum at $\bar{x}_2 = 0$. Furthermore we find that we are unable to obtain any further improvement in $g(x_1, x_2)$, so the optimum solution is

$$x_1 = 0, \qquad x_2 = 0, \qquad g(x_1, x_2) = 0$$

This is, of course, the true optimum.

FIGURE 5–9 Sectional search (one variable at a time): two variables, six unidimensional searches

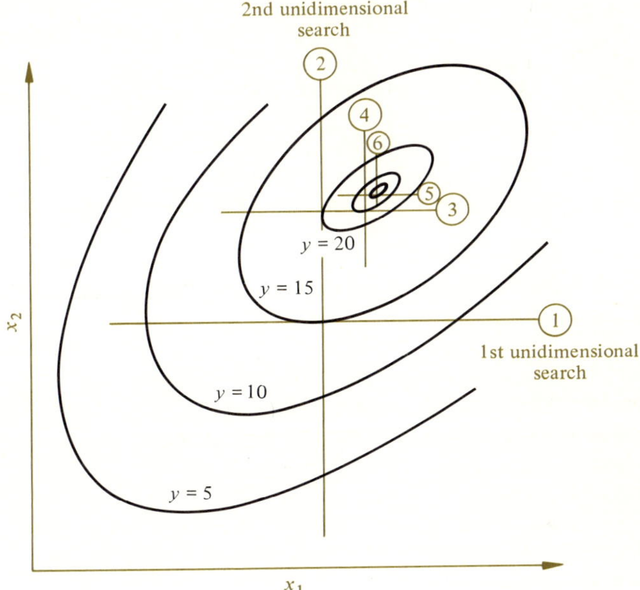

Now this example is quite trivial, but it illustrates the basic steps involved in a sectional search. Had the simplified criterion functions $g(x_1, \ldots)$ and $g(\ldots, x_2)$ required it, we could have applied one of the unidimensional search procedures described earlier. For instance, we could have minimized the simplified criterion $g(\ldots, x_i, \ldots)$ by employing a golden-section search.

SIMPLICIAL SEARCH

Another multidimensional search procedure is *simplicial search.* This scheme involves placing $n + 1$ observations at the vertices of an n-dimensional simplex, which is an n-dimensional geometric figure having exactly n vertices. For instance, the three observations in $n = 2$ dimensions form the vertices of an equilateral triangle, while the four observations in $n = 3$ dimensions form a regular tetrahedron. To describe the locations of the $n + 1$ observations algebraically, define the quantities

$$p_n = \frac{\sqrt{n+1} - 1 + n}{n\sqrt{2}} \quad (5\text{–}19) \qquad\qquad q_n = \frac{\sqrt{n+1} - n}{n\sqrt{2}} \quad (5\text{–}20)$$

Then we place the $n + 1$ observations at point $X^0 = (x_1^0, x_2^0, \ldots, x_n^0)$, called the *starting point,* and at the n additional points

$$\begin{aligned} X^1 &= X^0 + a(p_n, q_n, \ldots, q_n) \\ X^2 &= X^0 + a(q_n, p_n, \ldots, q_n) \\ &\vdots \\ X^n &= X^0 + a\,(q_n, q_m, \ldots, p_n) \end{aligned} \tag{5–21}$$

Note that X^i is a point in n-dimensional space; its coordinates are $(x_1^i, \ldots, x_n^i)$. The quantity a is the length of each edge of the simplex. For example, for $n = 2$, the values of the quantities p_2 and q_2 are

$$p_2 = \frac{\sqrt{2+1} - 1 + 2}{2\sqrt{2}} = 0.966, \qquad q_2 = \frac{\sqrt{2+1} - 1}{2\sqrt{2}} = 0.259$$

If the starting point X^0 is taken at (0,0) and the length a of the edge is 1, the remaining points on the simplex are

$$\begin{aligned} X^1 &= X^0 + a(p_2, q_2) = (0,0) + (1)(0.966, 0.259) = (0.966, 0.259) \\ X^2 &= X^0 + a(q_2, p_2) = (0,0) + (1)(0.259, 0.966) = (0.259, 0.966) \end{aligned}$$

These three points form a regular triangle.

The next step in *simplicial search* is to evaluate the function $g(x_1, \ldots, x_n)$ at each vertex of the simplex, discard the worst point X^w, and create a new search point that is the reflection $X^{w'}$ of this worst point. For instance, Figure 5–10 illustrates a simplex for $n = 2$ and demonstrates how a new point is created by reflecting the worst point. The new point $X^{w'}$ and the n remaining points in the

FIGURE 5–10 Reflecting the worst point in simplicial search

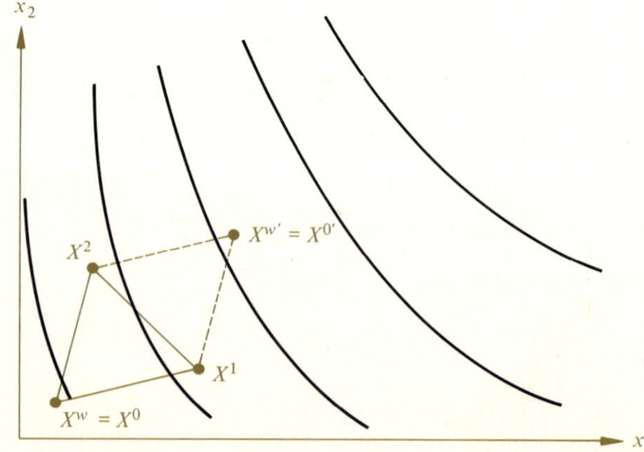

previous simplex together form a *new* simplex. We determine the placement of the reflected point by using

$$X^{w'} = \sum_{j=0}^{n} X^j - \frac{n+2}{n} X^w \qquad (5\text{-}22)$$

For example, suppose that in a particular problem the point $X^0 = (0,0)$ is the worst point X^w. Then the reflected point is

$$\begin{aligned} X^{0'} &= [X^0 + X^1 + X^2] - \frac{2+2}{2} X^0 \\ &= [(0,0) + (0.966, 0.259) + (0.259, 0.966)] - (2)(0,0) \\ &= (1.224, 1.224) \end{aligned}$$

We repeat this procedure until at some point the reflected point is one that we have already evaluated. Then we take the *second-worst point* as X^w and find its reflected point. Finally, when we cannot place a point without repeating a previous point, we terminate the search and select the best available point as the solution $\hat{X}$. Of course, the best point is the one that yields the maximum (or minimum) value of the criterion function $y = g(x_1, \ldots, x_n)$ among all points evaluated in the search. Figure 5-11 graphically illustrates a simplicial search in two dimensions.

FIGURE 5-11 Simplicial search in two dimensions

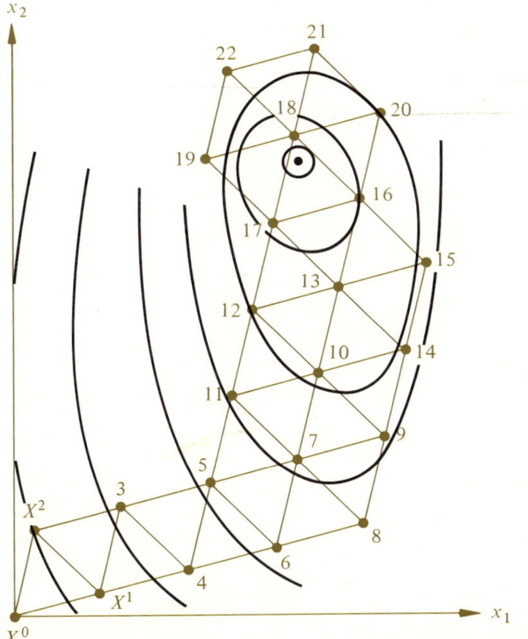

Exercises

5–4 Use a sectional search to minimize the criterion

$$g(x_1, x_2) = 2x_1^2 - 2x_1x_2 + 2x_2^2 - 6x_1 + 6$$

Start at $x_2 = 0$. If you use a unidimensional search, let $L = 0.2$.

5–5 Use a simplicial search to minimize the criterion given in Exercise 5–4. Use a side length $a = 0.5$ and a starting point $X = (0,0)$.

5–3 LINEAR PROGRAMMING

A tool of optimization that has been enormously successful in solving engineering problems, as well as problems in business, military, and social systems, is *linear programming*. Its origins date back more than 50 years to the developments in economic analysis by Leontief. However, not until the advent of the digital computer did linear programming blossom into the powerful tool that it has become in the last two decades. Almost coincident with the widespread use of the digital computer—and providing tremendous thrust to the evolution of linear programming—was the development of simplex algorithm† by Dantzig in 1947. This algorithm enables us to use the computer to solve large linear programming problems. More recent developments in linear programming have refined and improved the performance of the simplex algorithm.

Linear programming is a mathematical technique typically employed for finding the optimum use of limited resources. For instance, an industrial engineer might be interested in determining the combination of products manufactured in a plant that will both maximize profits and operate within constraints imposed by available equipment, workers, storage space, raw materials, and shipping capability. A metallurgical engineer might want to establish the optimum combination of ingredients to produce cast iron at minimum cost while satisfying various physical and chemical specifications for the product. The chemical engineer might need to find the optimum blend of different types of crude oil to maximize profits while satisfying demand for certain petroleum products, such as gasoline, heating oil, jet fuel, and lubrication oils. These are just a few of the technical problems for which engineers have used *linear programming* to obtain optimum solutions.

The most useful technique for solving a linear programming problem is the *simplex method*. Although the rules for applying the simplex method are not complex, a thorough development of its procedure is beyond the scope of this book. However, we can convey the rudiments of linear programming by using a *graphical* solution. But let us first examine the mathematical structure of the linear programming problem.

†The *simplex algorithm* referred to here is not to be confused with the "simplex" employed in simplicial search described in the previous section.

In the linear programming problem, we want to optimize a criterion function of the form

$$y = \sum_{i=1}^{n} c_i x_i = c_1 x_1 + c_2 x_2 + \cdots + c_n x_n \tag{5-23}$$

subject to restrictions or constraints of the form

$$\sum_{i=1}^{n} a_{ij} x_i \begin{Bmatrix} \leq \\ = \\ \geq \end{Bmatrix} b_j, \qquad j = 1, \ldots, m \tag{5-24}$$

and the *nonnegativity restrictions*

$$x_i \geq 0, \qquad i = 1, \ldots, n \tag{5-25}$$

For example, consider the following linear programming problem:

$$\text{Maximize } y = 3x_1 + 5x_2$$

subject to the constraints

$$0 \leq x_1 \leq 4, \qquad 0 \leq x_2 \leq 6, \qquad 3x_1 + 2x_2 \leq 18$$

This problem has the basic structure stated in relations (5-23) through (5-25), with $n = 2$ and $m = 3$.

A graphical solution to the linear programming problem is, of course, limited to two- and three-variable problems. The graphical solution of two-variable problems requires the following steps.

Step 1. Plot the constraints expressed by the inequalities (5-24) and (5-25) on a graph of x_2 versus x_1.

Step 2. Plot the *slope* of the criterion stated by Equation (5-23) on the same graph. This slope describes a family of parallel lines, increasing in the value of y in one direction and decreasing in the other.

Step 3. Identify the vertices of the region enclosed by the constraints. A vertex lies at the intersection of two or more constraints.

Step 4. Evaluate Equation (5-23) at each vertex. Select as a solution the vertex that has the maximum (or minimum) value of y. Of the family of lines with the slope determined in Step 2, the line having the greatest (or least) value of y passes through the vertex that is the optimum solution.

To illustrate this graphical procedure, suppose that Dynamic Automotive Corporation manufactures two classes of automobiles, a four-cylinder compact sedan and a six-cylinder compact station wagon. This company earns a profit of \$240 on each sedan and \$300 on each wagon. DAC can earn \$1.2 million profit annually by manufacturing 5000 sedans or 4000 station wagons. Moreover, certain *combinations* of sedans and wagons will also yield profits of \$1.2 million. The function that describes the annual profit earned by DAC is

$$y = \$240x_1 + \$300x_2$$

where x_1 is the number of sedans and x_2 is the number of station wagons. Figure 5–12 illustrates some "isoprofit" lines for the DAC operation. An isoprofit line has the slope of the criterion function and describes the combination of x_1 and x_2 that produces a given profit y. The prefix *iso* means "constant"; thus an *isoprofit line* is a line of constant annual profit y and is a specific type of criterion function.

DAC would, of course, want to manufacture all the automobiles it could sell, especially all the six-cylinder station wagons that yield higher profits. But DAC must meet some *constraints*. These constraints arise because the three major departments in the manufacturing process—the engine department, the body department, and the assembly department—each have limited capacity.

The engine department can produce as many as 7000 four-cylinder engines (sedans) each year if they build only four-cylinder engines, or up to 4000 six-cylinder engines (station wagons) each year if they build only six-cylinder engines. This department can handle both engines at the same time. We can write this constraint as

$$\frac{x_1}{7000} + \frac{x_2}{4000} \leqslant 1 \qquad \text{or} \qquad \frac{1}{7}x_1 + \frac{1}{4}x_2 \leqslant 1000$$

To verify this relation, let $x_2 = 0$ and restate the inequality in terms of x_1. Then let $x_1 = 0$ and restate the inequality in terms of x_2. Figure 5–13 depicts the region bounded by this constraint. Any point within the shaded region satisfies this constraint; that is, the engine department can produce any combination of engines that falls within this region.

The body department can manufacture up to 6000 sedan bodies each year, or 5000

FIGURE 5–12 Isoprofit lines for illustrative problem

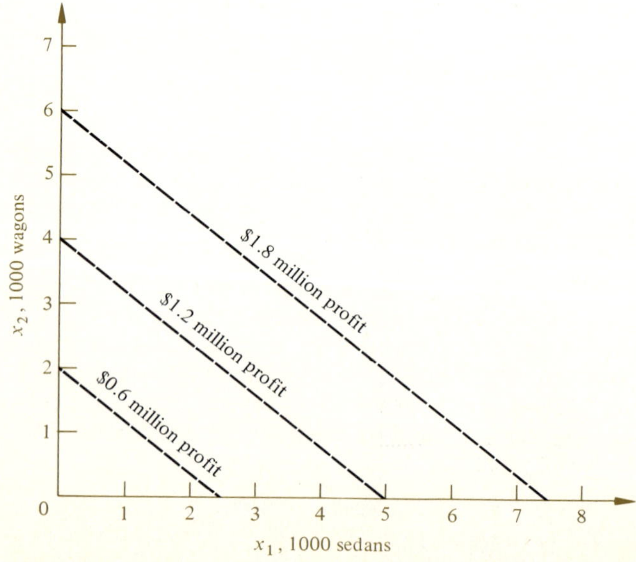

FIGURE 5–13 Feasible region resulting from the constraint on the engine department's capacity

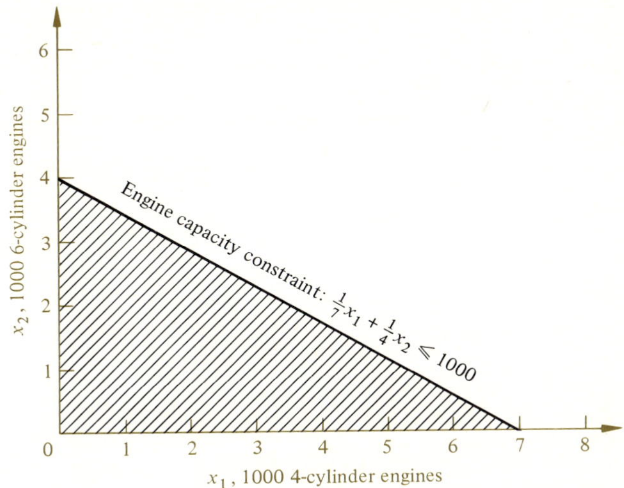

station wagon bodies, or combinations that lie between these values. The constraint arising from the body department's capacity is

$$\frac{1}{6}x_1 + \frac{1}{5}x_2 \leqslant 1000$$

Figure 5-14 depicts this constraint.

The assembly department handles the two types of automobiles on separate assembly lines. Each year it can handle up to 4500 sedans *and* up to 3500 station wagons. The constraints expressing this capacity are

$$x_1 \leqslant 4500, \qquad x_2 \leqslant 3500$$

Thus sedans and wagons do not compete for resources in the assembly department, as they do in the engine and body departments. The feasible region enclosed within these two assembly-department constraints is a rectangle, as shown in Figure 5–15.

The mathematical statement of the linear-programming problem for optimizing the DAC production is then

$$\text{Maximize } y = \$240\,x_1 + \$300\,x_2$$
$$\text{subject to } 0 \leqslant x_1 \leqslant 4500, \qquad 0 \leqslant x_2 \leqslant 3500$$
$$\frac{1}{7}x_1 + \frac{1}{4}x_2 \leqslant 1000$$
$$\frac{1}{6}x_1 + \frac{1}{5}x_2 \leqslant 1000$$

An important and universal characteristic of linear-programming problems is the *nonnegativity restriction* stated in (5–25). In this example, this restriction states that

FIGURE 5–14 Feasible region resulting from the constraint on the body department's capacity

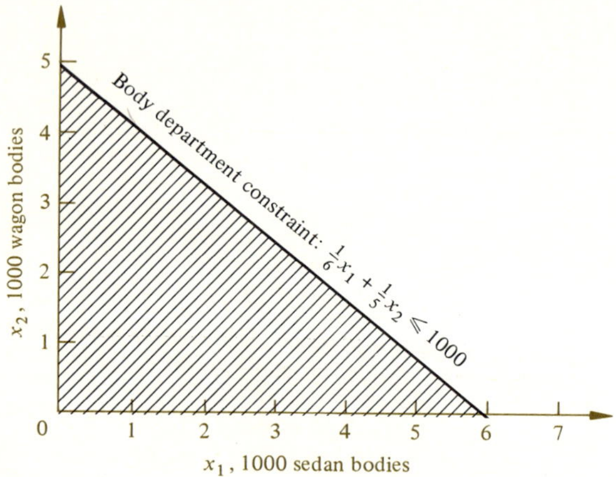

DAC can either manufacture no sedans (wagons) or a positive number of them, but it cannot produce a *negative* number of sedans (wagons).

Figure 5–16 is the graphical representation of this linear-programming problem. We observe that the isoprofit line of greatest value is the one that passes through the intersection of the engine-department constraint and the body-department constraint. Hence it satisfies these two constraints at their equalities and can be determined by solving two equations in two unknowns:

FIGURE 5–15 Feasible region resulting from the constraint on the assembly department's capacity

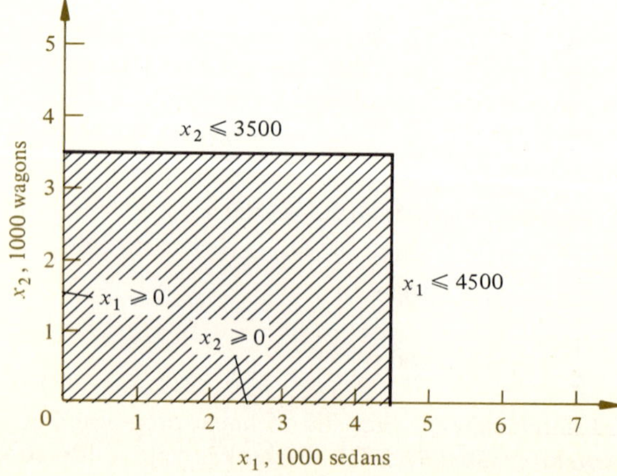

FIGURE 5-16 Graphical solution to the DAC linear programming problem

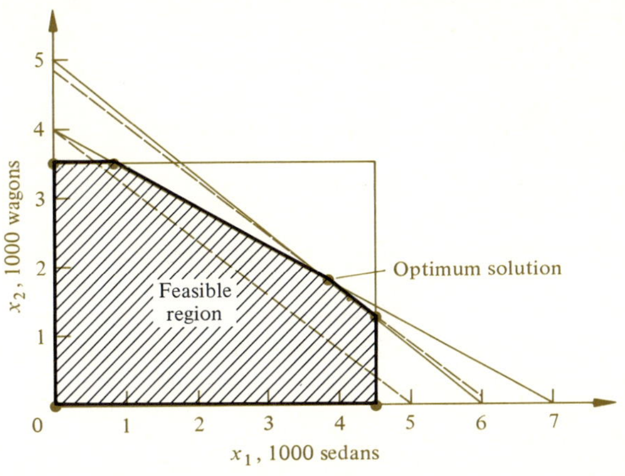

$$\frac{1}{7}x_1 + \frac{1}{4}x_2 = 1000, \qquad \frac{1}{6}x_1 + \frac{1}{5}x_2 = 1000$$

The algebraic solution of these equations yields the optimum solution,

$x_1^* = 3818$ four-cylinder compact sedans
$x_2^* = 1818$ six-cylinder compact station wagons
$y* = \$1,461,720$ profit

We can also obtain this solution by evaluating each vertex of the feasible region and selecting the vertex that yields the greatest annual profit. Proceeding in this way, we obtain the following results:

Vertex	*Annual profit*
$x_1 = 0;\ x_2 = 0$	0
$x_1 = 4500;\ x_2 = 0$	\$1,080,000
$x_1 = 0;\ x_2 = 3500$	\$1,050,000
$x_1 = 875;\ x_2 = 3500$	\$1,260,000
$x_1 = 4500;\ x_2 = 1250$	\$1,455,000
$x_1 = 3818;\ x_2 = 1818$	\$1,461,720

Of course, we could have obtained this solution simply by observing the graphical coordinates of the vertex, provided that the graph is sufficiently precise.

Exercises

5-6 Develop a graphical solution for the following linear-programming problem.

Maximize $y = 3x_1 + 5x_2$
subject to $0 \leqslant x_1 \leqslant 4, \quad 0 \leqslant x_2 \leqslant 6, \quad 3x_1 + 2x_2 \leqslant 18$

5–7 A manufacturer of television sets wants to produce the combination of black-and-white sets and color sets that will maximize profit and meet the restrictions imposed by plant capacity. Unit contributions to profit for black-and-white and color television sets are \$20 and \$50, respectively. The manufacturing process consists of four steps: (1) fabrication of major components, (2) chassis assembly, (3) set assembly, and (4) inspection and testing. The plant's daily capacity for components is 60 color tubes and 240 black-and-white tubes. In chassis assembly, each black-and-white set requires 5 work hours and each color set 15 work hours. The plant employs 195 workers in the chassis-assembly department. Each of these persons works 8 hours per day. In set assembly, each black-and-white set requires 1.2 work hours and each color set, 2.0 work hours. This department has 36 workers per 8-hour day. In final inspection and testing, each black-and-white set requires 0.6 work hour and each color set 1.8 work hours. The plant employs 18 inspectors per 8-hour day. Determine the optimum number of each class of television set produced each day, as well as the total daily profit.

5–4 DERIVATIVE METHODS

The term *classical optimization* refers to optimization methods that employ the *derivative* to obtain a solution, as expressed in relation (5–2). A *stationary point* is any point x that satisfies Equation (5–2),

$$y' = \frac{d[g(x)]}{dx} = g'(x) = 0$$

FIGURE 5–17 Different types of stationary points

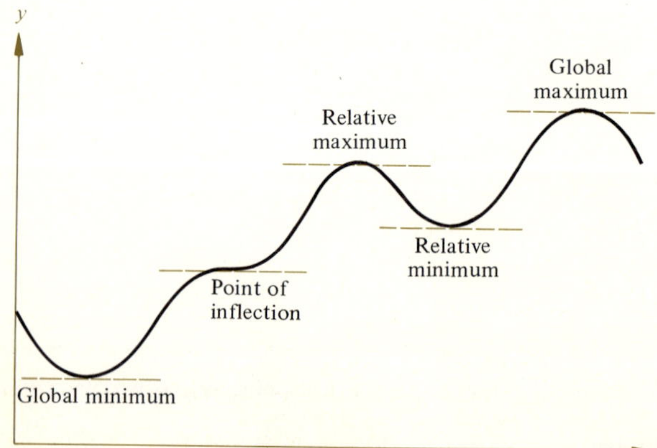

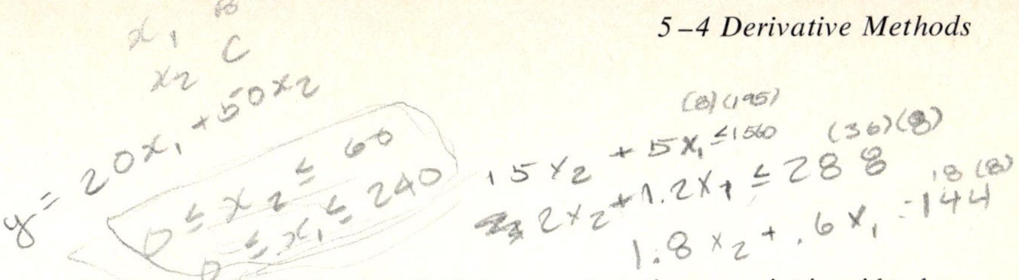

that is, a point at which the derivative dy/dx is zero. A stationary point is said to be a *relative maximum* if the second derivative evaluated at x is negative, and a *relative minimum* if the second derivative evaluated at x is positive. If the second derivative evaluated at x is zero, the stationary point is called a point of *inflection*. If a relative maximum is also the *best* relative maximum found for the function, it is called a *global maximum*. The same is true for the minimum. The points are illustrated in Figure 5-17.

For example, consider the function

$$y = g(x) = \sin x, \qquad 0 \leq x \leq 2\pi$$

The derivative of y with respect to x is

$$y' = g'(x) = \cos x$$

The values of x for which $g'(x)$ is zero are $\pi/2$ and $3\pi/2$, as Figure 5-18 shows. Evaluating the second derivative at these points gives

$$y'' = g''(x) = -\sin x$$
$$g''\left(\frac{\pi}{2}\right) = -\sin\frac{\pi}{2} = -1$$
$$g''\left(\frac{3\pi}{2}\right) = -\sin\frac{3\pi}{2} = +1$$

Thus the function has a maximum at $\pi/2$ and a minimum at $3\pi/2$.

Exercises

5-8 Plot values of the function $y = g(x)$, its first derivative, and its second derivative, to determine the optimum value as well as the type of optimum. Compare this result with that obtained by solving the first derivative for the stationary point.

(a) $g(x) = x^3 + 6x^2 + 3x - 10$

(b) $g(x) = e^x - 5x^2 + 3$, where $e = 2.71828$. (The derivative of e^x is equal to e^x.)

FIGURE 5-18 Stationary points for the function $y = \sin x$ in the interval $0 \leq x \leq 2\pi$

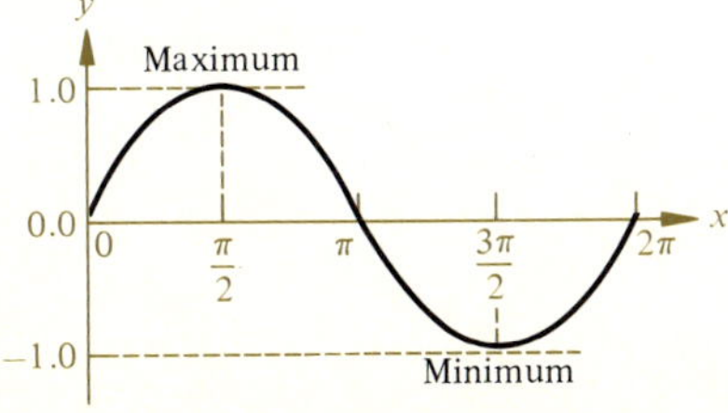

5–5 NEWTON–RAPHSON METHOD

The *root* of a function $g(x)$ is the value of x for which the function $g(x)$ has the value zero. There are numerous computational methods for finding the root of a function $g(x)$. Perhaps the best known is the *Newton–Raphson method.* It consists of the following steps.

Step 1. Start at an *estimate* x_0 of the root. Obtain this estimate by evaluating $g(x)$ at various values of x until a sign change (either from a positive $g(x)$ to a negative $g(x)$ or vice versa) occurs. Since $g(x)$ must be zero at some point between these two successive values of x, take the midpoint between them as the starting point x_0.

Step 2. Determine the $(n + 1)^{\text{th}}$ estimate of the root from the relation

$$x_{n+1} = x_n - \frac{g(x_n)}{g'(x_n)} \tag{5–26}$$

where $g'(x_n)$ is the first derivative of the function $g(x)$ evaluated at the point x_n.

Step 3. Repeat Step 2 until at some point $x*$, $g(x*)$ is sufficiently close to zero. That is, stop when

$$|g(x*)| \leqslant \epsilon$$

where ϵ is some preselected small value, say 10^{-3}.

Figure 5–19 graphically illustrates the mechanism of the Newton–Raphson procedure for finding a root.

FIGURE 5–19 The Newton–Raphson method for determining a root of a function $g(x)$. See Equation (5–26).

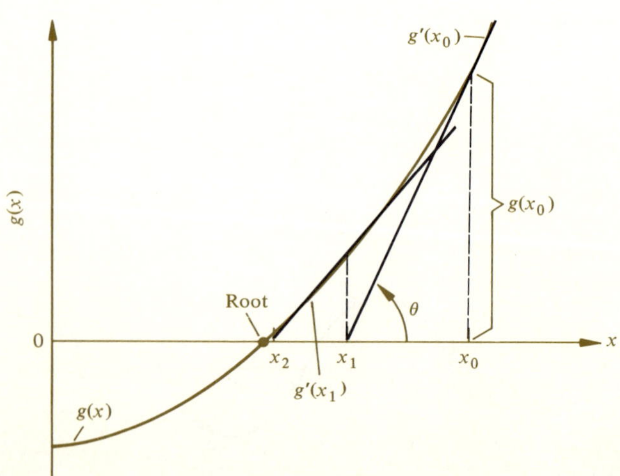

The Newton–Raphson procedure can also be used to determine the *stationary point* of a function $g(x)$; that is, the point at which the first derivative equals zero, or

$$y' = \frac{dy}{dx} = g'(x) = 0$$

Thus the Newton–Raphson procedure can be used to find the *root* of the first derivative $g'(x)$ and hence the stationary point of the function $g(x)$. The procedure is as follows.

Step 1. Estimate the *stationary point* at x_0 evaluating the first derivative $g'(x)$ for a sign change, as we described in Step 1 of the Newton–Raphson procedure for finding roots.

Step 2. Determine an $(n + 1)^{th}$ estimate from the relation

$$x_{n+1} = x_n - \frac{g'(x_n)}{g''(x_n)} \tag{5–27}$$

where $g'(x_n)$ is the *first derivative* of $g(x)$ evaluated at x_n, and $g''(x_n)$ is the *second derivative* of $g(x)$ evaluated at x_n. (The second derivative is simply the derivative of the first derivative.)

Step 3. Stop when the first derivative of $g(x)$ evaluated at some point $x*$ is very nearly zero. That is, stop at the point $x*$ such that $|g'(x*)| \leq \epsilon$.

As an example of the Newton-Raphson procedure for finding a stationary point, consider the function

$$y = g(x) = \frac{x^4}{4} - \frac{5x^2}{2} - 7x + 12$$

The first and second derivatives of the function $g(x)$ are, respectively,

$$g'(x) = x^3 - 5x - 7, \qquad g''(x) = 3x^2 - 5$$

To determine a starting point x_0, observe the value that changes the sign of the first derivative. In this example,

$$g'(1) = (1)^3 - 5(1) - 7 = 1 - 5 - 7 = -11$$
$$g'(2) = (2)^3 - 5(2) - 7 = 8 - 10 - 7 = -9$$
$$g'(3) = (3)^3 - 5(3) - 7 = 27 - 15 - 7 = 5$$

Thus a root of the first derivative $g'(x)$, and hence a stationary point of the function $g(x)$, lies between the values $x = 2$ and $x = 3$. Letting $x_0 = 2.5$, we perform the Newton–Raphson procedure:

$$\begin{aligned} x_1 &= x_0 - \frac{g'(x_0)}{g''(x_0)} \\ &= 2.5 - \frac{[(2.5)^3 - 5(2.5) - 7]}{[3(2.5)^2 - 5]} \\ &= 2.5 - \frac{-3.875}{13.75} = 2.5 + 0.282 \\ &= 2.782 \end{aligned}$$

TABLE 5–4. ILLUSTRATION OF THE NEWTON–RAPHSON METHOD

Iteration	x_n	$g(x_n)$	$g'(x_n)$	$g''(x_n)$
0	2.5	−11.36	−3.875	13.75
1	2.782	−11.85	−0.621	18.22
2	2.748	−11.86	−0.0115	17.65
3	2.749	−11.86	−0.0292	17.67

Table 5–4 gives the results of three iterations with the Newton–Raphson procedure with this example. The solution appears after two iterations, and the third iteration confirms it. Observe that the value of the slope $g'(x_3)$ is approximately, but not exactly, equal to zero. The positive value of the second derivative $g''(x_3)$ indicates that the solution is a *minimum*.

The Newton–Raphson procedure is very sensitive to the starting point x_0. If x_0 is chosen poorly, the procedure will *diverge* from the solution rather than converging to it. Taking the midpoint between values of x that change the sign of $g(x)$ (or $g'(x)$ for the stationary-point procedure) usually circumvents this difficulty.

Exercises

5–9 Using the Newton–Raphson method, determine a stationary point of the functions
(a) $g(x) = x^3 + 6x^2 + 3x - 10$, (b) $g(x) = e^x - 5x^2 + 3$
where $e = 2.718$. The derivative of e^x is e^x.

5–10 Develop a FORTRAN program for the Newton–Raphson method, solving the problems given in Exercise 5–9.

5–6 SUMMARY

This chapter has introduced some of the fundamental concepts in *optimization*. In earlier chapters we saw that one of the essential functions of the engineer was to develop *models* of physical systems. Now we see that an equally important aspect of the engineer's work is to *manipulate* these models to obtain an optimum solution to a problem.

Although a thorough understanding of optimization must wait until you know more mathematics, at least second-year calculus, these basic concepts and techniques should prove valuable to you as you progress through your program of study in engineering.

PROBLEMS

P5–1 Given the unidimensional search problem to minimize

$$g(x) = 2.8^{-0.01x^2} \cos(0.5x)$$

in the interval $0 \leqslant x \leqslant 10$, determine a solution within a final interval of uncertainty of 0.2 by (a) exhaustive search, (b) dichotomous search, and (c) golden section search.

P5–2 Given the unidimensional search problem to maximize $g(x) = x^2 \sin x$ in the interval $0 \leqslant x \leqslant \pi$, determine a solution within a final interval of uncertainty of 0.1 by (a) exhaustive search, (b) dichotomous search, and (c) golden-section search.

P5–3 Consider the function

$$g(x) = 12x_1 - x_1^2 - 5x_2 + 2.5 \ln x_2$$

With $(x_1 = 0, x_2 = 2)$ as a starting point, locate a relative maximum by the following two methods.

(a) Sectional (one-variable-at-a-time) search, using an interval of uncertainty $L = 0.2$ if unidimensional search is applied
(b) Simplicial search, using a side length of 0.5

P5–4 An electrical engineer wishes to design a set of electrical transmission lines. The cost of constructing this set of lines is given by the equation

$$C_1 = n(7500 + 800d^2)$$

where n is the number of electrical lines and d is the diameter of each line in centimeters. The cost of electrical power over the life of this system is given by

$$C_2 = \frac{90{,}000}{nd^2}$$

Minimize the total cost of this system. Start with the point $(n = 5, d = 2.0)$ and determine the number n and diameter d of these electrical lines by the following two methods:

(a) Sectional (one-variable-at-a-time) search using an interval of uncertainty of 0.2 if unidimensional search is applied
(b) Simplicial search with a side length of 0.2

[*Note:* Since n must be an integer, simply round your solution to the nearest integer.]

P5–5 Use linear programming to maximize $y = x_1 + 2x_2$ subject to the constraints

$$x_1 \geqslant 0, \qquad x_2 \geqslant 0$$
$$x_1 + x_2 \leqslant u, \qquad 2x_1 + x_2 \leqslant w$$

Solve graphically for the following values of $(u, w,)$: (a) (1,4), (b) (4,4), and (c) (4,1).

P5–6 Use linear programming to minimize $y = ux_1 + wx_2$ subject to the constraints

$$x_1 \geqslant 0, \qquad x_2 \geqslant 0$$
$$x_1 + 2x_2 \geqslant 6, \qquad 2x_1 + x_2 \geqslant 8$$

Solve graphically for the following values of (u, w): (a) (0.2, 1.2), (b) (1.2, 1.2), and (c) (1.2, 0.2).

P5–7 Tiny Dynamo Company manufactures two types of dynamos, A and B. Each unit of dynamo A requires 1 hour of engineering service, 10 hours of direct labor, and 0.3 kilogram of material. Each unit of dynamo B requires 2 hours of engineering, 4 hours of direct labor, and 0.2 kilogram of material. There are 40 hours of engineering, 280 hours of direct labor, and 20 kilograms of material available each week for manufacturing dynamos. Each unit of dynamo A yields a profit of $12, while each unit of dynamo B returns a profit of $9.50. Determine the most profitable production mix of dynamos.

P5–8 A closed cylindrical tank is to have a volume of 50 cubic meters. If the cost of the material per square meter in the top and bottom of the tank is twice that used for the curved sides, find the ratio of height to diameter that yields the minimum cost.

P5–9 Figure P5–1 shows two towns, located at points A and B. A water purification plant is to be located at some point C along the river to furnish water to both towns. Where must the purification plant be located to minimize the amount of pipeline?

FIGURE P5–1

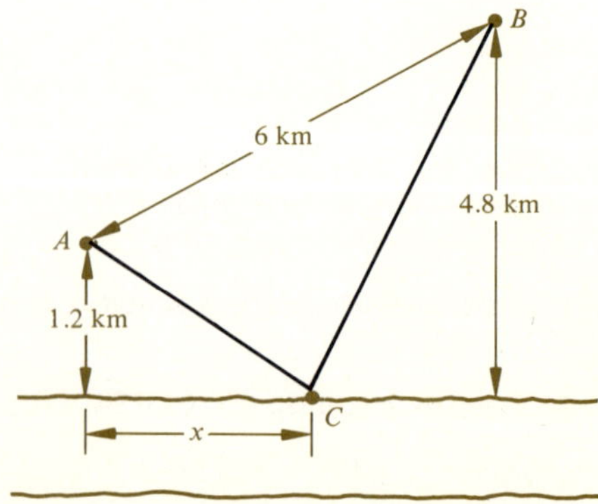

P5–10 The speed of signaling in a submarine telegraph cable is proportional to

$$x^2 \ln(1/x),$$

where x is the ratio of the radius of the core to the thickness of the sheathing. Find the value of x that maximizes the signaling speed.

(a) Use the principle stated in Equation (5–2).

(b) Using five iterations of the Newton–Raphson method, search from a suitable starting point. The derivative of $\ln(1/x)$ is $-1/x$.

P5–11 The total annual cost of operating an electric motor in a particular irrigation system is a function of x, the size (horsepower) of the motor. This relationship is

$$C = \$225 + \$0.7x + \frac{\$0.01}{x}(125{,}000)$$

Using the Newton–Raphson method, find the motor size that minimizes total annual cost.

EXERCISE 5-1 (PG 178) DUE
EXERCISE 5-2 (PG 179) OCTOBER 9
PROBLEM 5-1 (PG 195)

EXERCISE 5-6 (PG 189)
EXERCISE 5-7 (PG 190)
PROBLEM 5-1 (PG 195)

EXERCISE 5-9 (PG 194)
EXERCISE 5-10 (PG 194) DUE
PROBLEM 5-10 (PG 197) OCTOBER 16
PROBLEM 5-11 (PG 197)

DO NOT USE MULTIDIMENSIONAL SEARCH TECHNIQUE
ONLY $y=f(x)$

CHAPTER SIX
ADDITIONAL TOPICS IN FORTRAN

In Chapter 3 we introduced the elements of FORTRAN and applied them many times throughout Chapters 4 and 5. The examples in these chapters illustrated that the digital computer and the FORTRAN language, when used together efficiently, save vast amounts of an engineer's time. Several additional types of higher-level statements in FORTRAN greatly increase the usefulness of the digital computer. This chapter will discuss several of these statements. In subsequent chapters, we shall use them to analyze problems involving various engineering concepts.

The basic techniques of Chapter 3 will enable you to accomplish anything that a person can accomplish using the methods that we are about to discuss in this chapter, but the purpose of this chapter is to simplify FORTRAN programming. With these new statements and concepts, you can write more compact programs, and compose them more easily. In some situations, these new approaches will greatly reduce the length of programs and the amounts of keypunching you have to do. When you read this chapter, you should keep these thoughts in mind and constantly evaluate the added convenience provided by the new types of statement.

6–1 DØ LOOPS

Central to FORTRAN programming are loops within a program. With such loops, you can repetitively and easily evaluate complex formulas for different values of one or more independent variables. Several problems and exercises in the preceding chapters demonstrated the usefulness of loops. Before we introduce the DØ loop, let's review the essential elements used in Chapter 3 to generate a loop within a FORTRAN program.

The first statement was the unconditional GØ TØ used to branch "back" through a sequence of instructions. Without such a branching instruction, or its equivalent, we could not establish loops within programs. After we introduced this statement, we discussed the need to terminate the execution of the loop after the desired number of cycles had been completed. The simple use of a GØ TØ results in a loop that never ends. To solve this problem, we introduced a counting variable, along with a suitable test. The following program segment illustrates the statements that are typically used to generate a loop.

```
  J = 0
1bREAD (5, * ) A,B
      .
      .
      .
  J = J + 1
  IF (J .LE. 10) GØ TØ 1
```

Three statements are essential. The first sets the counter variable (J) at some suitable initial value, zero in this case. The second (J = J + 1) increases the counter by one each time the loop is executed. The last is a LØGICAL IF. This statement tests the counter; if the specified test is satisfied, the GØ TØ is executed. This branching statement establishes the cyclic behavior required.

A set of three statements like these will produce any loop desired within a program. However, a much simpler method for generating loops within FORTRAN programs is available. Rather than three statements, one will suffice—a DØ statement. The general form is

$$\text{bbbbbbDØb}\begin{matrix}\text{statement}\\\text{number}\end{matrix}\text{ b }\begin{matrix}\text{integer}\\\text{variable}\end{matrix} = \begin{matrix}\text{initial}\\\text{value}\end{matrix}, \begin{matrix}\text{final}\\\text{value}\end{matrix}, \text{increment} \qquad (6\text{–}1)$$

This statement always begins with the word DØ followed by at least one blank space and a statement number. (Remember that statement numbers are integers used as address labels for particular lines in a given program.) After this, at least one more blank is required, then an integer variable, an equals sign, and finally a set of three integer quantities separated by commas.

The following example illustrates the use of a DØ statement.

```
      DØ 10   J = 1,3,1
      READ (5,*) A,B

      SUM = A + B
10bWRITE (6,*) SUM

      STØP
      END
EXECUTE CARD
1.b2.
2.b3.
3.b4.
FINISH CARD
```

This program does not have the user identification and compiler selection cards. As usual, we have left blank spaces beneath the READ and WRITE statements for you to insert any necessary local modifications. When you study this program, assume that compilation is complete and that execution is beginning. The first statement is a DØ; it causes the following sequence of events within the computer system. The variable specified in the DØ statement (J) is set equal to the first integer value appearing on the right-hand side of the equals sign in the statement. In this case, the initial value is 1. Then the system executes all the statements from the DØ up to and including the statement labeled 10, that is, the statement number after the letters DØ. Once this has been done, the relevant section of the computer's memory is as follows.

Once statement 10 has been executed, the system checks the current value of J (1) and compares it to the final value specified in the DØ statement (in this case, 3). If the current value is less than the final value, the system returns to the DØ, increases

J by 1, and continues execution. This results in a second execution of the statements from the DØ up to and including the WRITE. Once this second pass has been completed, the computer memory looks like this.

J [2] A [2.] B [3.] SUM [5.]

At this point, the computer again compares the current value of J (2) with the final value (3). Since the current value does not exceed or equal the final value, the computer generates another branch back instruction and repeats the loop. The new values of the variables of immediate interest once this third pass is completed are:

J [3] A [3.] B [4.] SUM [7.]

At this time, the current value of the loop index J equals the final value and, under the rules of FORTRAN, control passes to the statement after the WRITE. Although we have not mentioned it explicitly, the third integer specified on the right-hand side of the equals sign in the DØ statement is the increment. The increment determines the amount by which the loop index (J) is increased each time the loop is executed.

Before we consider additional examples of DØ loops, we should mention that this FORTRAN tool motivates our consideration of integer quantities. You may have thought, in the preceding chapters, that integers were more trouble than they were worth. After all, using integer quantities at the wrong time can result in significant computational errors, and requires the novice programmer to memorize the segment of the alphabet (I–N) allocated to integer variables. Despite all this difficulty, we did nothing with integer variables that we could not have done just as well with real quantities. It is reasonable to wonder why we simply did not ignore the integer concept and spare ourselves the difficulties. One reason is that the DØ statement depends on integer quantities. The DIMENSIØN statement that we shall introduce in the next section will provide yet a second justification for considering integers.

After examining the execution paths resulting from a DØ statement, you can see that the following two programs segments will produce the same output. The segment on the left is more compact because it uses only one statement (the DØ) to establish the necessary loop.

```
  DØ 1  J = 1,3,1              J = 1
  READ (5,*) A,B              1bREAD (5,*) A,B

  SUM = A + B                  SUM = A + B
1bWRITE (6,*) SUM              WRITE (6,*) SUM

                               J = J + 1
                               IF (J .LT. 3) GØ TØ 1
```

The grammar of FORTRAN allows a DØ statement to be written without an increment, that is, the third integer quantity on the right side. In such cases, the compiler assumes that the increment is 1. This feature is solely for the convenience of the programmer—it saves a few strokes at the keypunch. We encourage you to include the increment in your first few programs containing DØ loops. The following statements are equivalent under this rule.

DØ 10 J = 1,51,1 DØ 10 J = 1,51

In addition, the initial value of the loop's variable does not have to be 1. The following are valid expressions.

DØ 7 K = 3,5 DØ 122 L1 = 7,18,1

The increment need not have the value 1. In some situations, it is quite convenient to use other values for increments. The following are acceptable statements.

DØ 2 K = 3,9,3 DØ 18 MN = 5,15,2 DØ 27 N3 = 7,2,2

The last example is somewhat puzzling. In fact, such a statement should never be used because it will not accomplish the primary purpose of a DØ, the repetitive execution of some sequence of FORTRAN instructions. Note that, in this example, the initial value of the loop variable (N3) is greater than the final value specified. In such a situation, the statements within the range of the DØ (the range includes all the instructions from the DØ to the statement number, 27 in this example) are executed only once, and the number 7 is stored in the memory slot labeled N3. This statement has little value and should be avoided. It is a misuse of the DØ concept.

Before we make some general comments on DØ loops, we shall consider the nuances of their execution. We can do this most easily by considering a program segment containing a DØ statement with an increment other than 1.

```
   DØ 11 I = 2,5,2
   L = 2 * I - 3
   M = 3 * I - 2
11bWRITE (6,*) L,M
```

What numerical output will this set of FORTRAN instructions produce? To answer this question, let's consider the execution step by step. Initially, the number 2 is stored in the memory slot labeled I; the second and third instructions calculated numerical values and store them in L and M. When statement 11 is executed for the first time, the numbers stored in memory are

I [2] L [1] M [4]

At this time, the current value of the loop index I (2) is less than the final value (5). The index is incremented (by 2), and the program branches back to the top of the segment. The value stored in I is now 4. L and M are calculated, and when statement 11 is executed for the second time, the numbers stored in I, L, and M are:

I [4] L [5] M [10]

The loop index is still less than the final value. Now we might expect the computer system to increase the loop index by the increment (in fact, it does) and

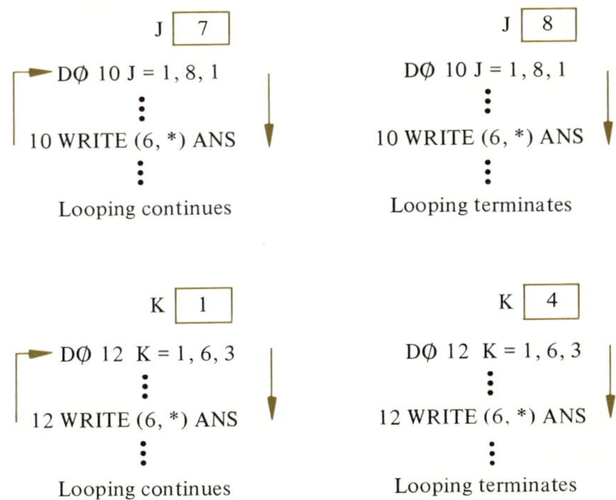

FIGURE 6–1 Examples of DØ loop execution. The examples on the left continue looping because the current values of the loop index are less than the maximums specified in the DØ statements.

then branch back to the DØ statement (in fact, it does not). By the rules of FORTRAN, the current value of the loop index is again checked to determine whether it exceeds the maximum value. If it does, control passes to the statement following the end of the range of the DØ loop. As a consequence, these statements would be executed only twice, once with I equal to 2 and once with I equal to 4.

We can summarize the branching behavior of a DØ loop in the following way. Each time the computer system executes the last statement within the range of a DØ loop, it compares the current value of the loop index to the maximum specified value. If these two quantities are equal, control passes to the first statement outside the range. If the current value is less than the specified maximum, the increment is added to the loop index. If the resulting value is less than or equal to the maximum, the computer system branches back and executes the statements within the DØ loop again. If, on the other hand, this value exceeds the maximum, control passes out of the DØ loop. In short, the computer system must make two decisions each time the last statement in the range is executed. Figure 6–1 illustrates this process.

Exercises

6–1 How many times will the following DØ loop be executed? What values of I3 will be printed?

```
      DØ 127   I3 = 1,6,1
127bWRITE (6,*) I3
```

6–2 How many times will the following DØ loop be executed? What values of L will be printed?

```
 DØ 4  L = 1,6,2
4bWRITE (6,*) L
```

6–3 How many times will the following DØ loop be executed? What values of L will be printed?

```
  DØ 91  J = 3,9,4
  L = J * J - 2
91bWRITE (6,*) L
```

6–4 How many times will the following DØ loop be executed? What values of J will be printed?

```
 I = 4
 DØ 6  J = 1,I
6bWRITE (6,*) J
```

Exercise 6–4 should have caused you some concern, since it introduced a new concept which we had not discussed. Refer to the general form of the DØ statement (Equation 6–1). You will observe that we made no definitive comment about the allowed forms of the integer quantities appearing on the right-hand side of the DØ statement. We did this intentionally to avoid any unnecessary confusion in our initial discussion of this topic. The examples we chose indicated that all these quantities (the initial value of the loop index, the final value, and the increment) were FORTRAN integer constants. In addition, Exercise 6–4 indicates that FORTRAN integer variables are also allowed. For example, the following program segments are valid.

```
I = 4
J = 15
L = 2
DØ 4  M = I,J,L
```

```
J1 = 3
J2 = 8
DØ 5  I = J1, J2
```

Exercises

6–5 What values of M will be printed when the following program segment is executed?

```
 I = 2
 J = 4
 L = I + 2 * J
 DØ 4  M = I,L
4bWRITE (6,*) M
```

6–6 What values of I will be printed when the following program segment is executed?

```
        N = 10
        M = N - 7
        L = N - 8
        DØ 5  I = M,N,L
      5bWRITE (6,*) I
```

To further illustrate the usefulness of DØ statements, we shall generate a set of multiplication tables using integers only. To construct a set of such tables, one must multiply a given integer by a sequence of integers. For example, to construct a "five times" table, one must multiply the integer 5 successively by the integers 1–12. This is the type of problem for which a DØ loop was created. The following segment illustrates one possible solution.

```
        I = 0
      2bDØ 1 J = 1,12,1
        MULT = I * J
      1bWRITE (6,*) I,J,MULT

        I = I + 1
        IF (I .LT. 12) GØ TØ 2
        STØP
        END
```

Examine this program carefully. It uses a DØ loop to generate one of the two product factors, and a method from Chapter 3 to generate the other. In a moment, we shall examine a possible solution that uses two DØ loops.

Before we consider this, observe that the index variable used in a loop may also be used in subsequent calculations. In this example, the loop index J is used as one of the factors. This use of a loop index is always permitted provided the variable occurs only on the right-hand side of a FORTRAN statement. An index used in a DØ loop must never appear on the left-hand side of a statement within the range of the loop. If it does, its value will change from that assigned by the loop execution, and this will result in an error when the program is executed.

We can print the desired set of multiplication tables by executing the following program segment which contains two DØ loops.

```
        DØ 1  I = 1,12,1
        DØ 1  J = 1,12,1
        MULT = I * J
      1bWRITE (6,*) I,J, MULT

        STØP
        END
```

FIGURE 6–2 Horizontal and vertical line notation used to denote DØ loop range

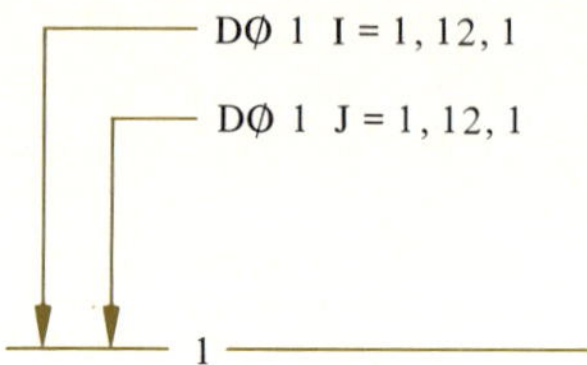

Because it uses two DØ loops, this segment is two statements shorter than the first solution. When one DØ loop is contained within another, the loops are said to be *nested*. To explain nesting, we must discuss the rules governing the execution of program segments containing nested DØ loops. We can do this most easily by studying the schematic illustration of the ranges of the two loops shown in Figure 6–2. This figure represents the loops in our sample program segment. In this example, both loops terminate on the same statement, and the execution sequence is as follows.

The first DØ statement is encountered, and the value of I is set equal to 1. Then the second DØ is executed, and the value of J is also set equal to 1. Execution continues until the terminal statement of both loops (statement 1) is reached. At this point, the computer's memory looks like this:

What will the computer do next? Will it execute the inside loop again, or will it execute the outside loop again? In fact, all FORTRAN compilers are designed to execute the inside loop (with the index J in our example) completely (12 times in our example) before it repeats the outside loop a second time.

This rule implies that control will return to the second DØ statement after the computer has executed the WRITE for the first time. When the computer reaches the WRITE the second time, the computer's memory will have changed to

This sequence of events will continue 10 more times until the memory looks like this:

At this point, the inner loop will have been completely finished (the current value of the loop index J is equal to the maximum value specified in the DØ statement), and control will return to the outer loop where the value of I will increase by 1 to a value of 2. Next, the inside DØ statement is executed again, and the value of J is reset to 1. When the computer again reaches the WRITE statement, the memory looks like this:

Execution will continue for a while because this simple six-statement program will generate 144 products. The inner loop (index J) will be executed 12 times for each value of the outer loop variable (I). There are 12 values of I, and the program will execute the WRITE statement 144 times. When two loops are nested so that they end on the same statement, the inner loop is executed completely (over the entire range of the loop index) before the outer loop is repeated.

When one loop is completely inside another, the inner loop is executed completely before the outer loop is incremented. The following program segment gives an example of this.

```
   DØ 1  I = 1,3,1
   READ (5,*) A,B

   DØ 2  J = 1,5,1
   READ (5,*) C

   SUM = A + B + C
2bWRITE (6,*) SUM

   DIFF = A − B
1bWRITE (6,*) DIFF
```

If the proper data cards are present, this program will produce 15 values of SUM (there are 5 values of J for each of 3 values of I) and 3 values of DIFF.

One DØ loop contained within another is said to be a *two-level nest*. It is possible to nest to three, four, and higher levels. The maximum level of nesting permitted depends on the computer system, but the number of available levels is rarely a programming limitation.

Exercises

6–7 What output will the following program produce?

```
   DØ 1  I = 1,4,1
   DØ 2  J = 1,3,1
   M = J * J
   N = J * J * J
   K = J
2bWRITE (6,*) J,M,N

   L = I ** K
1bWRITE (6,*) L
```

6–8 What output will the following program produce?

```
DØ 1   I = 1,6,1
DØ 1   J = 1,4,1
DØ 1   K = 1,3,1
1bWRITE (6,*) I,J,K
```

Not all conceivable nests of DØ loops are allowed in FORTRAN. To clearly illustrate the differences between valid and invalid nests, we shall use a simple line representation of DØ loops. A short horizontal line represents the position of the DØ statement within a given program, and a slightly longer horizontal line represents the final statement. A vertical line connecting the two horizontal lines represents the entire range of the DØ. The following example illustrates this convention.

```
DØ 1   I = 1,3
J = I * I
DØ 2   K = 1,4
L = J * K − 4
2bWRITE (6,*) L

M = J − L
1bWRITE (6,*) M
```

Using this form of representation, Figure 6–3 shows acceptable and Figure 6–4 unacceptable types of DØ loop nests. In all the unacceptable forms, the range of one DØ loop extends beyond the range of the DØ loop within which it began. In terms of the schematic diagrams, a vertical line is crossing a horizontal. Whenever this occurs, the program will not run successfully because there is no consistent way for the computer to carry out the two conflicting sequences of instructions. The nests shown in Figure 6–3 do not have overlapping ranges; no vertical line crosses a horizontal line. All these nests are thus acceptable.

FIGURE 6–3 Examples of valid nested DØ loops. The ranges of the inside loops do not extend beyond the ranges of the outside loops.

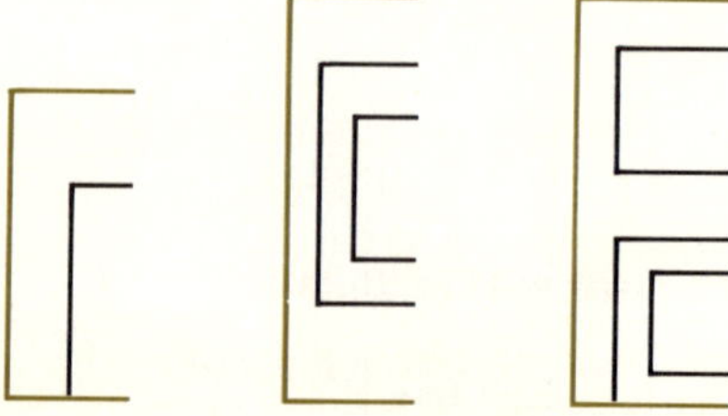

FIGURE 6–4 Invalid nested loops. There are three illegal crossings of horizontal and vertical lines.

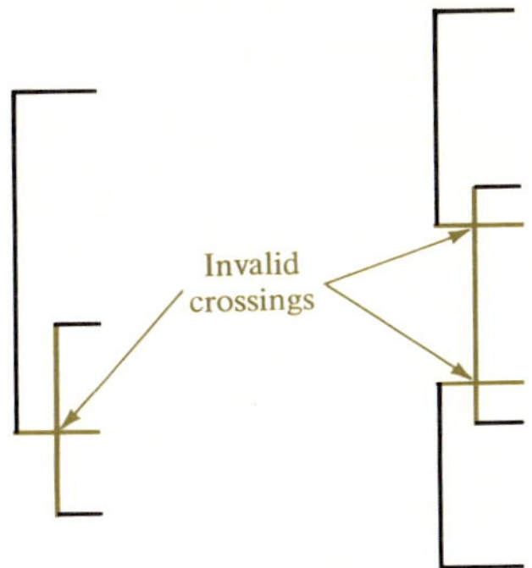

We shall close this section by briefly considering another new FORTRAN statement, the CØNTINUE. It causes no changes in memory, no numerical calculations, nor any input–output operations. It simply tells the computer system to continue to the next statement in the sequence. It does consume a small amount of time. The exact amount depends on the computer being used.

The CØNTINUE is valuable primarily as a convenience to the programmer. It is often useful, as in the following example, to terminate DØ loops.

```
         DØ 10  L = 1,13,1
         READ (5,*) A,B,C

         SUM = A + B + C
         PRØD = A * B * C
         WRITE (6,*) SUM, PRØD

      10bCØNTINUE
```

In such simple program segments, there is no convincing way to make the CØNTINUE statement appear useful, but more complex programs will demonstrate the value of this statement.

The next section will introduce a new class of FORTRAN variables, variables that are often useful in formulating and solving engineering problems.

Exercises

6–9 Write and execute a FORTRAN program that uses an appropriate DØ loop and prints a table for converting temperatures in degrees Celsius to temperatures in degrees Fahrenheit. The program should produce two columns of equivalent temperatures and should label each properly. Celsius temperatures should range from −200° to 200° in increments of 5°. The equation that relates Celsius and Fahrenheit temperatures is

$$T(°F) = \tfrac{9}{5}\, T(°C) + 32$$

6–10 A thermocouple is a simple device often used to measure temperatures. For example, the thermocouple is used to monitor the temperatures at which metals are being heat-treated to optimize their strength and resistance to fracture. Figure 6–5 illustrates the basic thermocouple. It consists of two wires of different materials welded together. The temperature at their junction results in an electrical voltage E as shown. The actual use of this measuring tool is a bit more complex, but we can safely ignore these details. For a thermocouple in which one wire is copper and the other is constantan, the relationship between the voltage produced E and the temperature T is

$$E = 2.5(10^{-5})T - 0.88(10^{-3})$$

where E is in volts and T is in degrees Fahrenheit.

Write and execute a FORTRAN program that produces a table of voltage versus temperature for this thermocouple. The temperature should range from −100° to 350° in increments of 25°. The two columns of numbers produced by the program should be appropriately labeled. Use the computer output to prepare a graph of voltage as a function of temperature.

6–11 An engineer is designing a diving board for a swimming pool. The critical design factor is the amount of deflection y that will occur when a person of weight W stands at the end of the board. The geometry of interest is sketched in Figure 6–6. The engineer refers to a standard text on strength of materials and finds that the equation connecting the deflection and applied weight is

$$W = \frac{3Ebh^3}{12L^3}\, y$$

In this equation, E is Young's modulus, a quantity related to the material used. It has the dimensions of N/m². L is the length of the board, b the width, and h the thickness; all three quantities are measured in meters. For this

FIGURE 6–5 A simplified view of a thermocouple. This device generates a voltage E proportional to the temperature T at the junction.

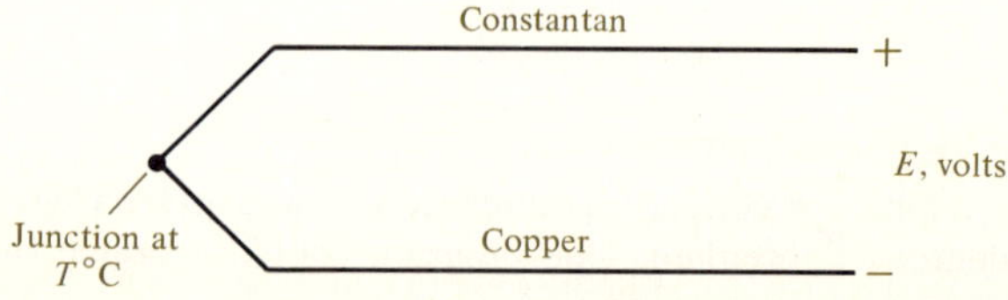

FIGURE 6–6 Diving board design. (a) The three dimensions characterize the board: length L, width b, and thickness h. (b) The quantity y is the deflection caused by a diver of weight W.

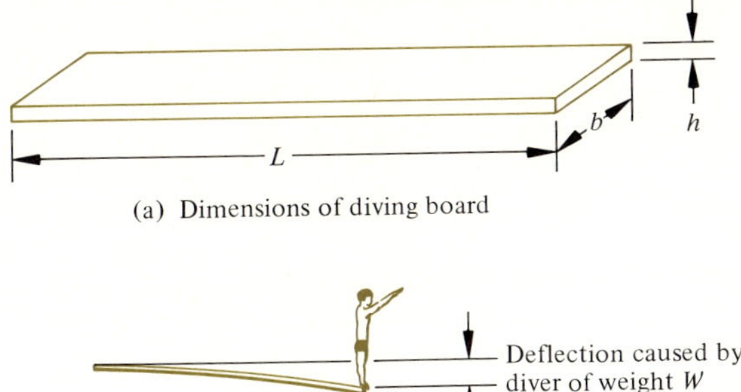

(a) Dimensions of diving board

(b)

problem, we assume the board is made from a piece of wood having a value of $E = 4.8(10^9)\ \text{N/m}^2$.

Write and execute a FORTRAN program that uses nested DØ loops and calculates and prints the deflections y that result in diving boards with the following dimensions and applied weights.

$$L = 2\ \text{m in all cases,} \qquad b = 0.4\ \text{m in all cases}$$
$$h = 0.50,\ 1.2,\ 1.9,\ 2.6,\ 3.3,\ 4.0\ \text{cm}$$
$$W = 200,\ 350,\ 500,\ 650,\ 800,\ 950\ \text{N.}$$

Your program should produce 36 values of deflection—6 values for each of 6 different thicknesses. Be sure you check dimensional homogeneity.

6–2 THE DIMENSIØN STATEMENT AND DIMENSIONED VARIABLES

Quantities with subscripts are important in the formulation and solution of many engineering problems. They naturally occur in many kinds of problems, but one of their most natural uses is for labeling different designs. For example, in Exercise 6–11, six different diving board designs were examined. In referring to these designs, it is convenient to label each in the following way.

design 1	h = 0.50 cm
design 2	h = 1.20 cm
design 3	h = 1.90 cm
design 4	h = 2.60 cm
design 5	h = 3.30 cm
design 6	h = 4.00 cm

The assignment of integers to each of the designs is arbitrary and depends only on the user's choice. With a 200-newton child standing on the diving board, a deflection y is associated with each of the six designs.

deflection 1 or y_1

.

.

.

deflection 6 or y_6

The use of a numerical subscript on the quantity y allows a compact representation of the six deflections. In computer analyses, it is convenient to have FORTRAN variables that may be used to represent each subscripted quantity. One obvious choice for the six deflections is the following.

Deflection	FORTRAN *variable*
y_1	Y1
y_2	Y2
y_3	Y3
y_4	Y4
y_5	Y5
y_6	Y6

Such FORTRAN variables are sometimes—but not always—useful and convenient, as the following considerations indicate. Imagine that we need to calculate the average deflection of the six designs under a 200-newton load. We use the variable AVE to denote arithmetic average and write the following FORTRAN statement to calculate the desired quantity.

```
AVE = (Y1 + Y2 + Y3 + Y4 + Y5 + Y6)/6.
```

We can easily write a compact statement in this case.

Now imagine that we were considering 1000 diving boards! To write the necessary FORTRAN statement would require an apallingly large amount of keypunching. We could do it, but such an exercise is not conducive to either spiritual or physical well-being.

We can greatly simplify the solution to our problem by using a different set of FORTRAN variables to represent the deflections of the diving boards. FORTRAN enables us to define a large array of variables in one statement, the DIMENSIØN. In our example, we could write

```
DIMENSIØN Y(6)
```

This single statement defines six different FORTRAN labels attached to six memory slots. Figure 6–7 shows the appropriate section of the computer's memory after the DIMENSIØN statement has been compiled.

With these new variables, we can solve the problem of calculating the average in a different way, as the following segment illustrates.

FIGURE 6–7 A one-dimensional array of six FORTRAN variables

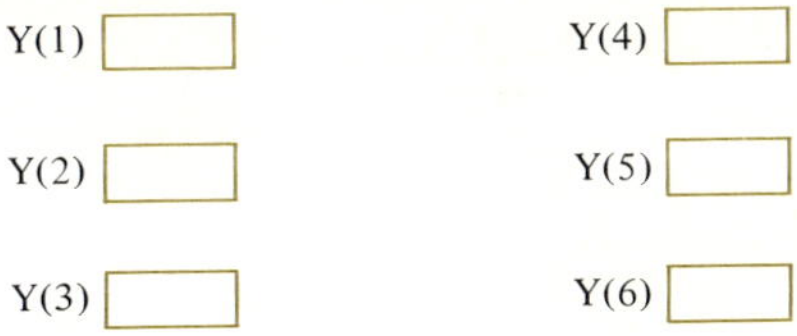

Y(1) Y(4)

Y(2) Y(5)

Y(3) Y(6)

```
   DIMENSIØN  Y(6)
   SUM = 0.
   DØ 1  I = 1,6
   READ (5,*) Y(I)

1bSUM = SUM + Y(I)
   AVE = SUM/6.
```

This solution is less than optimum, but it emphasizes the value of properly choosing FORTRAN variables. To expand this program segment to the case of 1000 diving-board deflections is not difficult. All we have to do is change the number 6 in three statements (the DIMENSIØN, the DØ, and the last) to the number 1000. This is not difficult and does not require a significant increase in keypunching. The use of the DIMENSIØN statement has greatly simplified the overall task of calculating the average of a large set of numbers.

The DIMENSIØN statement uses a *dimensioned variable*. A valid dimensioned FORTRAN variable is any FORTRAN variable followed by a pair of parentheses containing an allowed integer quantity. Allowed integer quantities include integer constants, integer variables, and some simple integer expressions. We shall not specify all acceptable integer expressions, since these vary from one computer to another. Again, you should check local ground rules. The following columns give examples of valid and invalid dimensioned variables.

Valid	*Invalid*
Y(6)	Y(6.)
Y(I)	Y(A)
I(J)	I(Z)
Ø12(L)	012(L)

The first three expressions in the right-hand column are invalid because the parentheses contain real quantities. FORTRAN compilers do not allow such expressions. The last expression is invalid because 012 is not a valid FORTRAN variable.

The compiler will recognize a dimensioned quantity only when it has been defined in a DIMENSIØN statement. This statement must occur before any other FORTRAN statements. It has the general form

DIMENSIØN valid variable (integer constant), another valid variable (another integer constant), and so forth

The hallmark of this statement is the word DIMENSIØN followed by a list of valid variables (either real or integer) separated by commas. After each variable is a pair of parentheses containing integer constants; these constants specify the number of memory slots to be reserved for each subscripted variable. The integers within the parentheses must be positive. The following are examples of valid and invalid DIMENSIØN statements.

```
DIMENSIØN  G(15), M(4), AL(8), IL(36)     valid
DIMENSIØN  G(15) M(4) AL(8) IL(36)        invalid
DIMENSIØN  1A(3), I2(10)                  invalid
DIMENSIØN  A(10), B(10), C(20)            valid
DIMENSIØN  Z(20), Y(I)                    invalid
```

The second statement is invalid because commas are missing, while the third is invalid because 1A is not a valid FORTRAN variable. The fifth DIMENSIØN is also invalid because one of the quantities within the parentheses is not an integer constant.

You should avoid the promiscuous use of DIMENSIØN statements and subscripted variables because memory must be assigned to each subscripted variable called for in a DIMENSIØN whether or not it is used in a program. If your computer system is small, you can run into difficulty by allocating memory needlessly.

Exercises

6–12 Are the following statements valid or invalid? If invalid, explain why.

```
DIMENSIØN  A(15), B(27.)
DIMENSIØN  I(12), Ø(14), K(18)
DIMENSIØN  ALP(7), SEAS(7.)
DIMENSIØN  A12(1024)
```

6–13 Each of the following DIMENSIØN statements is incorrect. Explain why.

```
DIMENSIØN  U(12 + 6), V(11 + 7)
DIMENSIØN  A(100000)
DIMENSIØN  S(10), T(10), FØRCE (I)
```

The DIMENSIØN statement cannot be placed at any point in a FORTRAN program. It must appear before any dimensioned variable is used in another statement, or else the compiler will generate an error message when it encounters the variable. However, the rules are even stricter than this. Most FORTRAN compilers require all DIMENSIØN statements to appear right after the compiler selection card. This means that the order of cards must be

ACCOUNTING CARD
COMPILER CARD
DIMENSIØN STATEMENT(S) (if any)
OTHER FORTRAN STATEMENTS
END
EXECUTION CARD
DATA CARDS (if any)
FINISH CARD

We can further extend the concept of subscripted quantities and dimensioned FORTRAN variables by reexamining the results of the diving-board analysis. In this section, we have considered only the set of 6 deflections ($y_1, y_2, \ldots, y_6$) that result from the application of a 200-N force (a small child) at the end of the diving board. However, in Exercise 6–11, you calculated the deflections resulting from 6 different applied forces. In all, you calculated 36 deflections, 6 for each of the 6 different forces. We may represent these 36 deflections symbolically in the following way.

		Thickness, cm					
		0.50	*1.20*	*1.90*	*2.60*	*3.30*	*4.00*
	200	y_1	y_2	y_3	y_4	y_5	y_6
	350	y_7	y_8	—	—	—	—
Weight,	500	—	—	—	—	—	—
N	650	—	—	—	—	—	—
	800	—	—	—	—	y_{29}	y_{30}
	950	y_{31}	y_{32}	y_{33}	y_{34}	y_{35}	y_{36}

This array contains 36 entries, one for each deflection. An alternative method that may be used to label the y's involves double subscripting.

		Thickness, cm					
		0.50	*1.20*	*1.90*	*2.60*	*3.30*	*4.00*
	200	$y_{1,1}$	$y_{1,2}$	$y_{1,3}$	$y_{1,4}$	$y_{1,5}$	$y_{1,6}$
	350	$y_{2,1}$	$y_{2,2}$	$y_{2,3}$	$y_{2,4}$	$y_{2,5}$	$y_{2,6}$
Weight,	500	$y_{3,1}$	—	—	—	—	$y_{3,6}$
N	650	$y_{4,1}$	—	—	—	—	$y_{4,6}$
	800	$y_{5,1}$	$y_{5,2}$	$y_{5,3}$	$y_{5,4}$	$y_{5,5}$	$y_{5,6}$
	950	$y_{6,1}$	$y_{6,2}$	$y_{6,3}$	$y_{6,4}$	$y_{6,5}$	$y_{6,6}$

Each entry in the array has two subscripts. The first denotes the force applied to the board (1 for 200 N, 2 for 350 N, and so forth), and the second indicates the thickness of the board. FORTRAN also has doubly subscripted variables; thus a one-to-one correspondence between the 36 deflections and 36 FORTRAN variables is possible.

$$y_{1,1} \quad y_{1,2} \quad \ldots \quad y_{6,6}$$

Y(1,1) Y(1,2) . . . Y(6,6)

The statement required to define these 36 variables is

```
DIMENSIØN  Y(6,6)
```

This statement allocates 36 slots in memory, as shown in Figure 6–8. Once the values of the deflections have been calculated and stored in the memory locations, it is possible to calculate the average of the 36 deflections as well as the average deflection for each of the six different loads.

We can further extend this concept by considering three and four subscripts. The number of subscripts that you can use depends on local ground rules, but in a first course, you will rarely need to use more than two subscripts. Our warning about wasting the computer's memory applies even more stringently to multiple subscripts. The number of memory slots allocated by a given DIMENSIØN statement is the product of the integers contained within the parentheses. As a result, innocuous-looking statements can use large amounts of computer memory. For example, this one statement reserves 37,500 memory slots!

```
DIMENSIØN  G(250, 10, 10), X (250, 50)
```

Dimensioned variables are commonly used to analyze problems like our example of tabulating board deflections. Later chapters will include additional examples of engineering problems in which subscripted variables are both natural and helpful. Defining many FORTRAN variables is much simpler with a DIMENSIØN statement. Remember that a dimensioned variable must always be a valid FORTRAN variable followed by a pair of parentheses containing a valid integer quantity.

FIGURE 6–8 A two-dimensional array of 36 FORTRAN variables

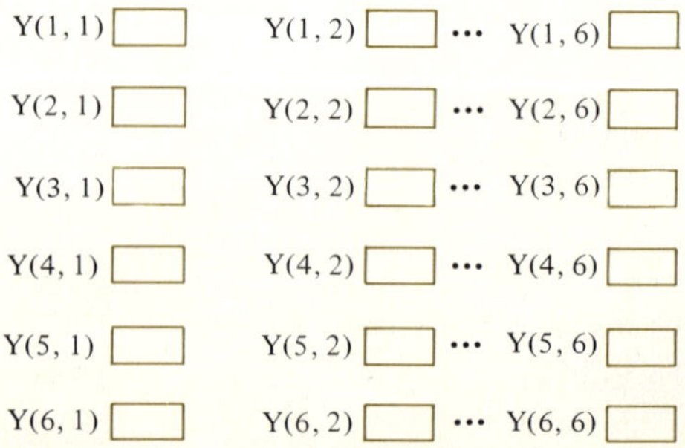

Y(1, 1) [ ]	Y(1, 2) [ ]	···	Y(1, 6) [ ]
Y(2, 1) [ ]	Y(2, 2) [ ]	···	Y(2, 6) [ ]
Y(3, 1) [ ]	Y(3, 2) [ ]	···	Y(3, 6) [ ]
Y(4, 1) [ ]	Y(4, 2) [ ]	···	Y(4, 6) [ ]
Y(5, 1) [ ]	Y(5, 2) [ ]	···	Y(5, 6) [ ]
Y(6, 1) [ ]	Y(6, 2) [ ]	···	Y(6, 6) [ ]

6–3 IMPLIED DØ LOOPS

The input or output of numerical values assigned to subscripted variables is often necessary. As shown in Section 6–2, this transfer of data can be efficiently accomplished by means of DØ loops. This need occurs so frequently that special forms of both READ and WRITE statements exist in FORTRAN. These are the *implied* DØ READ and the *implied* DØ WRITE. The general forms of these statements are as follows. As usual, we have allowed space for you to record any differences between this and the computer system available to you.

READ (5,*) (variable (integer variable), integer variable = initial value, final value, increment)

WRITE (6,*) (variable (integer variable), integer variable = initial value, final value, increment)

To understand these general forms, examine the following specific examples.

```
READ (5,*) (G(I),   I = 1,10,1)

WRITE (6,*) (G(J),   J = 1,10,1)
```

These two lines are nearly exact FORTRAN equivalents of the following program segment.

```
  DØ 1   I = 1,10,1
1bREAD (5,*) G(I)

  DØ 2   J = 1,10,1
2bWRITE (6,*) G(J)
```

We say *nearly* because there are differences in the precise manner in which data are read and written.

More than one variable can be either read or written using one implied DØ loop, as the following example illustrates.

```
READ (5,*)   (G(I), L(I),   I = 1,10,1)
```

is "almost" equivalent to

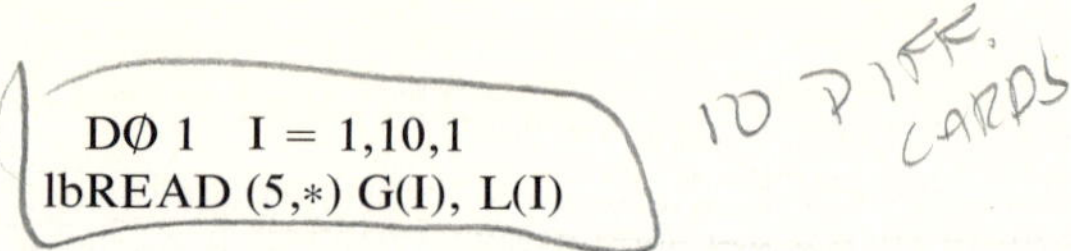

```
 DØ 1   I = 1,10,1
1bREAD (5,*) G(I), L(I)
```

One caveat concerning the arrangment of numerical values on data cards is essential, since these examples use free FØRMAT READ and WRITE. This last segment contains an explicit DØ statement. The first statement (the READ statement) will be encountered 10 times as the program is executed. Thus at least 10 separate data cards are needed, since a data card will be pulled through the card reader each time the READ statement is executed. Hence 10 executions of the READ require 10 separate data cards with two pieces of data on each.

This is not true for the implied DØ loop, since the READ is executed only once. This one execution causes the computer to go to the card reader and "look" for 20 numbers. The first two are assigned to the memory slots labeled G(1), L(1), the second two to G(2), L(2), . . . , and the last pair to G(10), L(10). In this way, all 20 required numbers may be placed on the same data card (assuming they fit). You should keep this distinction in mind when you use the implied DØ READ, since it simplifies keypunching considerably. When a free FØRMAT input or output is not used, the specific set of FØRMAT statements being used still determines the arrangement of data.

Implied DØ loops may also be nested. This is accomplished through the use of multiple parentheses, as shown in the following example.

```
READ (5,*) ((Y(I,J), J = 1,3), I = 1,3)
```

is "almost" equivalent to

```
 DØ 1   I = 1,3
 DØ 1   J = 1,3
1bREAD (5,*) Y(I,J)
```

The increment is assumed equal to 1 if it is not explicitly stated in implied DØ statements. Nesting to more than two levels proceeds in a fashion analogous to the details shown in this example.

In summary, the implied DØ loop READ and WRITE statements add little that is new to the arsenal of FORTRAN tools, but they do provide techniques for reducing the labor required in some input–output operations.

6–4 ARITHMETIC-STATEMENT FUNCTIONS

All FORTRAN compilers contain a built-in set of library functions. These are equivalent to the function keys found on advanced scientific calculators (log *x*, sin *x*, etc.). For example, the FORTRAN statement

```
Y = SIN(X)
```

means that the trigonometric sine of the value found in the memory slot labeled X (assumed to be in radians) will be calculated and stored in Y. This statement is valuable, since it allows the user to refer to the function sine as needed. The argument of such a so-called library function must be contained within parentheses, exactly as in algebraic expressions. Some commonly available library functions are as follows.

Algebraic	FORTRAN
$y = \sqrt{x}$	Y = SQRT(X)
$y = \sin(x)$	Y = SIN(X)
$y = \cos(x)$	Y = CØS(X)
$y = \tan^{-1}(x)$	Y = ATAN(X)
$y = \exp(x) = e^x$	Y = EXP(X)
$y = \ln(x)$	Y = ALØG(X)
$y = \lvert x \rvert$	Y = ABS(X)

These functions are frequently useful, but invariably some functions that are repeatedly needed in a program are not available in the computer system's library. An example of this is the function

$$y(x) = 18x^3 + 7x + 8 + 6x^{-2}$$

This function has no physical meaning that we know of; we selected it at random to introduce a new concept. Imagine that we need to evaluate this function several times in the course of executing a sequence of instructions. We could write

```
Y = 18. * X ** 3 + 7. * X + 8. + 6./X ** 2
```

each time the function was needed. This could lead to cumbersome expressions and large amounts of keypunching.

Fortunately, we can define this function by specifying an arithmetic-statement function. Once we have done this, we can use the function as we would use any library function. The required specification has the form:

```
Y(X) = 18. * X ** 3 + 7. * X + 8. + 6./X ** 2
```

The left-hand side of this statement is not simply the variable *Y* but rather the quantity Y(X). This alerts the compiler to the existence of a "new" function.

Caution is necessary in handling arithmetic-statement functions. The compiler could easily confuse the left-hand side of an arithmetic-statement function with a subscripted variable, since they both involve parentheses. For this reason, all compilers rigidly specify the sequence in which definitions of DIMENSIØN, REAL, INTEGER, and arithmetic functions must appear at the beginning of all programs. One common sequence is given below (and will be used hereafter) along with space to record any local deviations.

All DIMENSIØN statements
All REAL and INTEGER statements

```
All arithmetic-statement functions
Rest of FORTRAN instructions
```

This sequence enables the computer to "see" all subscripted variables before any arithmetic statement functions are defined. Thus, the system is "aware" that any variables followed by parentheses that did not appear in the DIMENSIØN, REAL, or INTEGER statements must be arithmetic statement functions. The one class of exceptions to this statement is the set of all defined library functions.

The need to properly order these statements at the beginning of a set of FORTRAN instructions will become very clear after you have compiled a few improperly ordered programs. We have found that this mistake generates an embarrassingly large number of error messages.

In arithmetic-statement functions, the mode (integer or real) of the function name (Y in our example) and the argument (X above) do not need to agree. To illustrate this point, consider the following examples.

```
Y(X) = (4. * X ** 2 - 11.)/6.
Z(I) = (4 * I ** 2 - 11)/6
W(X) = SIN(X) - SQRT(X)
```

If the first two functions were used in the following program segment,

```
Y(X) = (4. * X ** 2 - 11.)/6.
Z(I) = (4 * I ** 2 - 11)/6
A = Y(2.)
B = Z(2)
```

the state of memory would be

A [0.8333]   B [0.]

The third example of an arithmetic-statement function illustrates that library functions may be used in definitions. In fact, one statement function may be used in the definition of another, as the following example illustrates.

```
A(X) = X ** 2 + 8. * X - SQRT(X)
B(T) = SIN (A(T)) - 3. * A(T)
```

These two statements illustrate the use of dummy variables. That is, an arithmetic-statement function defined in terms of the argument X can be referred to, without error, as a function of T. In fact, the argument of the function A(X) can be any real variable. The following segment further illustrates the use of dummy variables.

```
Z(A) = 8. * A ** 2 - 3. * A + 7.
READ (5,*)  P,R,S,T

P1 = Z(P)
R1 = Z(R)
S1 = Z(S)
T1 = Z(T)
```

is equivalent to

```
READ (5,*)  P,R,S,T

P1 = 8. * P ** 2 - 3. * P + 7.
R1 = 8. * R ** 2 - 3. * R + 7.
S1 = 8. * S ** 2 - 3. * S + 7.
T1 = 8. * T ** 2 - 3. * T + 7.
```

Once an arithmetic-statement function has been defined in a program, it may be used throughout that program in the same manner as any valid library function. Defining an arithmetic-statement function often significantly reduces the time for writing and keypunching a program as the last example demonstrates.

Arithmetic-statement functions may be defined as functions of more than one variable. For example, to solve a quadratic equation

$$ax^2 + bx + cx = 0$$

one must calculate the discriminant

$$b^2 - 4ac$$

An arithmetic-statement function that may be used to calculate this quantity for any set of coefficients (a, b, c) is

```
DISC(A,B,C) = B * B - 4. * A * C
```

Once the function DISC has been defined, it may be referred to in any subsequent part of the program without redefinition. Such statement functions are not difficult to construct, and they reduce programming effort. Any complex program requiring the use of involved mathematical expressions in several different places is an excellent candidate for arithmetic-statement functions. The rules governing their formation are quite straightforward.

One problem with arithmetic-statement functions is that such functions must consist of only one expression. This expression can be quite long; it can be continued on more than one FORTRAN card if necessary. But no LOGICAL IFs are permitted. For example, it would not be possible to define one arithmetic-statement function as being equivalent to the following algebraic function.

$$y(x) = \begin{cases} 5x, & x \geqslant 0 \\ -2x, & x < 0 \end{cases}$$

Section 6–5 defines a more powerful technique to overcome this defect.

6–5 FUNCTION SUBPROGRAMS

We can vividly illustrate the limitations of arithmetic-statement functions. Consider the problems of implementing a FORTRAN program to calculate the function $y(x)$ given at the end of Section 6–4. To calculate this function, we have to consider the sign of the argument. If the argument is negative, we evaluate the function as

$$y(x) = -2x$$

If the argument is positive, we must use the function

$$y(x) = 5x$$

There is no way to carry out these calculations using only one statement. Calculating $y(x)$ requires more than one FORTRAN statement, since some logic has

FIGURE 6–9 An arithmetic statement function cannot realize this flow chart because it contains a logical operation (Is $x < 0$?).

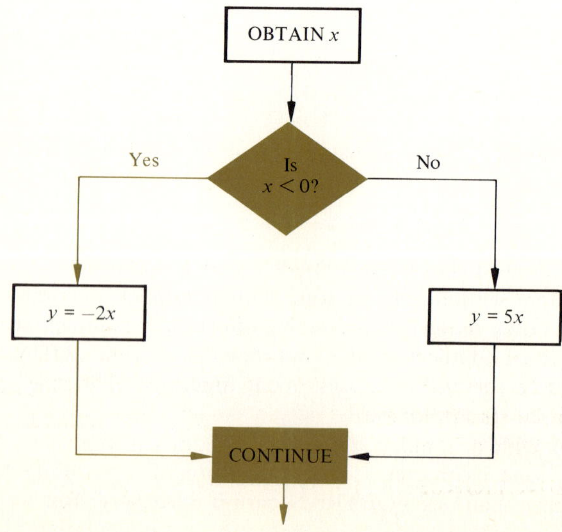

to be carried out. In particular, it is essential to determine whether the variable y is positive or negative, and this requires a LØGICAL IF statement. The following program segment illustrates the evaluation of the function.

```
   IF (X .LT. 0.)  GØ TØ 3
   Y = 5. * X
   GØ TØ 2
3bY = -2. * X
2bCØNTINUE
```

This program segment is shown as a block diagram in Figure 6–9.

As indicated above, an arithmetic-statement function cannot be used to calculate the function $y(x)$ because such a statement may consist of only one arithmetic expression, and evaluating $y(x)$ requires *two* algebraic expressions. But in FORTRAN, we can define more involved functions by means of *subprograms*. These are precisely what the name implies. They are program segments defined by a set of suitable statements. When a subprogram is compiled along with another program (we shall call this other program the *main* program), the subprogram may be invoked very simply. To see how subprograms work, let's consider an example.

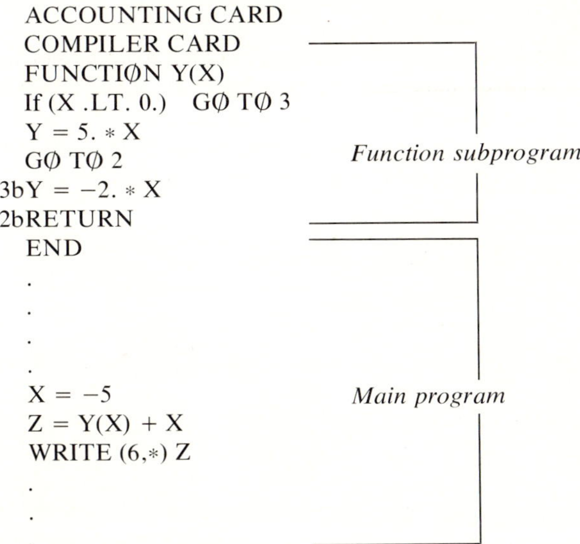

As usual, the first and second cards identify the user and the desired language. The third card is a new statement; it tells the computer that the set of cards between it and the first END card define a function with the name Y and the argument X. The general form of this statement is

```
                  valid
bbbbbbFUNCTIØNbFORTRAN(argument list, separated by commas)
                  label
```

The function name (in this case, Y) determines the mode of the numerical values calculated. This name is followed by a list of variables enclosed in parentheses and separated by commas. In this example, the list consists of only one variable X.

The statements following the FUNCTIØN card follow the algorithm for calculating the function $y(x)$. (The variable X must be assigned a numerical value by some mechanism outside the FUNCTIØN subprogram. We shall discuss this later.) We can follow through the next four cards in the subprogram with no difficulty and arrive at statement 2. Instead of a STØP, we come to a new FORTRAN statement, RETURN. The FUNCTIØN itself terminates in an END card that indicates that the subprogram is complete for purposes of compilation.

Next consider the main program. It may consist of many cards, but in this example only three are relevant. The first of these assigns a numerical value (−5.) to the variable X. The next statement refers to the defined function Y(X). When the computer executes this statement, it branches to the set of statements that define the function Y(X), that is, the function subprogram. Then it calculates the value of Y according to the recipe detailed in the subprogram. The following are the memory slots just before the RETURN statement is executed.

X [−5.] Y [10.] Z [ ]

The RETURN statement returns the system to the point at which it left the main program. Thus the value of the variable Z will next be calculated. This will be 5. The process we have just described is shown schematically in Figure 6–10.

Using the expression Y(X) transfers control from the main program to the subprogram. Execution remains within the subprogram until the RETURN state-

FIGURE 6–10 Execution of a function subprogram

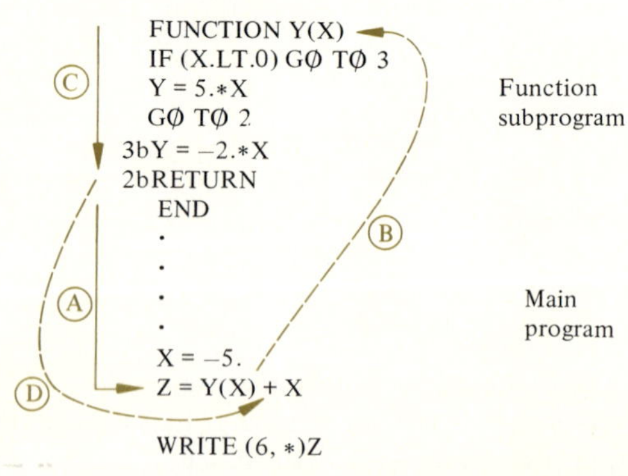

(A) First set of statements executed

(B) Branch caused by use of Y(X) in main program

(C) Execution of function subprogram Y(X)

(D) Branch caused by execution of return

ment is encountered. At this point, control returns to the main program at the point at which the expression Y(X) was first encountered.

The variable or variables appearing within the parentheses following the name of a function are truly dummy variables in the sense discussed in Section 6–4. For example, the following main segment of a program would successfully execute using the FUNCTIØN subprogram defined in our example.

```
.
.
.
T = -5.
Z = Y(T) + T
WRITE (6,*) Z
.
.
.
X = 15.
Q = -3.
A = X/Q
B = Y(A) + A
WRITE (6,*) B
.
.
.
```

After these two segments of the main program have been executed, the values stored in Z and B are

Z [5.]   B [5.]

It is also often desirable to define functions of several variables. For example, in calculating the roots of quadratic equations, it is helpful to have a function that will produce the absolute value of the discriminant of a given set of coefficients (*a, b, c,*). The following FUNCTIØN subprogram does this.

```
  FUNCTIØN ADISC (A,B,C)
  DISC = B * B - 4. * A * C
  IF (DISC .LT. 0.) GØ TØ 1
  ADISC = DISC
  GØ TØ 2
1bADISC = -DISC
2bRETURN
```

Exercise

6–14 Assume that the arithmetic-statement function Y(X) and the FUNCTIØN subprogram ADISC (A,B,C) discussed above have been defined. What values of Z will the following two main program segments produce?

```
A = 10.                    X = -3.
R = 11.                    A = Y(X)
C = 15.                    B = 3.
Z = ADISC (A,R,C)          C = 4.
WRITE (6,*) Z              Z = ADISC (A,B,C)
                           WRITE (6,*) Z
```

Several general features of FUNCTIØN subprograms must be considered each time such a function is constructed. The first card within a subprogram must be the FUNCTIØN card containing the name of the function and its argument or arguments within parentheses. Any subprogram must contain at least one RETURN card to direct the system back to the main program. Finally, the mode of the variables used in the main program as arguments of the function must agree with the modes used in the subprogram. The following example illustrates this last point.

```
   FUNCTIØN Z(X,Y,I)
   IF (I .LT. 0)   GØ TØ 2
   Z = X ** I - Y
   GØ TØ 1
 2bZ = X ** (-I) - Y
 1bRETURN
```

Suppose now that the following segment of a main program invoked this function

```
   .
   .
   .
A = 2.
B = 0.
I = 2
Y = Z(A,B,I)
   .
   .
   .
```

Then memory would look like this after the execution of the program.

Y | 4. |

On the other hand, the execution of the following main program segment

```
   .
   .
   .
A = 2.
```

```
B = 0.
C = 2.
Y = Z(A,B,C)
  .
  .
  .
```

would give an error, since the third variable used in the parentheses following the function name Z is real in the main program and an integer in the defining subprogram. The modes must agree.

Although we did not say so explicitly, it is possible to use constants as arguments of functions. Again, the modes in the main program must agree with those in the defining subprogram. For example, the following are valid expressions using the function Z.

```
Y = Z(2., 0., 2)
Q = Z(3., 7., −4) + Z(7., −1., 2)
```

We can now summarize the capabilities and limitations of arithmetic-statement functions and FUNCTIØN subprograms. Arithmetic-statement functions are necessarily limited to one expression, and they do not provide for any logical processing. In addition, each time the name of such a function is used in a FORTRAN statement, the computer branches to the defining statement and calculates the one functional value that corresponds to the current values of the arguments. A FUNCTIØN subprogram, on the other hand, can consist of more than one statement and can easily accommodate logical processing. Thus FUNCTIØN subprograms are more powerful than arithmetic-statement functions because they allow the programmer to handle more complex functions. However, a FUNCTIØN subprogram calculates only one numerical value each time it is invoked, and that value is the functional value corresponding to the current values of the arguments.

Both arithmetic-statement functions and FUNCTIØN subprograms are useful tools. However, it is often desirable to calculate more than one functional value at a time. This need is satisfied by a FORTRAN concept called a SUBRØUTINE.

6–6 SUBRØUTINES

Let us consider an example involving elementary principles of mechanics. Consider a body that is initially at rest and that undergoes a constant acceleration a for a time t. We can calculate the distance traveled x and the final velocity attained v by using the formulas

$$x = \frac{1}{2}at^2, \qquad v = at$$

If we needed to evaluate these expressions for different values of acceleration and time, a natural FORTRAN approach would be to define two arithmetic-statement functions.

```
X(A, T) = .5 * A * T * T
V(A, T) = A * T
```

The problem becomes complicated when only the magnitudes of the velocity and distance are desired, that is, when the signs are not necessary. We can rewrite the formulas as

$$\text{Distance} = \frac{1}{2}|a|t^2, \qquad v = |a|t.$$

Although we could compose new arithmetic-statement functions to evaluate these two expressions, it is more convenient for our purposes here to evaluate these quantities using FUNCTIØN subprograms. Two subprograms that would do the job are as follows:

```
FUNCTIØN DIST(A,T)
IF (A .LT. 0.) A = -A
DIST = 0.5 * A * T * T
RETURN
END

FUNCTIØN SPEED(A,T)
IF (A .LT. 0.) A = -A
SPEED = A * T
RETURN
END
```

Note that a separate FUNCTIØN subprogram is needed for each function. In this example, one subprogram is required to evaluate the distance traveled, and another to calculate the final speed. In both cases, the main program would define the values of time and acceleration.

This example illustrates a limitation of both arithmetic-statement functions and FUNCTIØN subprograms. Each time either is executed, only one numerical value is calculated. To calculate values for several quantities, we have to define several arithmetic-statement functions (if we can write the quantities in a single line) or several FUNCTIØN subprograms.

To overcome this limitation and provide more program flexibility, FORTRAN allows the definition of a *subroutine*. This is precisely what the name implies. It is a program segment composed of suitable statements. When a subroutine is compiled along with a main program, a single statement in the main program will evoke the subroutine. Let's consider a specific example using the same mechanics problem.

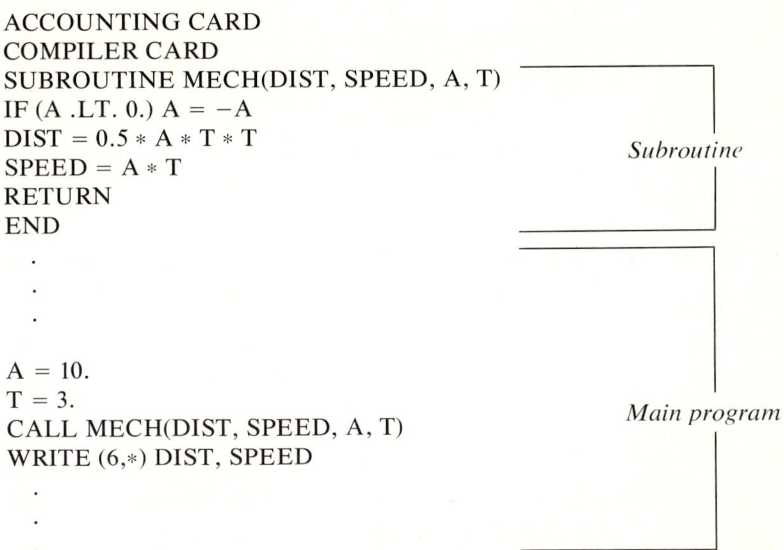

The first two cards serve the usual purpose, while the third card identifies the set of statements as a subroutine. The general form of this statement is

bbbbbbSUBRØUTINEbname (variable list, separated by commas)

The subroutine name is only a reference label; its mode has no importance in any calculation. It is often helpful to use a descriptive name, as in our example. The name is followed by a list of variables, separated by commas, within parentheses. You should carefully consider the nature of the variables in this example. Note that two (A, T) are input values to the subroutine. That is to say, they must be specified by some means external to the subroutine. They appear only on the right-hand side of statements occurring within the subroutine itself. This means that the quantities to be stored in the memory locations A and T must be supplied by the main program. The other two variables (DIST, SPEED) are derived quantities; they appear on the left-hand side of statements within the subroutine. Later we shall have more to say about this difference between the variables within the parentheses following the subroutine name.

The statements following the subroutine card implement the algorithms that define the two FUNCTIØN subprograms presented earlier in this section. These three statements enable one to calculate the quantities DIST and SPEED in the desired manner. The RETURN statement has the same meaning as in a FUNCTIØN subprogram. The subroutine itself terminates in an END card, which indicates that the subroutine is complete for purposes of compilation.

Next let us consider the new element added to the main program. The first two statements in this segment assign numerical values to the variables A (10.) and T

(3.). Then a totally new statement occurs. This is the CALL statement, which is required to invoke a subroutine.

CALL MECH (DIST, SPEED, A T)

It consists of the word CALL followed by the name of the subroutine and the four variables within parentheses. To understand the effect of this statement, we shall consider the execution of the main program. First, the values 10. and 3. are stored in memory locations A and T. Then the CALL statement initiates the subroutine MECH, and control within the computer goes to the first statement after the subroutine card. Since A is greater than zero, the statement after the parentheses of the logical IF is not executed and values for DIST (45.) and SPEED (30.) are calculated. The RETURN statement returns the system to the main program at the statement following the CALL. However, values for DIST and SPEED are now stored within the memory. This is the effect of executing the subroutine.

In summary, the CALL statement transfers control from the main program to the subroutine. Execution remains within the subroutine until the RETURN statement is encountered. At this point, control returns to the main program at the statement following the CALL. This sequence is similar to that occurring with a FUNCTIØN subprogram, except that a CALL statement is not required for a subprogram.

The variables appearing within the parentheses following the name of a subroutine are dummy variables. For example, the following two main program segments would both work as well as the example given above.

```
A1 = 10.
T1 = 3.
CALL MECH (DIST, SPEED, A1, T1)
```

or

```
A1 = 10.
T1 = 3.
CALL MECH (X, V, A1, T1)
```

On the other hand, the main program segment

```
.
.
.
A = 10.
T = 3.
CALL MECH (X, V, A, T, Q)
.
.
.
```

would not work at all, since five variables occur in the CALL statement, while only four occur in the SUBRØUTINE expression. The number of variables within the parentheses of both the CALL and the SUBRØUTINE statements must agree.

Although the mode of the subroutine name itself is unimportant, the modes of the variables in the calling statement and the subroutine statement must be the same. For example, the following would be an invalid CALL for the subroutine MECH.

```
I = 10
T = 3.
CALL MECH (X, V, I, T)
```

We shall consider the formal rules governing subroutines after a brief philosophical digression. We have introduced new concepts by using simple examples. In general, this approach has significant value, but in the case of subroutines, it runs the risk of obscuring one significant point.

Engineering problems, even on the undergraduate level, are often quite complex. There are many ways of attacking involved problems, and one of the most useful and natural is to resolve a complex issue into a number of simple parts. You may think that we are simply stating the obvious, but we have observed that many people do not approach complex problems in such a methodical fashion. They may pay lip service to this approach, but if one carefully examines their methods of attack, the failure to resolve the problem into a number of simpler parts is evident.

The subroutine provides a natural vehicle for decomposing a long program into a number of simpler segments. Indeed, once you have successfully constructed these segments, you may use them to help solve other problems. For this reason, we encourage you to use subroutines, since using them will force you to reduce complex issues into simple constituents. Few would argue with the proposition that engineers take complex problems and—by intuition and, in some cases, dumb luck—resolve them into simpler pieces. Using subroutines frequently helps you develop this habit, as the problems in subsequent chapters will demonstrate.

Now let us consider the formal rules for constructing subroutines. The first card is a title card of the general form shown on page 229. The subroutine name itself has no use in the compilation or execution of the program other than as a label. Consequently, the mode of this name is unimportant. Any name may be used (students in the past have proved to be very creative in this area), but there is value in choosing a name with an obvious relationship to the purpose of the subroutine.

The variables in parentheses are divided into two classes: input and output. No particular ordering within the parentheses is required. That is, the output variables do not necessarily have to precede the input variables, as has been the case in our examples. However, at least one of the variables must occur on the left-hand side of at least one statement within the subroutine. This is, of course, the output variable.

The flow of execution through the subroutine must terminate in a RETURN, and the last card in the FORTRAN deck composing the subroutine must be an END card.

If dimensioned variables are used in a subroutine, there *must* be an appropriate DIMENSIØN statement *in* the subroutine. If dimensioned arguments are used, they *must* be dimensioned in *both* the main program *and* the subroutine. The usefulness of subroutines will become apparent when we study several examples in the following chapters.

Exercises

6–15 A basic equation of mechanics states that the velocity v attained by a body undergoing a constant acceleration a over a distance x is given by the equation

$$v^2 = v_0^2 + 2ax$$

where v_0 is the initial velocity of the body. Write an arithmetic-statement function to calculate the quantity v as a function of a, x, and v_0. Write a segment of a main program that will use this arithmetic-statement function to evaluate v for

$$\begin{aligned} v_0 &= 10 \text{ meters/sec} \\ a &= 5 \text{ meters/sec}^2 \\ x &= 100 \text{ meters} \end{aligned}$$

6–16 Use a FUNCTIØN subprogram to solve the problem in Exercise 6–15.

6–17 Use a subroutine to solve the problem in Exercise 6–15.

6–18 If two bodies undergo different constant accelerations (a_1, a_2) for different times (t_1, t_2), their respective velocities are given by the equations

$$v_1 = a_1t_1, \qquad v_2 = a_2t_2, \qquad \Delta v = v_1 - v_2$$

The last expression is the difference between the two velocities. Write a subroutine that will use the two accelerations and two times as input variables and produce the two velocities and the difference between them as outputs.

6–7 MORE ON OUTPUTS

In Chapter 3, we discussed techniques to produce literal output. The method used apostrophes embedded within a FØRMAT statement, since the compiler instructs the output device to reproduce literally whatever symbols are contained within the apostrophes. This method of producing literal output is not universally available now, although it is becoming more widespread. All FORTRAN compilers, however, do provide another technique for producing such output. This is based on the H specification.† The following two-line program segment illustrates the use of this specification.

Program segment	*Output*
WRITE (6, 101)	THEbANSWERbIS
101bFØRMAT (14HbTHEbANSWERbIS)	

†H is used to honor Herman Hollerith, who originated the idea of punched cards in data processing.

The specification is written as 14H, and it directs the printer to reproduce the 14 symbols appearing after the H in the FØRMAT statement.

You will undoubtedly conclude that the use of apostrophes is preferable to the H specification, since the latter involves carefully counting symbols and blank spaces to ensure the correct output, while the former does not. This is indeed true, and if you are using a system that lets you use apostrophes to produce literal output, you can completely avoid the H specification.

Now let us take a closer look at the details of producing output on the printer. Consider the following program segment and its output.

Program segment	*Output*
WRITE (6, 101) 101bFØRMAT (14HTHEbANSWERbISb)	HEbANSWERbIS

Compare the output produced by this segment with the output produced by the preceding segment. In one case, the phrase appears as desired; in the other, the T in the first word is omitted. But what is the difference between the two program segments themselves? It is only an apparently insignificant shift of a blank from before the T to after the S. But this very small change has important consequences; it illustrates a concept known as *carriage control.*

In general, a computer produces output in the following way. During execution, the system encounters a WRITE statement containing directions for printing a set of data and, in some cases, literal output. The system proceeds from left to right through the output statement and accumulates the desired information in a buffer memory. It is helpful to think of this buffer memory as a scratch pad for output. In many cowboy movies, the grizzled clerk in the Western Union office wrote the hero's message on a scratch pad before he transmitted it. In fact, the impression left on the next sheet of paper in the pad often helped the villain obtain information concerning the hero's next move. (Of course, in some movies, it is necessary to interchange the words hero and villain.)

In any event, all computer systems have this kind of output scratch pad. Basically, it accumulates one line of output data. The first character in this line is significant, since it instructs the output printer, to space, double-space, or even skip to the top of the next sheet of paper, and so forth. This leading symbol in any line of output in the buffer memory is called the *carriage-control symbol.*

If the first character is a zero, the printer spaces twice before printing the line in the buffer memory. If the first symbol is a 1, the printer skips to the first line of the new sheet of paper before printing the line. If the first symbol is a plus sign, the printer does not advance at all, but rather prints on the same line. Local ground rules may be different.

If the first symbol is a blank or any character other than a 0, 1, or +, the printer advances one line before printing. Carriage-control symbols are used only with the printer, not with any other input or output device. The carriage-control symbol is never printed.

One last symbol often useful in specifying output is the slash (/). When it appears in a FØRMAT statement, the slash says, "Fill the remainder of the output buffer memory with zeros, print the resulting memory, and then continue with the rest of the FØRMAT specification." The following example illustrates this effect.

Program segment	*Output*
WRITE (6, 102)	THEbbANSWER
102bFØRMAT ('bTHE ANSWER'/'bbbbIS')	bbbIS

Exercise

6–19 Write and execute a suitable program to determine the effect of the following statement on output.

FØRMAT (/////)

Many computer systems have a FORTRAN subroutine to plot graphs of equations on the printer. This is often a useful option, since it can save much drafting time. The subroutines used to plot numerical data vary considerably from one computer to another, so we shall not try to describe them. But we encourage you to investigate the techniques for such subroutines.

6–8 SUMMARY

The end of our formal discussion of FORTRAN is near. After devoting two chapters to this topic, we think it would be helpful to review our motivation and general approach. FORTRAN should be viewed as an engineering tool that offers significant computational advantages to both the student and the practicing engineer. Novices will perceive these advantages only when they make a concerted effort to solve problems on a digital computer using FORTRAN. There are no short cuts to this goal, and the material within this text provides only a broad framework for learning FORTRAN. We have consciously omitted many FORTRAN topics and specific details from Chapter 3 and from this chapter. In this section, we shall mention a few additional topics briefly. Many detailed computer manuals are readily available; they provide much more information about the details of FORTRAN and particular computer systems. Let us now consider some additional useful statements.

The first of these is the arithmetic IF. It has the general form

IF (expression) SN1, SN2, SN3

It consists of the letters IF followed by a set of parentheses containing a FORTRAN expression. This expression may be simply one variable or a more complicated expression involving defined variables, constants, algebraic operators, and functions. Three statement numbers separated by commas follow the expression in parentheses. If the value of this expression is zero when the IF is encountered, control branches to statement 2 (SN2). If its value is negative, the computer branches to the first statement (SN1). And if its value is positive, the system branches to the third statement. The arithmetic IF was widely used when FORTRAN was a younger language and LOGICAL IF statements were not available.

Another useful statement that we have not discussed is a computed GØ TØ with the following general form.

```
GØ TØ  (SN1, SN2, . . . , SNN), integer variable
```

It consists of the words GØ TØ followed by a set of parentheses containing a set of statement numbers separated by commas. The parentheses themselves are followed by another comma and a single integer variable. Consider a specific example.

```
GØ TØ  (7, 14, 1, 9), I
```

When the computer encounters this statement during execution, the value stored in the memory slot labeled I controls branching. If I contains the value 1, control passes to the first statement number within the parentheses, 7. If I contains the value 2, control passes to statement number 14. This pattern continues for the remaining statement numbers within the parentheses. If, as in this example, there are only four statement numbers within the parentheses of the GØ TØ and the value stored in I is other than 1, 2, 3, or 4 when the statement is executed, the system will generate an error message and terminate execution.

Two library functions that we have not mentioned yet in this text are widely used. They are FLØAT and IFIX.

The FLØAT function converts a number stored in a memory slot labeled with an integer variable to a floating point quantity. In simple terms, this library function adds a decimal point to the integer. The function is most commonly used when the programmer wishes to use integer variables in calculations. The following example illustrates this.

```
  DØ 1  I = 1, 12
  DØ 1  J = 1, 12
  PRØD = FLØAT (I) * FLØAT (J)
1bWRITE (6,*) PRØD
```

This segment will produce a multiplication table in which all the entries (the products) have decimal points.

This program segment illustrates one additional concept. The function FLØAT may appear unnecessary, since the following segment generates the same set of answers.

```
  DØ 1  I = 1, 12
  DØ 1  J = 1, 12
  A = I
  B = J
  PRØD = A * B
1bWRITE (6,*) PRØD
```

It does do so, but there is an important difference between these two. The second segment requires two additional FORTRAN variables (A and B), and this uses more memory. Of course, worrying about this in such a simple program is like worrying about Jupiter's gravitational effect on an automobile. But in more complicated programs, the use of FLØAT can save computer memory and hence computer time. Since computer time costs money, this fine difference becomes important in some engineering jobs.

You may have already guessed the meaning of the second mode-conversion function IFIX. This function converts a number stored in a memory slot labeled by a real FORTRAN variable into an integer. In simple terms, IFIX truncates the decimal point and all digits to the right of the decimal point.

The following program segment produces the memory storage shown.

Program segment	*Memory*
Z = 5.47	
J = 3	Y [3.]
Y = FLØAT (J)	
I = IFIX (Z)	I [5]
M = IFIX (Z) * J	M [15]

One warning. IFIX can result in a loss of numerical information and should always be used with caution.

Before concluding this chapter, we want to consider one last point about DØ loops that may confuse the beginner. We could have mentioned it in the first section of this chapter, but we felt there was some value in deferring this topic to avoid unnecessary complications. In general, a program will execute properly if a transfer out of a DØ loop is made before the loop index (the integer variable that controls the looping) has reached its terminal value. This transfer could be accomplished, for example, by means of a logical IF or a GØ TØ. Transfers into a DØ loop from some other part of a program are possible, but only under special conditions. A transfer in can only occur *after* a transfer out, and it will execute without error only if the DØ loop index has the same value as it did when the transfer out occurred. You can see why these restrictions are necessary by considering the logical problems that would otherwise result.

This completes our brief introduction to computer systems and FORTRAN. At the risk of boring you, we repeat that the computer and FORTRAN are a pair of useful tools for solving problems. To be sure, the design and organization of computer systems and languages are ends in themselves for some engineers. But our purpose is to convey the foundations of engineering—modeling and computation.

In principle, with patience and lots of time, one could calculate all the solutions to the exercises and problems in this book by hand. The digital computer provides little that is new. You already know how to add, subtract, multiply, divide, raise numbers to powers, find trigonometric functions, and find logarithms. You certainly know how to perform repetitive calculations. You can add decimal points or truncate decimal points. You can store intermediate answers on a piece of scrap paper. In short, you have already learned, somewhere in your academic career, to carry out all the functions that a digital computer can. The computer has two significant advantages. It is orders of magnitude faster than you are and orders of magnitude more accurate. As such, it is a useful tool, and the wise engineer will master the technique of using it. These techniques will stand you in good stead throughout your professional career.

In Chapter 7, we begin a systematic study of the models that are essential for formulating and solving engineering problems. Indeed, models are used to solve just about all problems that human beings address.

PROBLEMS

P6–1 Which of the following DØ statements are incorrect? Why?

```
DØ 1  X = 1,10.3
DØ 2  J = 1,5,2
DØ 3  L = 1,10,1.
DØ 4  L1 = 2,5
DØ 5  L2 = I,J,K
DØ 6  L3 = I,J,2
```

P6–2 For each of the following program segments, list the values of L produced by execution.

```
   DØ 11  I = 1,4,1
   L = I + 5
11bWRITE (6,*) L
```

```
   DØ 12  J = 2,8,1
   L = J * J
12bWRITE (6,*) L
```

```
   DØ 13  K = 2,8,3
   L = K - 5
13bWRITE (6,*) L
```

```
   DØ 14  M = 3,2,1
   L = M/3
14bWRITE (6,*) L
```

P6–3 For each of the following program segments, determine the output produced.

```
   J = 5
   K = 2 * J
   DØ 31  I = J,K
   M = I * I + 2 * I
31bWRITE (6,*) M
```

```
   I = 5
   J = I/3
   K = I/4
   DØ 23  M1 = J,I,K
23bWRITE (6,*) M1
```

```
   L = 14
   M = 2.
   DØ 32  I = M,L
   N = 15 * I
32bWRITE (6,*) N
```

```
   N1 = 2
   N2 = 5
   DØ 12  K = N1, N2
   N3 = K - 8
12bWRITE (6,*) N3
```

P6–4 In each of the these DØ statements, one of the loop parameters is missing. In the parentheses above each statement is the number of times the loop should be executed. Fill in the missing integer.

```
      (10 times)                  (7 times)
DØ 15  I = 1,   ,1        DØ 17  J = 2,   ,2

      (8 times)                   (3 times)
DØ 19  K =     ,26,2      DØ 21  L = 1,20,

      (4 times)                   (6 times)
DØ 23  M = 5,   ,3        DØ 25  N = 8,   ,2
```

P6–5 How many times will statement 1 be executed for each of the nested DØ loops?

```
DØ 1  I = 1,5,1       DØ 1  I = 1,6,1
DØ 1  J = 1,5,1       DØ 1  J = 1,4

DØ 1  L = 1,6,2       DØ 1  M1 = 3,6
DØ 1  J = 1,5         DØ 1  M2 = 5,10,2

DØ 1  II = 1,7,3      DØ 1  N = 3,7,5
DØ 1  III = 2,6       DØ 1  NØ = 1,4,1
```

P6–6 Show the output that the following program segments will produce.

```
  DØ 1  I = 1,5,1            DØ 1  I = 1,5,1
  DØ 1  J = 1,I,1            DØ 1  J = I,5,1
1bWRITE (6,*) I,J          1bWRITE (6,*) I,J

  DØ 1  I = 1,5,1            DØ 1  I = 1,5,1
  DØ 1  J = 1,5,I            DØ 1  J = I,5,I
1bWRITE (6,*) I,J          1bWRITE (6,*) I,J
```

P6–7 Show the output for each of the following segments.

```
  DØ 1  I = 1,4              DØ 1  I = 1,4
  DØ 2  J = 1,4              DØ 2  J = I,4
  M = J - I                  M = J + I/2
2bWRITE (6,*) M,J          2bWRITE (6,*) M,J

  N = J + 2 * I              N = I - 3 * J
1bWRITE (6,*) N,I          1bWRITE (6,*) N,I
  DØ 1  I = 1,4              DØ 1  I = 1,5
  DØ 2  J = I,4              DØ 2  J = 1,I
2bWRITE (6,*) J            2bWRITE (6,*) J

1bWRITE (6,*) I            1bWRITE (6,*) I
```

P6–8 Show the output produced by the execution of the following program. Briefly describe the function of this program.

```
      DIMENSIØN  K(100)
      READ (5,*) N
      READ (5,*) (K(J), J = 1,N)
      DØ 10  L = 1,N
      DØ 10  J = L,N
      IF (K(L) .LE. K(J))  GØ TØ 10
      M = K(L)
      K(L) = K(J)
      K(J) = M
   10bCØNTINUE
      WRITE (6,*) (K(J), J = 1,N)

      STØP
      END
```

Data cards

```
5
5b2b7b1b4
```

P6–9 Indicate whether each of the following DIMENSIØN statements is valid or invalid. Give a reason for each invalid statement.

```
DIMENSIØN  X(15), Y(10), Z(5)
DIMENSIØN  S(10), I(12), K4
DIMENSIØN   (4) P(7) Q(10)
DIMENSIØN  AA(10), AAA(10.)
DIMENSIØN  I34(10), I35(10), I36(10)
DIMENSIØN  AI(J), K(L)
```

P6–10 Determine whether each of the following DIMENSIØN statements is valid. If valid, determine the total number of memory slots reserved. If invalid, give a reason.

```
DIMENSIØN  X(10), Y(15,7), Z(10,8,6)
DIMENSIØN  XY(15,15,3), XYZ(8,12,2)
DIMENSIØN  I(12,12.), J(12,12,3)
DIMENSIØN  M(11 + 12,3), J(N,2)
DIMENSIØN  A15(10,10,), A16(5,5), A17(10)
```

P6–11 For each of the following program segments, show the numbers stored in the array Q, that is, sketch the memory slots labeled Q(1), Q(2), . . . and their contents.

```
   DIMENSIØN  Q(10)
   DØ 10  I = 1,10
10bQ(I) = I * I - 7 * I
```

```
   DIMENSIØN  Q(12)
   DØ 12  I = 2,11
12bQ(I) = 5 * I - 10
```

P6–12 The following storage is in part of a computer's memory.

Y(1) 10.	Y(2) 11.	Y(3) 12.	Y(4) 18.
Y(5) 17.	Y(6) 27.	Y(7) 24.	Y(8) 21.

Show the printout that each of the following segments will produce.

(a)
```
      DØ 1   J = 1,2
    1bWRITE (6,*) Y(J), Y(J + 2), Y(J + 4), Y(J + 6)
```
(b)
```
      DØ 2   J = 1,4
      WRITE (6,*) Y(J)
      M = J + 4
    2bWRITE (6,*) Y(M)
```
(c)
```
      WRITE (6,*) (Y(J), J = 1,8)
```
(d)
```
      WRITE (6,*) (Y(J), J = 1,4)
      WRITE (6,*) (Y(K), K = 5,8)
```

P6–13 Determine the numbers stored in the memory locations labeled with dimensioned variables when each of the following two program segments is executed.

```
      DIMENSIØN   X(4)
      Z = 10.
      DØ 1   I = 1,4
      X(I) = Z
    1bZ = Z - 2.
```

```
      DIMENSIØN   S(6)
      T = 10.
      DØ 1   I = 1,6
      J = I/4
      IF (J. EQ. 0)   GØ TØ 2
      S(I) = T
      GØ TØ 1
    2bS(I) = 0.
    1bCØNTINUE
```

P6–14 Determine the numbers stored in the memory locations labeled with dimensioned variables when each of the following two program segments is executed.

```
      DIMENSIØN   T(4,4)
      L = 14
      DØ 1 I = 1,4
      DØ 1 J = 1,4
    1bT(I,J) = L - I - J
```

```
      DIMENSIØN   S(6,6)
      DØ 1 I = 1,6
      DØ 1 J = 1,6
    1bS(I,J) = I + 2 * J
```

P6–15 Write the FORTRAN statement needed to reserve the required number of memory locations for the dimensioned variable in the following program segment. N is 100 or less.

```
      DØ 50   JACK = 2,N,2
      CI(2 * JACK + 2) = JACK
   50bCØNTINUE
```

P6–16 Determine the numbers stored in the memory location D after each of the following program segments has been executed.

```
  A = 2.0
  B = 1.0
  N = 4
  C = 0.0
 1bC = C + A + B
  N = N + 1
  IF (N .LT. 6) GØ TØ 1
  D = FLØAT (N) * C
```

```
  CAT(Z,K) = Z ** K + FLØAT(K)
  W = 0.
  D = 1.
  DØ 25  K = 1,2
  W = W + CAT(D,K)
25bD = D + W
```

P6–17 Each of the following FORTRAN statements contains at most one error. Indicate whether each is a valid statement for the local computer system. If any statement is invalid, write one or more statements to carry out the intended operation.

```
DØ 10  J = 1, K + 1
READ (5,*) X(I), Y(I), Z(I)
DIMENSIØN  A(10), B(10), C(I)
WRITE (6,*) (X(K), K = 1(M)
C = SQRT (A * A + B * B)
DØ 56  I = 0, 10
```

P6–18 Determine the value or values of X printed by each of the following short programs.

```
  DIMENSIØN  T(5)
  DØ 25  I = 1,3
25bT(I) = I ** 2 - 1
  X = T(3) * T(2) * T(1)
  WRITE (6,*) X

  STØP
```

```
  X = 0.0
  DØ 50  I = 2,4
  X = X + FLØAT(I)
50bCØNTINUE
  X = SQRT(X)
  WRITE (6,*) X

  STØP
```

P6–19 Give the sequence of numbers that the following program will print.

```
FUNCTIØN JUMP(I,J)
JUMP = I - I * J
RETURN
END
SUBRØUTINE SALT(I,J,K,L)
K = I - J
L = I + J
RETURN
END
L(K) = K * K + 7 * K
N = 4
```

```
M = 9
I = L(N)
WRITE (6,*) I
J = JUMP(M,N)
WRITE (6,*) J
CALL SALT (M,N,LL,II)
WRITE (6,* ) LL,II
I = JUMP(I,M) + 302
J = L(I)
WRITE (6,*) J
STØP
END
```

P6–20 Consider a computer program that is evaluating a large number M of possible designs for a bridge; the designs are numbered 1, 2, 3, . . . ,M. Each design has an associated cost ($cost_1$, $cost_2$, $cost_3$, . . . , $cost_M$) and these costs are available in the main program as CØST (1), CØST (2), . . . , CØST (M).

Write a subroutine that will find both the lowest cost and the associated design number. The subroutine should return the answers to the main program. Assume that $1 \leq M \leq 1000$.

P6–21 Given N values each of x and y, write a FORTRAN program to read these values in pairs, compute

$$S = [(x_1 - y_1)^2 + (x_2 - y_2)^2 + \cdots + (x_N - y_N)^2]^{1/2}$$

and print out S. Execute your program using the following data.

x	0.0	0.4	0.8	1.2	1.6	2.0	2.4
y	1.2	1.1	1.4	1.7	1.8	1.9	2.2

P6–22 Write a subroutine to perform the following evaluations.

$$\text{PRICE} = 4.25\text{X}^2 + 5.69\text{Y}^2$$
$$\text{CØST} = 2.95\text{X}^2 + 4.03\text{Y}^2$$

Write a main program to read in values of X and Y and call the subroutine to find CØST and PRICE. The main program should then print the results.

CHAPTER SEVEN
DETERMINISTIC MODELS

Chapter 1 provided a general introduction to the different categories of models and their uses, and Chapter 2 discussed in detail the different techniques used to classify them. Chapters 3 and 6 introduced the elements of FORTRAN, a convenient language for using the digital computer in formulating models and obtaining information from analyzing them. Chapter 4 examined the physical units associated with models, and Chapter 5 developed optimization techniques.

Of the many viewpoints we developed earlier, the distinction between deterministic and probabilistic situations is important. A deterministic model describes or predicts the behavior of a system in which the outcome of an experiment under a specified set of conditions occurs with certainty. A probabilistic model is necessary when all we can say (after studying many experiments) is that a given outcome occurs in some fraction of the total number of trials conducted. Uncertainty is always present in physical situations but we may sometimes safely ignore it. No set of unambiguous rules specifies when we can ignore uncertainty in a model. Even experienced engineers sometimes have difficulty making this decision. This chapter develops deterministic concepts, while Chapter 8 examines probabilistic ideas. In Chapter 9 we introduce a third major theme: We develop the economic models that often determine whether a given engineering project will be initiated. These three aspects—deterministic, probabilistic, and economic—are often present simultaneously in a single engineering problem. To illustrate how they dovetail, let us consider a concrete example.

7–1 THE BASIC TYPES OF MODELS

Let us examine the task of designing a rust-resistant children's swing set. The most natural first step is to determine exactly what rust is. Until we understand the nature and mechanism of rusting, we cannot formulate any solutions. Rust is the chemical compound that results from the reaction between steel and the oxygen in the air. Next, we must find materials that are immune to this oxidation reaction. Stainless steel is such a material. Once we have selected it, we must determine its mechanical strength. We can do this by examining published data about the material and its strength under various mechanical loads. This information appears in research journal articles, textbooks, and design handbooks.

Once we have this information, we must apply the laws of statics and dynamics (engineering mechanics) to the design of a structure that will resist the anticipated forces resulting from children at play. The word design means several things: a sketch of the main structure of the swing set, the number of supporting posts, the number and positions of braces, the sizes of the various elements, the sizes of the bolts and nuts used to secure the entire structure, and so forth. We can answer the question of the total loads imposed only probabilistically. The germane questions are: How many children will play on the swing set at any one time? What is the average weight of a child? And so forth. Once we have completed a suitable probabilistic analysis to determine the average loads imposed and the maximum likely loads, we apply the deterministic laws of mechanics and perform calculations to ensure that the supporting beams and bolts can carry the anticipated loads.

Imagine that you have completed this analysis and that you then show your design to your supervisor. The reaction will be negative, since stainless steel is prohibitively expensive. To be sure, it is rust resistant, but few (if any) people would pay the price involved. Most swing sets are made of lower-grade (and less-expensive) steels, painted to prevent rust. This is not a perfect solution, since the paint will be nicked and the steel will rust, but it appears to be a suitable compromise between the cost of the finished product and its serviceability. It is possible to improve, in technical terms, the performance of almost any consumer product presently available. However, economic factors often prevent us from doing so.

This analysis went wrong because it did not consider an economic model. People who work in industry consistently criticize universities for the lack of economics in engineering education. This text develops economic models to enable us to evaluate alternative engineering designs in economic terms. After all, any company must make a profit if it is to survive. Any government program must show greater benefits than cost if government officials are to responsibly continue funding these programs. And consequently, engineers must include economic considerations in their designs.

To summarize, any problem can and often must be approached from three perspectives: deterministic, probabilistic, and economic. These viewpoints are complementary, but in some cases, we shall focus attention on just one to the exclusion of the others, to facilitate the communication of ideas. To introduce all the ramifications of problems in a first course in engineering would be unfair to you. You need several years of experience before you can successfully formulate a total picture. This chapter examines deterministic models. Chapter 8 develops probabilistic models, and Chapter 9 introduces economic considerations.

7–2 THE ELEMENTS OF DETERMINISTIC MODELS

We begin by restating a basic definition. A model is a simplified description of some real-world phenomenon or device; its purpose is to aid analysis, understanding, and design ability. Models come in many forms, but the ruthless suppression of uncertainties distinguishes deterministic models from probabilistic models. The deterministic model always implies absolute precision. As an example, consider Newton's second law,

$$F = Ma \qquad (7\text{–}1)$$

where F is the force applied to a body of mass M and a is the resulting acceleration. This equation must, of course, be used with a consistent set of units, and it constitutes a basic deterministic model for a large set of mechanical phenomena. Normally we use it in the following way. Given a body with a mass of 1 kilogram to which a force of 1 newton is applied, what acceleration results? We find the answer by direct substitution.

$$a = \frac{F}{M} = \frac{1\text{ N}}{1\text{ kg}} = \frac{1\text{ m-kg/sec}^2}{1\text{ kg}} = 1\text{ meter/sec}^2$$

All physical quantities are considered to be known with absolute accuracy.

Analysis of deterministic models is simpler than that of probabilistic models, since we can more readily obtain answers that are useful in design when we use deterministic models. To be sure, there is a trade-off. We eliminate detailed probabilistic information about a physical system to gain computational ease. Often this is of benefit, but sometimes the loss of accuracy is unacceptable.

Consider a familiar example—15 pocket billiard balls and a cue ball set on a table. How does a pool player make a shot? What factors does the player consider? The player is normally trying to solve Newton's laws of motion mentally, worrying about velocities, momenta, forces, and angles of impact, although usually in a very intuitive way. These aspects of play are deterministic, and all the balls are assumed to be identical. However, the problem clearly has a probabilistic facet. Every pool player almost always selects the shot that has the highest probability of success. To do otherwise is usually to invite defeat.

We shall say more about the uses of deterministic and probabilistic models later in this chapter and in Chapter 8. Now we shall consider several examples of deterministic models in engineering settings.

7–3 DETERMINISTIC MODELS IN ENGINEERING

Deterministic models are well suited to representation by mathematical equations and graphs. For this reason, we shall devote the remainder of this chapter to a consideration of symbolic and graphical analog forms of deterministic models. This section will consider a number of physical processes and phenomena for which deterministic models are well known.

The first concerns the fuel economy† (in miles per gallon) of an "average" automobile as a function of its speed (in miles per hour). It is on such deterministic models that the U.S. government based its policies with regard to automobile fuel conservation during the 1974 oil embargo. Figure 7-1 shows the model. This graph indicates that the "average" automobile uses gasoline most efficiently when it is traveling at a speed of 40 mph. We can also represent this information approximately by the following equation.

†Fuel consumption (gallons per mile) is commonly used to characterize aircraft, but is rarely used in connection with automobiles.

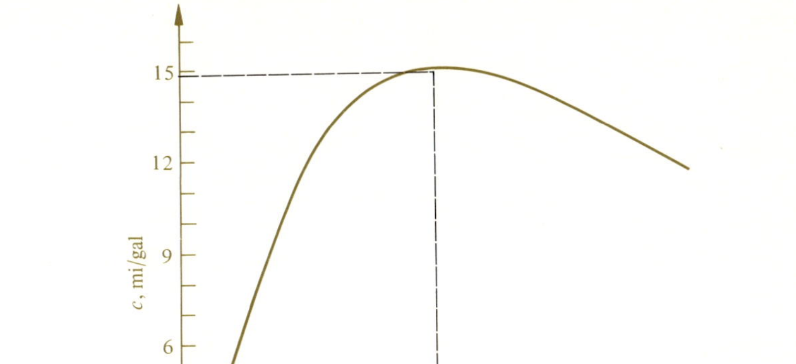

FIGURE 7–1 Fuel economy versus speed for an "average" automobile

$$c = 1200\left(\frac{v}{v^2 + 1600}\right) \qquad (7\text{–}2)$$

The importance of this model (in either of the two forms) is that it allows individuals considering automobile fuel economy to focus on essentials and eliminate confusing details. The model clearly indicates that fuel will be saved if speed limits are reduced and if the government enforces the reduction. Figure 7–2 displays the effects on fuel economy that result from reducing vehicle speeds. By reducing their speeds from 70 to 55, drivers increase their fuel economy from 12.9 to 14.3 mpg. On a national basis, this reduction results in a significant savings of gasoline. A purely engineering approach to this problem indicates that the optimum national speed limit is 40 mph. But political realities and the nature of people are important factors. The 55 mph speed limit has not been popular, and it would be difficult for any politician to support a 40 mph speed limit. In addition, even if a law were passed requiring the lower maximum speed, it is unlikely that it would be widely obeyed. Even the 55 mph speed limit has been disregarded in some areas.

A curve showing fuel economy c versus speed v could be drawn for each automobile in the country. However, we can apply suitable averaging methods to obtain a single curve that represents the "average" performance of the entire fleet of automobiles. So our deterministic model depends on probabilistic techniques of averaging.

FIGURE 7–2 Effects of changes in speed limit on fuel economy. Reducing speed from 70 mph to 40 mph improves fuel economy, while reducing speeds below 40 mph increases fuel consumption.

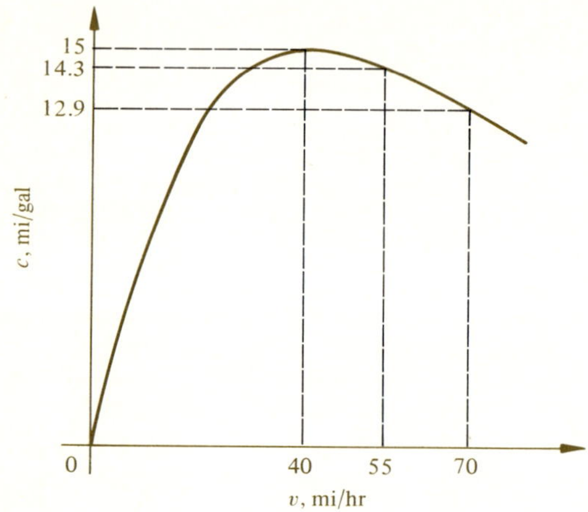

Exercises

7–1 Assume that automobile manufacturers followed a different course over the past decade and that the "average" c versus v model for the automobile population is

$$c = 2800\left(\frac{v}{v^2 + 4900}\right)$$

Plot this symbolic model in graphical form.

7–2 If the model in Exercise 7–1 were more accurate than that in Equation (7–2), would reducing the national speed limit from 70 mph to some lower value to conserve gasoline make sense? Why or why not?

The next example of a deterministic model concerns a popular topic—money. Suppose that an initial amount of money (P) is invested at 8% interest, compounded annually. The total amount (F) accumulated after n years is given by the formula

$$F = P(1.08)^n \tag{7–3}$$

We derive this and related expressions in Chapter 9.

For example, if $1000 were invested at 8% for five years, we could find the future value by substituting these values into Equation (7–3).

$$F = 1000(1.08)^5 = \$1469.$$

This represents an overall increase of nearly 47%.

We shall draw our third example of a deterministic model from mechanical engineering. Air may be compressed by decreasing the total volume of a fixed mass of air in a piston–cylinder apparatus. Extensive laboratory studies of this process have revealed that the relationship between the pressure P and the volume V at any time during the process is given by the following symbolic model,

$$PV^{1.3} = C \tag{7-4}$$

where C is a constant. Figure 7-3 shows a graph of this physical process. A manufacturer of compressed-gas cylinders would use this equation to determine the pressures to which cylinders will be subjected.

The following specific example gives some insight into the magnitudes of the physical quantities. A manufacturer wishes to compress 35 m^3 of air at atmospheric pressure (10^5 N/m^2) to a volume of 1 m^3. We can evaluate the value of the constant C by using the initial state of the air.

$$C = P_1V_1^{1.3} = (10^5)(35)^{1.3} = 1.02(10^7)$$

The pressure P_2 in the compressed state is given by the equation

$$P_2V_2^{1.3} = C$$

where V_2 is the compressed volume. In this case, $V_2 = 1$ m^3

Substituting numerical values results in

$$P_2(1)^{1.3} = 1.02(10^7) \qquad \text{or} \qquad P_2 = 1.02(10^7) \text{ N/m}^2$$

This means that, to provide a sufficient margin of safety, the tank holding the compressed air must be made of materials strong enough to withstand pressures in excess of 10^7 N/m^2.

FIGURE 7-3 Model of pressure P versus volume V for a fixed quantity of a gas

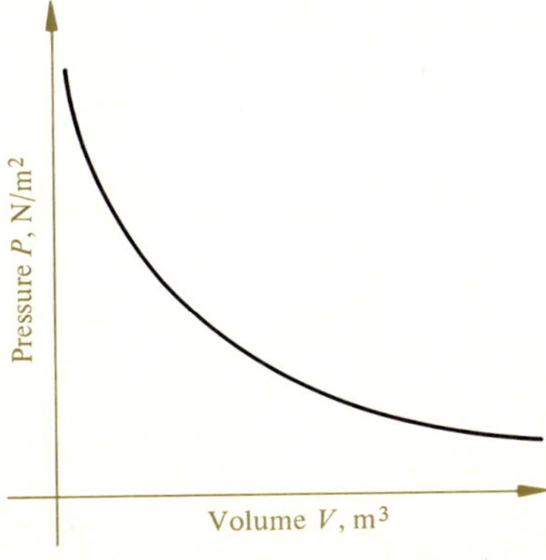

The last example in this section concerns a familiar topic—an oven cooling to room temperature once the heating elements have been turned off. Section 7–4 gives the mathematical model governing this phenomenon, but Figure 7–4 shows the cooling process by displaying oven temperature as a function of time. In this figure, the oven was turned off at time 0. The oven was initially at 200°C; after a period of time, it cooled to room temperature (20°C). This curve is an example of exponential decay; its shape occurs over and over again in physical systems. Section 7–4 examines exponentials in detail, but for now we can use Figure 7–4 to predict the oven's temperature at any time during the cooling cycle.

In steel mills, cooling curves for large ingots and slabs of metal must be known with extreme accuracy. In a typical processing operation, a slab that is removed from a gas-fired heating oven immediately begins to cool. The cooling curve, similar to that of Figure 7–4, governs the time available to shape the steel while it remains soft. In some mills, thick slabs are reduced to sheet steel by successively passing them through a series of large roller presses. This reduction in thickness must be completed before the steel has cooled and become too hard to be worked.

In summary, this section briefly examined typical deterministic models. The discussion demonstrated the usefulness of such models in predicting a system's behavior as changes occur. We extracted information from the models by simple mathematical manipulations or by reading numbers from the axes of graphs. We shall return to this concept of extracting information after we have developed some additional mathematical ideas.

FIGURE 7–4 Cooling curve, temperature T versus time t. The exponential curve is characteristic of the cooling of physical systems.

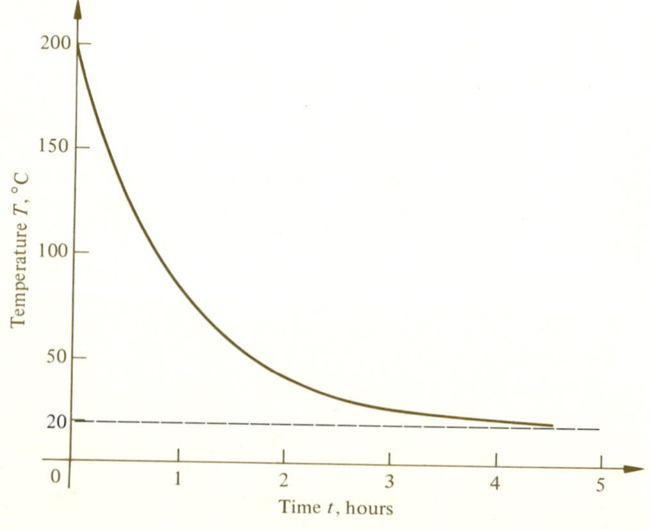

Exercises

7–3 An initial amount of money P is invested at an interest rate i (expressed as a fraction). After n years with compound interest, it grows to

$$F = P(1 + i)^n$$

For an initial investment of $5000, calculate the principal plus interest resulting after 10 years at 6% interest.

7–4 A well-known rule of thumb in financial circles is that the number of years required to double the value of an investment at a given interest rate may be estimated by dividing 72 by the interest rate (expressed as a percentage). For example, at 8%, we would need nine years to double the value of an investment. Using the model in Exercise 7–3, determine the accuracy of this rule for interest rates from 4% to 16%.

7–5 Consider a mass of air that was initially at a pressure of 10^5 N/m^2 and occupied a volume of 0.30 m^3. Calculate the pressure when the volume is reduced to 0.06 m^3.

7–6 Using Figure 7–4, estimate the time required for the oven to cool to 100°C. Also estimate the time required to cool to 50°C.

7–4 THE EXPONENTIAL FUNCTION

Consistent with our philosophy of throwing you directly into a new situation, we begin by presenting the mathematical model for the oven-cooling curve given in Figure 7–4.

$$T(t) = 180\, e^{-t} + 20\ [°\mathrm{C}] \tag{7–5}$$

In this equation, the temperature T is measured in degrees Celsius and the time t is measured in hours. The symbol e is commonly used in the engineering and scientific literature to denote a specific number:

$$e = 2.71828\ldots$$

This symbol denotes an irrational number; it can only be approximated by a rational quantity as indicated by the . . . on the right-hand side of the equation. It will help to consider e in the same category as the following symbol and equivalent number.

$$\pi = 3.14159\ldots$$

Both symbols are used to denote irrational numbers with important physical significance. The quantity π is, of course, the ratio of the circumference to the diameter of a circle. The physical significance of e is less obvious because it rests on some elementary concepts of calculus. As a prelude, let us examine a few specific functions.

Consider the meaning of

$$F(x) = x^2$$

This notation implies that, for any value of x, the square of x is the value of the function associated with that x. Now consider a somewhat less-familiar function:

$$g(x) = 10^x$$

We can construct a graph of this function by directly substituting different values of x. For example, for positive values,

$$\begin{aligned} g(0) &= 10^0 &&= 1 \\ g(0.25) &= 10^{0.25} &&= 1.78 \\ g(0.50) &= 10^{0.50} &&= 3.16 \\ g(0.75) &= 10^{0.75} &&= 5.62 \\ g(1.) &= 10^1 &&= 10 \end{aligned}$$

These results are plotted in Figure 7–5.

Now we may advance a step further and consider another function, closely related to the previous one:

$$h(x) = 2^x$$

FIGURE 7–5 The function $g(x) = 10^x$

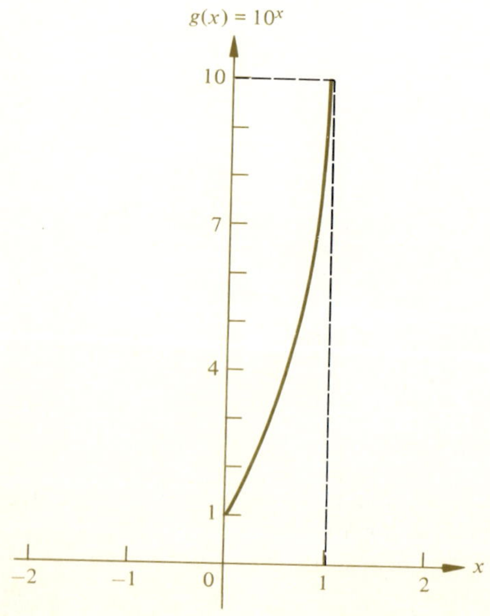

Again we may sketch the function by evaluating it for different values of the argument x.

$$\begin{aligned} h(0) &= 2^0 = 1. \\ h(0.25) &= 2^{0.25} = 1.19 \\ h(0.50) &= 2^{0.50} = 1.414 \\ h(0.75) &= 2^{0.75} = 1.682 \\ h(1.) &= 2^1 = 2. \\ h(2.) &= 2^2 = 4. \end{aligned}$$

Figure 7–6 displays a partial sketch of this function based on these values.

Exercises

7–7 Complete Figure 7–5 for negative values of x ($-2 \leq x \leq 0$).

7–8 Complete Figure 7–6 for negative values of x ($-2 \leq x \leq 0$).

The function

$$f(x) = e^x \tag{7–6}$$

is, in some sense, intermediate between $g(x)$ and $h(x)$. We can sketch it by using the techniques used to complete Figures 7–5 and 7–6.

$$\begin{aligned} f(0) &= 2.71828^0 = 1.00 \\ f(0.25) &= 2.71828^{0.25} = 1.28 \\ f(0.5) &= 2.71828^{0.50} = 1.65 \\ f(0.75) &= 2.71828^{0.75} = 2.12 \\ f(1) &= 2.71828^1 = 2.71828 \end{aligned}$$

FIGURE 7–6 The function $h(x) = 2^x$

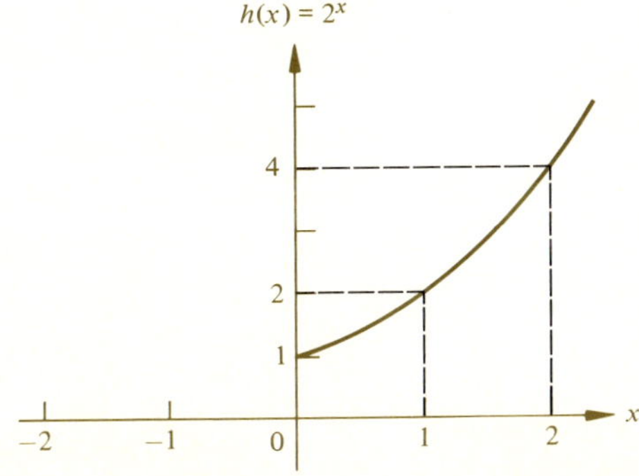

Note that we have had to limit the accuracy of our calculations to a finite number of digits because we are representing an irrational quantity by a rational approximation.

Exercises

7–9 Write and execute a FORTRAN program that will produce numerical values of 2.71828^x for $x = 0., \pm 0.125, \pm 0.25, \pm 0.375, \pm 0.5, \pm 0.625, \pm 0.75, \pm 0.875, \pm 01., \pm 1.25, \pm 1.5, \pm 1.75, \pm 2.$

7–10 Using the results of Exercise 7–9, sketch e^x on the axes of Figure 7–7.

In and of itself, the completed Figure 7–7 is unremarkable—not because of your sketching ability, but only because the function e^x is closely related to the two functions presented in Figures 7–5 and 7–6. The three figures, when considered together, do not imply any special significance for the function e^x. But the quantity e and the function e^x are important for two reasons: one mathematical and the other physical.

FIGURE 7–7 The function $f(x) = e^x$, do-it-yourself version

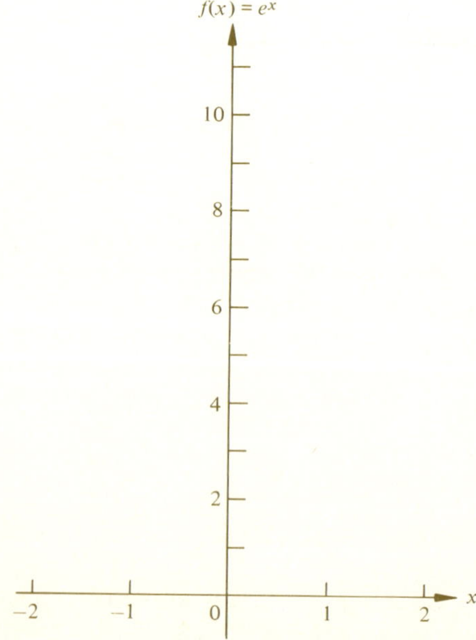

This function is mathematically important because it is unique. To see this, examine the properties of the following polynomial expansion.

$$f(x) = e^x = 1 + x + \frac{x^2}{2!} + \frac{x^3}{3!} + \frac{x^4}{4!} + \cdots \qquad (7\text{–}7)$$

The first year of calculus develops the methods used to construct such series expansions, but for now, we can only ask you to accept the correctness of this series. You can verify Equation (7–7) by writing and executing a FORTRAN program that will evaluate this polynomial expansion for different values of x and comparing the values produced by this program and the one written for Exercise 7–9.

Now we take the first derivative of e^x, using Equation (7–7):

$$\frac{df(x)}{dx} = \frac{d}{dx}(e^x) = 0 + 1 + \frac{2x}{2!} + \frac{3x^2}{3!} + \frac{4x^3}{4!} + \cdots$$

Simplified, this becomes

$$\frac{df(x)}{dx} = \frac{d}{dx}(e^x) = 1 + x + \frac{x^2}{2!} + \frac{x^3}{3!} + \cdots = e^x$$

or, remarkably enough,

$$\frac{df(x)}{dx} = \frac{d}{dx}(e^x) = e^x = f(x)$$

The mathematical function e^x is its own first derivative. Rigorous proof shows that no other function displays this property, with the trivial exception of

$$y(x) = 0$$

The function $f(x) = e^x$ is physically important because exponential functions describe many physical processes. Among these are the decay of radioactive materials, the heating and cooling of physical bodies, the weakening of radio and television signals with distance from the sending antenna, the decrease in density of air with height above the earth's surface, the emptying of liquids from tanks under the influence of gravity, and so forth. In a very meaningful sense, e is a "natural" number that arises in the mathematical descriptions of many physical processes. We shall give examples of exponential functions in physical and economic settings later in this chapter. First let us inject a note about terminology. Any function of the form

$$Ae^{\alpha x}, \qquad A,\ \alpha \text{ constants}$$

is said to be an exponential. This term will be used many times in our subsequent discussions.

To comfortably manipulate deterministic models that contain exponentials, we must be familiar with the concept of a logarithm.

Definition. The logarithm of a number x to the base b is the power p to which b must be raised to give x, or

$$b^p = x \qquad \text{is equivalent to} \qquad p = \log_b x$$

We may also state this as $b^{(\log_b x)} = x$.

Here are some examples of the evaluation of logarithms to the base 10 and the base 2.

$\log_{10}(100) = 2$	$\log_{10}(0.01) = -2$
$\log_2(8) = 3$	$\log_2(1/4) = -2$
$\log_{10}(2) = 0.301$	$\log_2(10) = 3.33$
$\log_2(1) = 0$	$\log_{10}(1) = 0$

Logarithms may be defined and calculated to any base.

$\log_7 49 = 2$	$\log_7(1/7) = -1$
$\log_{2.5} 6 = 1.956$	$\log_{2.5}(0.88) = -0.14$
$\log_e e = 1$	$\log_e(0.75) = -0.288$

When e is the base of the logarithm, it is customary to use the following shorthand notation.

$$\log_e x = \ln x$$

To understand logarithms to the base e (commonly called *natural logarithms*), note that the equation

$$\ln x = p$$

means that the quantity x is equal to e^p. In short, $\ln x = p$ means that $e^p = x$.

We shall often have to evaluate expressions containing natural logarithms and exponentials. Fortunately, all FORTRAN compilers have library functions to carry out these calculations. The appropriate forms are as follows.

Algebraic	FORTRAN
$y = e^x$	Y = EXP(X)
$z = \ln x$	Z = ALØG(X)

These tools make the evaluation of natural logs and exponentials comparatively simple. Scientific calculators can also handle these functions. Of course, the computer is preferred when many calculations must be made rapidly. Now let us consider how to extract information from deterministic models involving exponentials.

7–5 EXTRACTING INFORMATION FROM MODELS

The first example is the cooling of an oven, as sketched in Figure 7–4. Recall the symbolic model describing this process.

$$T(t) = 180\, e^{-t} + 20 \text{ [°C]}$$

Again, temperatures (T) are in degrees Celsius and times (t) are in hours.

The problem is to analytically determine the time at which the oven will have cooled to 66°C. Denote this time as t_1. Then, by definition,

$$180\ e^{-t_1} + 20 = 66$$

If you do not understand where this last equation came from, reread the discussion to this point and carefully consider Figure 7–4.

We can manipulate this last equation as follows.

$$180\ e^{-t_1} = 66 - 20 = 46, \qquad e^{-t_1} = 46/180$$

If any two numbers are equal, their natural logs must also be equal; thus we may write

$$\ln (e^{-t_1}) = \ln (46/180)$$

Now

$$\ln (e^{-t_1}) = -t_1$$

Finally, we can find a simple expression for the unknown time t_1.

$$t_1 = -\ln \left(\frac{46}{180}\right) = \ln \left(\frac{180}{46}\right)$$

We evaluate the right-hand side of this equation directly.

$$t_1 = 1.36 \text{ hours}$$

As a second example, consider the draining of the chemical storage tank sketched in Figure 7–8. The volume stored in the tank is symbolized by V, and the flow of

FIGURE 7–8 A storage tank draining under the influence of gravity. The flow is the negative time rate of change of the liquid volume.

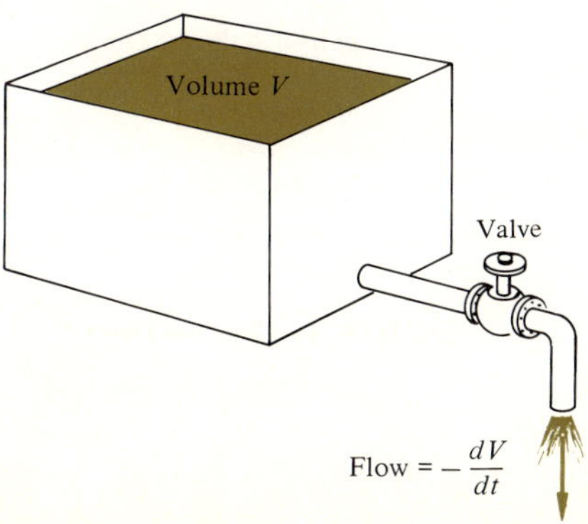

chemicals from the tank (in liters per hour) is simply the time rate of change of the volume V. The following equation relates the rate of flow to the volume in the tank at any time t,

$$\frac{dV}{dt} = KV(t) \tag{7–8}$$

where K is a constant.

The valve controlling the draining of the tank was opened at $t = 0$ and the tank's volume at $t = 0$ was

$$V(0) = 40{,}000 \text{ liters}$$

In addition, the initial flow rate was

$$\frac{dV}{dt} = -4000 \text{ liters/hour at } t = 0$$

This term is negative because the volume is decreasing.

Using this information and Equation (7–8), we may write the following equation to evaluate the constant K for this problem.

$$\text{At } t = 0, \qquad \frac{dV}{dt} = -4000 = KV(0) = K(40{,}000)$$

We may now find K directly.

$$K = -0.1$$

Equation (7–8) is called a first-order differential equation; its solution is given by the following equation:

$$V(t) = V(0)\, e^{Kt}.$$

Inserting the values known for $V(0)$ and K, we get

$$V(t) = 40{,}000\, e^{-0.1t}$$

This is the basic deterministic exponential model for the storage tank being drained solely under the influence of gravity. Note that no pumps are being used to remove the liquid chemical. For this example, let us now evaluate the time t_0 at which the tank will have been drained to one-half its initial volume, that is, 20,000 liters. The defining equation for t_0 is

$$V(t_0) = 20{,}000 = 40{,}000\, e^{-0.1t_0}$$

and we may manipulate it to the following form:

$$\ln 2 = 0.1t_0$$

Hence $t_0 = 6.93$ hours.

Both this section and Section 7–3 have pointed out the means to obtain information from deterministic models. We may classify these methods as mathematical or graphical. In either instance, we obtain numerical values that may be used to predict a system's behavior or make decisions.

Undergraduate engineering studies are aimed at developing suitable models for a wide range of physical phenomena. These models are essential to an engineer in any design problem. A civil engineer designing the concrete supports for a highway overpass must know the deterministic model relating the strength of concrete to the percentages of its constituents (cement, fine aggregate, and coarse aggregate). The engineer must also know the load-bearing ability of a concrete pillar of a given size. An electrical engineer designing a power-generating station must know the models that govern the behavior of electrical generators. For a given mechanical input, how much electrical energy will a generator produce? A chemical engineer must have a good grasp of the deterministic models that link temperature to rates of chemical reactions. This is essential to maintain control over the reactions occurring in a given process.

We could provide many more examples, but these suffice to illustrate the importance of deterministic models. It is not enough to have models. The engineer must also be able to manipulate these models and obtain information from them. These techniques commonly reduce to finding unknowns in equations. Graphical solutions are convenient, in some cases, while numerical solutions and the assistance of a digital computer are required in others. In the middle are problems that can be solved by algebraic and calculus methods; we have illustrated these in Section 7–2 and in this section.

Most of the time engineers use well-established deterministic models found in textbooks and handbooks. However, engineers sometimes need to formulate deterministic models. This interesting problem is the topic of Sections 7–6 and 7–7.

Exercises

7–11 When concrete is first poured, it has no mechanical strength because it is in a liquid state. As hydration of the elements (curing) proceeds, the concrete begins to harden and its strength increases. A knowledge of the model relating strength to curing time is essential if a construction project is to proceed rapidly. By knowing the model for strength versus curing time, an engineer will be able to remove braces and temporary supports at the proper time.

For this exercise, the compressive strength S_c of concrete will be measured in newtons per square meter. A pillar with cross-sectional area A will support a load of S_cA (newtons), where A is measured in square meters. For a certain mix of concrete, the curve of strength versus curing time is

$$S_c = 4.0(10^7)(1 - e^{-0.02t}) \text{ N/m}^2.$$

The time t is measured in days.

A concrete support with a circular cross section must withstand a load of $3.6(10^8)$ newtons when $t = 400$ days. Allowing a safety margin of 50%, what should the diameter of the pillar be? Five days after the pouring of the concrete, how strong will the pillar be?

7–12 For the concrete in Exercise 7–11, how much curing time is required before the strength is one-half of its ultimate strength?

7–13 How long must the concrete cure before it attains 99% of its ultimate strength?

7–6 FORMULATING MODELS

To be meaningful, deterministic models must be formulated on the basis of experimental data. In general, the more extensive the data, the better the model will be. The deterministic models we have considered so far have expressed a relationship between two experimentally related variables. To be sure, there is no intrinsic limitation to only two variables. But we shall continue to restrict our discussion to such simple models for the time being.

The first, often quite difficult, decision facing the model maker is the selection of variables. For example, to model the fuel economy of automobiles, one could relate miles per gallon to the phase of the moon or the time of the day. To model the cooling of an oven, one could relate oven temperature to the rate of rainfall outside the kitchen. But these choices of variables are unnatural; they simply do not seem correct. The model maker must select the variables intelligently. Further, the individuals collecting data for the models must take care to ensure that all experimental conditions—other than those being intentionally varied—remain the same.

To illustrate this, let us consider an experiment in which shields of various thicknesses are placed between a source of neutrons (a subatomic particle produced in nuclear reactors) and a neutron detector. The following experimental results give the detector reading as a function of shield thickness.

Detector reading (DR) (counts per minute)	*Thickness (t)* (cm)
91,202	25
80,028	50
63,119	75
51,892	100
46,188	150
30,222	200
20,410	300
11,738	400
7062	500
3879	600
1346	800
480	1000
142	1200

A question should naturally arise in light of our discussion about experimental conditions: If these data are to be useful in formulating a model, what conditions had to be maintained when the measurements were made? Figure 7–9 illustrates the basic experiment.

The neutron source had to emit neutrons at a constant rate. If this rate fluctuated as the shielding thickness was changed, variations in the detector readings could not be attributed entirely to changes in shielding thickness. Since the experiments are intended to investigate the relationship between the thickness and the effectiveness of the shield, the rate at which the source emits neutrons had to be constant. In addition, the characteristics of the neutron detector remain the same. Another necessary condition is that the shielding material had to remain the same throughout the experiments. Of course, the relative distance between the neutron source and detector had to remain unchanged.

If all these conditions were satisfied, then we may use these data to construct a model relating the shield's thickness t and its effectiveness (measured by the readings DR of the neutron detector). To begin, consider some of the possible relationships between the two experimental quantities of interest (DR, t).

Possible models

$$DR = at + b \qquad \text{(linear)} \qquad (7\text{–}9)$$

$$DR = \frac{c}{t} + d \qquad \text{(hyperbolic)} \qquad (7\text{–}10)$$

$$DR = Ae^{-Bt} \qquad \text{(exponential)} \qquad (7\text{–}11)$$

$$DR = Ct^{D} \qquad \text{(power law)} \qquad (7\text{–}12)$$

where $a,b,c,d,$ A,B,C,D are unknown constants.

FIGURE 7–9 A basic neutron-shielding experiment. The beam of neutrons is attenuated by the shielding of thickness t.

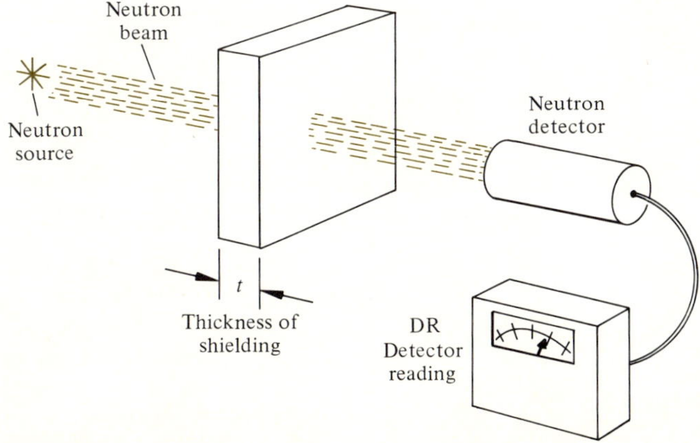

These four possibilities are not exhaustive (many other models exist), but they do provide a good starting point. Our task is to determine which, if any, of these four fit the experimental data accurately, and what numerical values of the two constants provide the best fit for that particular functional relationship. We shall discuss the words "accurately" and "best" in Section 7–7.

The process we are about to describe is a simplified version of commonly used methods of constructing deterministic models. Consider the essential elements of the problem. The experimental results give an implicit functional relationship between the two variables (DR, t). Our primary concern is to develop techniques for finding an explicit relationship—a specific deterministic model.

The first possible model—the linear relationship—says that equal changes in shielding thickness (t) must produce equal changes in detector reading (DR). One method of assessing whether the data closely fit this model is to construct a table of changes in thickness and corresponding changes in detector reading. We can do this simply by subtracting the value of each variable from the one below it in the chart. Carrying out this procedure is a lot of work. However, we can rapidly assess the accuracy of the linear model by realizing that the ratio of the change in detector reading to the change in thickness must be a constant. In mathematical terms,

$$\frac{d(DR)}{dt} = a = \text{constant} \qquad \text{(if the data are linear)}$$

A brief analysis of the top and bottom parts of the table on page 260 will demonstrate that this is not the case. Consequently, a linear model fits the data poorly.

The next possibility is the hyperbolic equation. This model will fit the data if and only if the detector reading is a linear function of $1/t$. To check this possibility, we construct a new pair of data columns showing DR versus $1/t$.

Detector reading (DR) (counts per minute)	*Inverse thickness (1/t)* (cm^{-1})
91,202	0.04
80,028	0.02
63,119	0.0133
51,892	0.01
46,188	0.00667
30,222	0.005
20,410	0.00333
11,738	0.0025
7062	0.002
3879	0.00167
1346	0.00125
480	0.001
142	0.000833

Again we can apply criteria similar to those used to evaluate the linear model. If the hyperbolic model fits the data, the ratio of the change in DR to the change in $1/t$ must be constant. We can see this by substituting two sets of values (DR_2, t_2), (DR_1, t_1) into Equation (7–10):

$$DR_2 = \frac{c}{t_2} + d, \qquad DR_1 = \frac{c}{t_1} + d$$

Subtracting the second from the first and rearranging, we obtain

$$DR_2 - DR_1 = \frac{c}{t_2} - \frac{c}{t_1}, \qquad \frac{DR_2 - DR_1}{(1/t_2) - (1/t_1)} = c$$

Because (DR_2, t_2) and (DR_1, t_1) are any two points, we may write

$$\frac{\Delta(DR)}{\Delta(1/t)} = c$$

Some brief calculations will show that this is not true, and we conclude that the hyperbolic function is not a good deterministic model for this experiment.

The third possibility is the exponential model. To see how to test the data for this case, we must begin with Equation (7–11) and perform some mathematical manipulations.

$$DR = Ae^{-Bt}$$

This equality implies that the natural logarithms of both sides must be equal.

$$\ln (DR) = \ln (Ae^{-Bt})$$

Since the log of a product of two factors is equal to the sum of the logs of both factors, we get

$$\ln (DR) = \ln (A) + \ln (e^{-Bt})$$

By definition of the natural logarithm, the second term on the right-hand side reduces to $-Bt$. So we have

$$\ln (DR) = \ln (A) - Bt$$

The meaning of this equation is clear. If the exponential deterministic model is to fit the data, the natural log of the detector readings must be a linear function of the shield's thickness. To evaluate this, we construct a third chart giving thickness in one column and the natural log of the detector readings in another.

ln (DR)	t
11.42	25
11.39	50
11.05	75
10.86	100
10.74	150
10.32	200
9.92	300
9.37	400
8.86	500
8.26	600
7.20	800
6.17	1000
4.96	1200

Applying the constant ratio test (evaluating the ratio of the change in ln (DR) to the corresponding change in t) finally yields success. The ratios are nearly constant over the entire range of experimental data. Thus the relationship between detector readings (DR) and shielding thicknesses (t) is exponential. We shall shortly consider the calculation of values for the two constants (A,B) in the exponential deterministic model.

For completeness, we check the validity of the fourth possible model, the power law of Equation (7–12).

$$DR = Ct^D$$

Some mathematical manipulation is required, and we begin by taking the natural logarithms of both sides of Equation (7–12).

$$\ln (DR) = \ln (Ct^D)$$

Using the additive properties of logarithms yields

$$\ln (DR) = \ln (C) + \ln (t^D)$$

and since

$$\ln (t^D) = D \ln (t)$$

the following result is obtained.

$$\ln (DR) = \ln (C) + D \ln (t)$$

This implies that the natural log of the detector reading is a linear function of the natural log of the shielding thickness. Now you should be able to develop a method of analysis to determine whether the experimental data follow this relationship.

In this lengthy discussion, we have developed the key concepts used to make an initial selection of a deterministic model. We have not yet attempted to evaluate the constants in the deterministic models. We have focused our initial attention on the task of choosing one of the four possibilities.

We chose a model by a laborious computational method. Fortunately, we can write a FORTRAN program to read a set of experimental data, use the data to prepare the necessary tables, and apply the constant ratio test in each case.

The problems at the end of this chapter give you the opportunity to write and use such a program. For now, we shall introduce another method for checking whether each of the four possible deterministic models fits a given collection of data. Before we do this, it is worth taking a moment to recall the considerations mentioned at the beginning of this section.

In any physical situation, there exist a host of experimental variables. In our discussion so far, we have assumed that we have properly chosen two related variables. If we have not, all the techniques of this section will be of absolutely no use. For example, the rate at which an oven cools has no relationship to the rate at which rain is falling. To look for such relationship is clearly folly. This simple example does not mean that the proper selection of variables is, in all cases, obvious. For example, the search for the variables that cause cancer continues without complete success.

As long as we have made a proper choice of related variables, we can test each of the four deterministic models. We can do this by carrying out a series of hand calculations or by writing a FORTRAN program to do so. The essentials are preparing tables of the experimental data recast into appropriate forms (reciprocals or logs) and applying the constant ratio test. In addition, there is a third way to carry out these tests.

Instead of performing a set of numerical calculations, we can test each model by plotting the data on an appropriate set of axes. A plot of the detector readings versus shielding thicknesses on a sheet of linear graph paper shows clearly that no linear relationship exists. Figure 7–10 shows this result. We can observe the same result by plotting detector readings versus $1/t$ on a sheet of linear graph paper. (This is not illustrated.)

We can test the exponential model by plotting logarithms of the detector readings versus thicknesses on a sheet of linear graph paper. To carry out this procedure, we have to perform some preliminary numerical analysis (finding the logs of the detector readings), but we can avoid this simply by using a sheet of semilogarithmic graph paper. Such paper has one linear scale and one logarithmic scale. The logarithmic scale lays out distances in proportion to the logarithms of the numbers displayed along the axes. Of course, linear graph paper lays out distances in proportion to the numbers on the axes. Figure 7–11 displays the plot of experimental data on a sheet of semilogarithmic paper. This figure indicates that the exponential model provides a good (but not perfect) fit to the data.

We can continue along these lines and test the power-law model as well. We can most conveniently test this model by using a sheet of log–log paper. On this kind of paper, both scales are logarithmic. Figure 7–12 shows the results of this procedure. Clearly, the power-law equation is not an adequate representation of the experimental data.

FIGURE 7–10 Linear plot, *DR* versus *t*

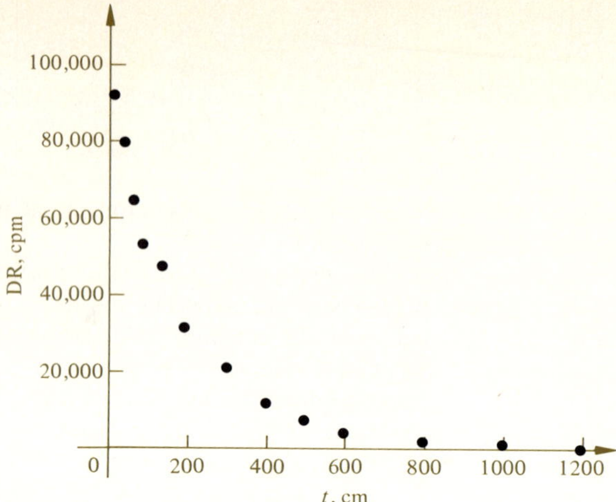

FIGURE 7–11 Semilogarithmic plot, ln (*DR*) versus *t*

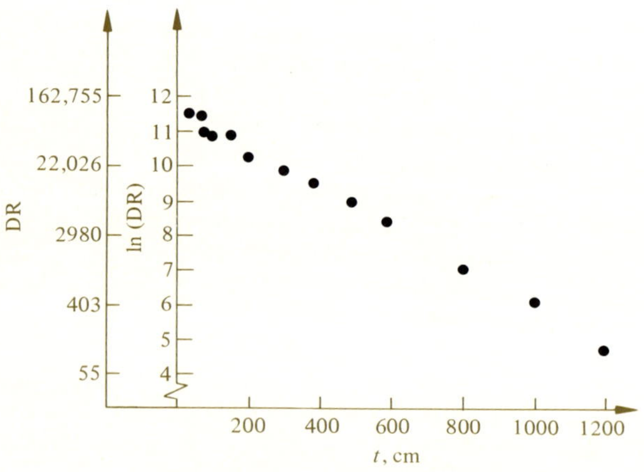

FIGURE 7–12 Logarithmic plot, ln (*DR*) versus ln (*t*) on log–log paper

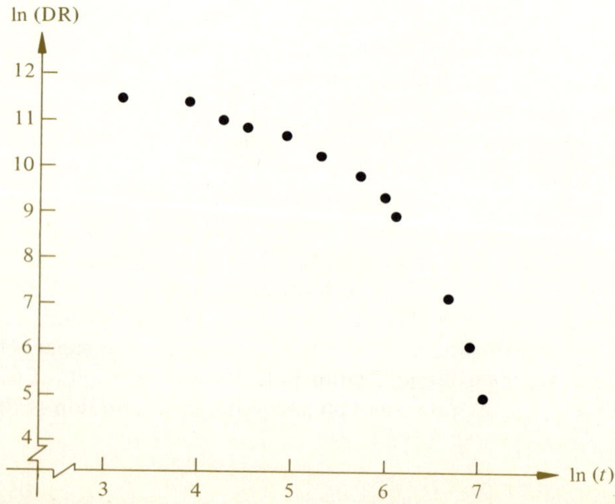

Exercises

7–14 Prepare four graphical plots [y versus x, y versus $1/x$, ln (y) versus x, ln (y) versus ln (x)] and determine which of the basic models is most appropriate for the following data.

y	−6	5	16	27	38	49	71	104	126	159
x	0	1	2	3	4	5	7	10	12	15

7–15 Repeat Exercise 7–14 for the following data.

y	0	2.86	3.5	5.7	7.	8.64	10.6	13.1	14.	14.8	16.7
x	0	0.005	0.01	0.05	0.1	0.2	0.4	0.8	1	1.2	1.8

7–16 Repeat Exercise 7–14 for the following data.

y	39.8	26.4	19.8	16.47	12.3	9.8	7.8	3.8	1.8	0.3	0.05	0
x	0.5	0.75	1	1.2	1.6	2.0	2.5	5	10	40	80	100

7–17 For the experimental data given in Exercise 7–15, prepare a set of four columns of the following ratios.

$$\frac{\Delta y}{\Delta x}, \quad \frac{\Delta y}{\Delta(1/x)}, \quad \frac{\Delta \ln (y)}{\Delta x}, \quad \frac{\Delta \ln (y)}{\Delta \ln (x)}$$

You should use an electronic calculator to do this exercise. Verify that this constant ratio test selects as most appropriate the same model as the graphical methods of Exercises 7–14 through 7–16 did.

Exercise 7–17 illustrates that tedious hand calculations are required to apply the constant ratio test to experimental data. A digital computer can, fortunately, easily solve this kind of problem. The flow chart of Figure 7–13 illustrates the essential elements of the necessary FORTRAN program.

Exercises

7–18 In a suitable set of FORTRAN statements, implement the algorithm presented in the flow chart of Figure 7–13.

7–19 Use the program you wrote in Exercise 7–18 to test the experimental data given in Exercise 7–15. Show that both the program and graphical methods select the same model as most appropriate.

FIGURE 7–13 Flow chart for selection of model. The ratios needed to test the four basic models are calculated and printed.

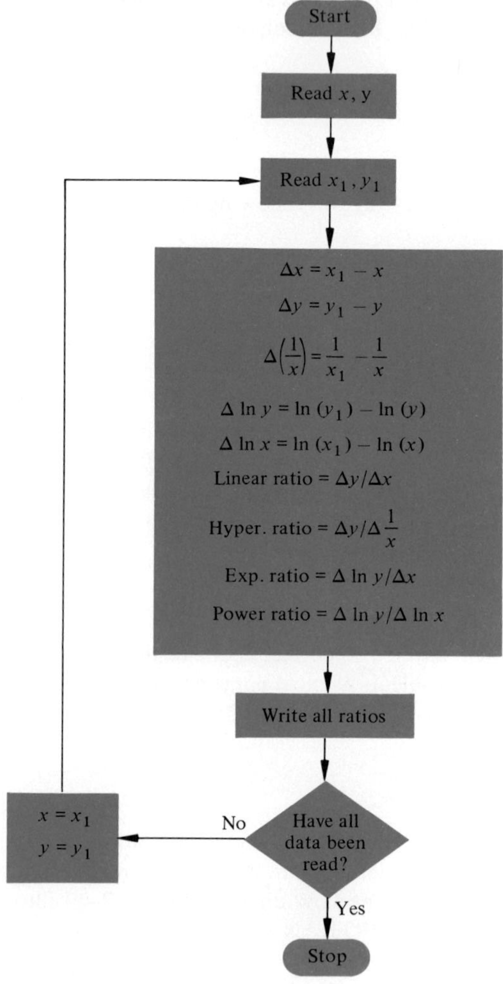

To set the stage for the next section, we return to the problem of relating neutron detector readings to shielding thicknesses. Consider the selection of the "best" pair of coefficients in the exponential deterministic model. The basic equation was

$$DR = Ae^{Bt}$$

We manipulated this into the form

$$\ln (DR) = \ln (A) + Bt$$

FIGURE 7–14 Three candidate straight lines. Two are clearly losers (a,c), and one is a possible winner (b). Plotted on semilogarithmic paper

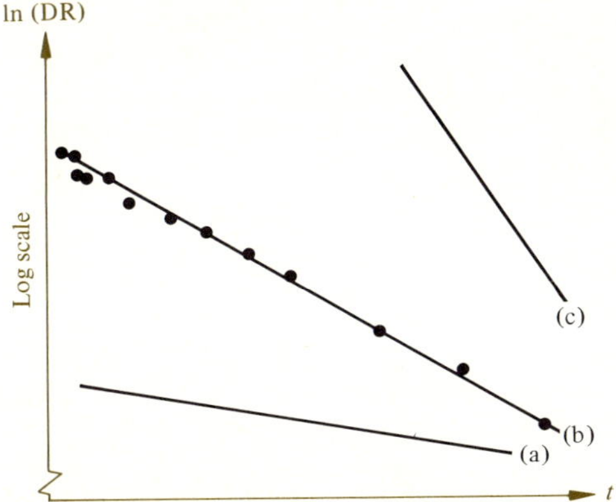

Figure 7–14 shows the experimental data plotted on semilogarithmic paper. By changing the parameters A and B, we can change the straight line given by the preceding equation. After all, the quantity B is the slope of the line and the quantity $\ln(A)$ is the intercept of the line on the vertical axis. The task is to determine which values of A and B give the "best" fit to the experimental data. Figure 7–14 shows three possible lines. The two straight lines labeled a and c are clearly losers, since they do not accurately match the data points. It is not difficult to select line b as the "best" fit in this case.

The choice is much less obvious for the two lines shown in Figure 7–15. Which is

FIGURE 7–15 Two possible close fits to a set of data. It is not possible to decide, by visual inspection, which is more accurate.

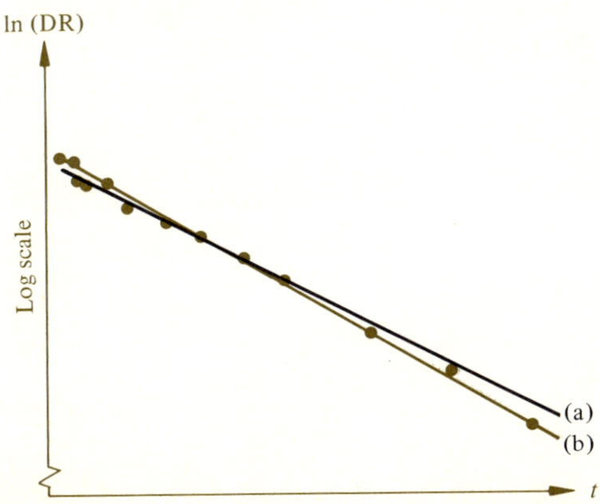

the best fit? The answer to this question is not clear. What do we really mean by the question: What straight line most accurately fits a collection of data points? We discuss the answer in Section 7–7.

In summary, we have devoted this section to techniques for selecting deterministic models. We have only gone part way in formulating a suitable model, since our approach has focused on selecting one equation from among four possibilities. No matter which we choose, we have to fix numerical values for the two parameters (a,b or c,d or A,B or C,D).

7–7 THE METHOD OF LEAST SQUARES

Section 7–6 concluded with a challenge to determine the criteria for judging whether a given straight line provides the best possible fit to a collection of data points. Figure 7–16 displays the sample data that we shall analyze in the following discussion. For simplicity, the figure shows axes labeled y and x and only four data points.

In mathematical terms, we seek the particular values of a and b that provide the best fit between the data and the straight line given by

$$y = ax + b$$

Figure 7–17 shows the four data points and a candidate straight line. Carefully examine the notation introduced in this figure. The quantities

$$x_1, \qquad x_2, \qquad x_3, \qquad x_4$$

are the four experimentally measured values of x. Associated with each of these are two values of y. For example, a straight line projected vertically upward from x_1 hits the candidate straight line at a value of y given by

$$y_{1M} = ax_1 + b$$

FIGURE 7–16 A sample set of data for which a linear model will be constructed

FIGURE 7–17 Notation to be used for least-squares analysis

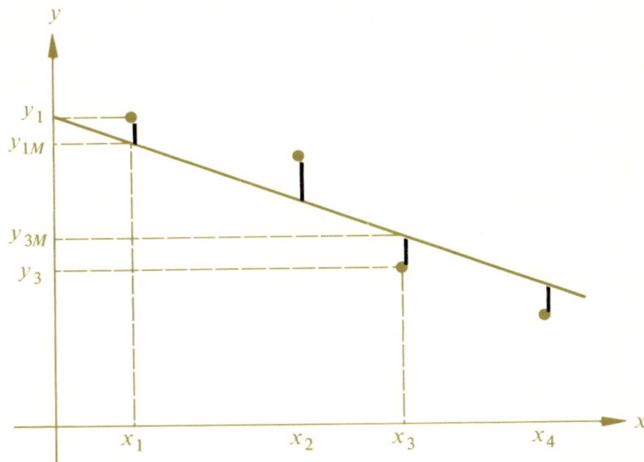

This same straight line then meets the first data point, denoted by the pair of coordinates

$$x_1, y_1$$

A natural measure of the degree of accuracy with which the candidate straight line "fits" this first data point is the difference between the two y values. Symbolically, we write

$$\text{Error (associated with first data point)} = y_1 - y_{1M}$$

In exactly the same fashion, we can identify errors with each of the three remaining data points.

$$\text{Error (second point)} = y_2 - y_{2M}$$

$$\text{Error (third point)} = y_3 - y_{3M}$$

$$\text{Error (fourth point)} = y_4 - y_{4M}$$

It is natural then to define the total error in "fit" as

$$\text{Total error} = \text{Error (1)} + \text{Error (2)} + \text{Error (3)} + \text{Error (4)}$$

In this expression, we have simplified the notation in a straightforward way.

To see this, consider a specific example. Calculate the total error directly for the three data points and the straight line labeled a in Figure 7–18.

$$\text{Error (1)} = (2.5 - 3.0) = -0.5$$
$$\text{Error (2)} = (2.0 - 2.0) = 0.0$$
$$\text{Error (3)} = (1.5 - 1.0) = 0.5$$

$$\text{Total error} = -0.5 + 0.0 + 0.5 = 0.$$

FIGURE 7–18 Two candidate straight lines. Line (a) misses the first and third points. Line (b) fits the data exactly.

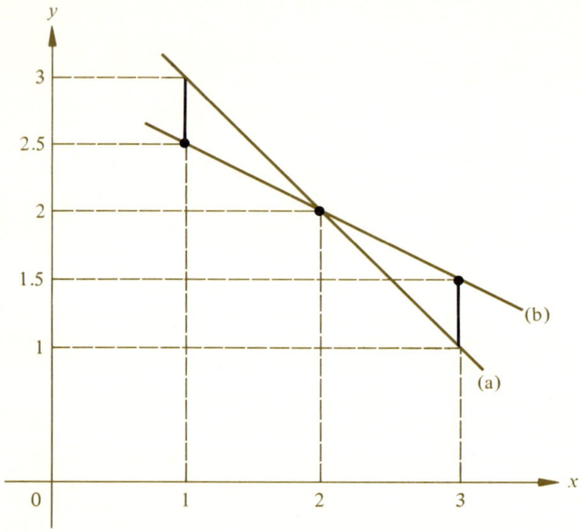

The total error turns out to be identically zero—and one certainly cannot hope to find any other straight line that will give a smaller error! But this result has to be incorrect because the straight line labeled b is a better fit to the experimental data.

This is a discouraging turn of events. We were careful to define total error in a way that appeared to make sense; yet we end up with an absurd conclusion.

To determine what went wrong, reconsider Figure 7–18. The line b is clearly an accurate fit to the data. The total error for this line is also zero. So things are beginning to improve. The criterion of total error indicates that the two lines are equally good fits to the three data points.

Let's look more closely at these two lines. Line a appears to be a good fit mathematically because the error associated with the first value of x cancels the error associated with the third value. This implies that the total error is zero. In fact, though, the straight line a misses the first and third data points by comparatively large margins. To get around this problem, we redefine the total error. One possibility is to define total error as the sum of the absolute values of the individual errors. This definition would lead to the conclusion that curve b is the best fit to the data.

Another choice is to define the total error as the sum of the squares of the individual errors. This definition gives greater weight to larger errors than the absolute value does. With this definition, curve b would again be the best fit.

Of these approaches, the sum of the squares is normally used. This not only produces a good fit to data, but also relates to other methods of estimating error in the analysis of data. Applying this concept to the data of Figure 7–18, we have

$$\text{Total error} = [\text{Error (1)}]^2 + [\text{Error (2)}]^2 + [\text{Error (3)}]^2$$

With this proper definition, we can recalculate the total errors associated with the two lines shown in Figure 7–18.

$$\text{Total error (line a)} = (-0.5)^2 + (0.0)^2 + (0.5)^2 = 0.5$$

$$\text{Total error (line b)} = (0.0)^2 + (0.0)^2 + (0.0)^2 = 0.0$$

This new definition of total error works well. Using it to calculate the error indicates that line b is a better fit to the experimental data than line a. An inspection of the figure verifies this assessment.

We can generalize this definition of total error to the case of N data points by introducing some new notation. We define the data points by giving their x and y values.

$$(x_1, y_1), (x_2, y_2), \cdots, (x_N, y_N)$$

Then the equation for the candidate straight line is, as before,

$$y = ax + b$$

This implies that the errors have the forms

$$\text{Error (1)} = y_1 - (ax_1 + b)$$

$$\text{Error (2)} = y_2 - (ax_2 + b)$$

$$\vdots$$

$$\text{Error } (N) = y_N - (ax_N + b)$$

We can write this more compactly as

$$\text{Error } (i) = y_i - (ax_i + b), \qquad i = 1, 2, \cdots, N$$

We then write the total error as

$$\text{Total error} = [\text{Error (1)}]^2 + [\text{Error (2)}]^2 + \cdots + [\text{Error } (N)]^2$$

or

$$\text{Total error} = [y_1 - (ax_1 + b)]^2 + [y_2 - (ax_2 + b)]^2 + \cdots + [y_N - (ax_N + b)]^2$$

Another, more compact, way to write this is

$$\text{Total error} = \sum_{i=1}^{N} [y_i - (ax_i + b)]^2 \qquad (7\text{–}13)$$

Equation (7–13) introduces a new notation that will be useful for reducing the complexity of equations. The Greek letter sigma (Σ) denotes *summation*. This notation means: Evaluate the term to the right of the Greek letter Σ with $i = 1$, add this result to the value of the expression with $i = 2$, and repeat this procedure for all the values of i up to and including N. This notation may seem strange at first, but its value will become apparent.

For a given set of data points, there exist a specific value of the slope a and a specific value of the intercept b which characterize the one straight line that "best" fits the experimental points. The word best implies the least total error. We can find the best slope a and the best intercept b, for a specific set of experimental values, by evaluating the following two equations.

$$a = \frac{N\Sigma x_i y_i - \Sigma x_i \Sigma y_i}{N\Sigma x_i^2 - (\Sigma x_i)^2} \qquad (7\text{–}14)$$

$$b = \frac{\Sigma y_i - a\Sigma x_i}{N} \qquad (7\text{–}15)$$

In these expressions, N represents the total number of experimental points. We have simplified these expressions by using the following shorthand notation.

$$\Sigma x_i = \sum_{i=1}^{N} x_i = x_1 + x_2 + \cdots + x_N$$

$$\Sigma y_i = \sum_{i=1}^{N} y_i = y_1 + y_2 + \cdots + y_N$$

$$\Sigma x_i y_i = \sum_{i=1}^{N} x_i y_i = x_1 y_1 + x_2 y_2 + \cdots + x_N y_N$$

$$\Sigma x_i^2 = \sum_{i=1}^{N} x_i^2 = x_1^2 + x_2^2 + \cdots + x_N^2$$

We shall now derive these two equations using basic calculus concepts. (If you have not yet had differential calculus, you can omit this derivation without losing the thread of the discussion.) The problems at the end of this section demonstrate numerically that Equations (7–14) and (7–15) result in minimum error. If you wish to omit this derivation, you may proceed to the example that starts on page 276.

A convenient starting point for this analysis is the expression for total error given in Equation (7–13).

$$\text{Total error} = e(a,b) = \sum_{i=1}^{N} [y_i - (ax_i + b)]^2$$

The functional representation $[e(a,b)]$ explicitly indicates that the total error is a function of two variables, the slope a and the intercept b of the line. This is a proper interpretation for a specific set of experimental points. If the points do not change, the total error is a function of both the slope and the intercept of the candidate line. As you may know from differential calculus, the function $e(a,b)$ must satisfy the following two conditions if it is to be either minimum or maximum.

$$\frac{\partial e(a,b)}{\partial a} = 0, \qquad \frac{\partial e(a,b)}{\partial b} = 0$$

It is important to explicitly recognize the structure of the expression that we are about to derive. We are seeking values for two unknowns, the slope a and the intercept b, that characterize the best straight line. To find two unknowns requires two equations. The two preceding conditions provide the required two equations.

To apply these, we shall calculate the derivatives step by step. First we calculate the partial derivative of e with respect to a.

$$\begin{aligned}\frac{\partial e}{\partial a} &= \frac{\partial}{\partial a}\Sigma[y_i - (ax_i + b)]^2 \\ &= \Sigma\frac{\partial}{\partial a}[y_i - (ax_i + b)]^2 = \Sigma[-2x_i(y_i - ax_i - b)] \\ &= -2\Sigma x_i(y_i - ax_i - b)\end{aligned}$$

Setting the last equation to zero and dividing both sides by -2 yields the following condition.

$$\Sigma(x_iy_i - ax_i^2 - bx_i) = 0$$

Similarly, we calculate the second equation.

$$\begin{aligned}\frac{\partial e}{\partial b} &= \frac{\partial}{\partial b}\Sigma[y_i - (ax_i + b)]^2 \\ &= \Sigma\frac{\partial}{\partial b}[y_i - ax_i - b]^2 = \Sigma[-2(y_i - ax_i - \mathrm{b})]\end{aligned}$$

Setting this expression equal to zero and dividing both sides by -2 gives

$$\Sigma(y_i - ax_i - b) = 0$$

Thus the two equations from which we can derive expressions for the slope a and the intercept b are

$$\Sigma(x_iy_i - ax_i^2 - bx_i) = 0 \tag{7–16}$$

$$\Sigma(y_i - ax_i - b) = 0 \tag{7–17}$$

Distributing the summation across the three terms in both equations results in

$$\Sigma x_iy_i - \Sigma ax_i^2 - \Sigma bx_i = 0$$

$$\Sigma y_i - \Sigma ax_i - \Sigma b = 0$$

In these two equations, the only unknowns are a and b. The other quantities are the experimental points on which the analysis is based. Now we are in the happy position of having two equations and two unknowns. Note that we can remove a and b from the summations in which they occur.

We next multiply the first equation by $-N$ and the second by Σx_i, and we get

$$-N\Sigma x_iy_i + aN\Sigma x_i^2 + Nb\Sigma x_i = 0$$

$$\Sigma x_i\Sigma y_i - a(\Sigma x_i)^2 - Nb\Sigma x_i = 0$$

Note that $\Sigma b = Nb$. Adding these two equations eliminates b and yields

$$\Sigma x_i\Sigma y_i - N\Sigma x_iy_i - a[(\Sigma x_i)^2 - N\Sigma x_i^2] = 0$$

Now we can find an expression for a.

$$a = \frac{N\Sigma x_i y_i - \Sigma x_i \Sigma y_i}{N\Sigma x_i^2 - (\Sigma x_i)^2}$$

Substituting this expression into either Equation (7–16) or (7–17) results in an explicit equation for b.

$$b = \frac{\Sigma y_i - a\Sigma x_i}{N}$$

These last two formulas are identical to Equations (7–14) and (7–15).

To illustrate the use of these expressions, we shall consider a three-point set of "experimental data."

$$\begin{array}{llll} x_1 = 1 & x_2 = 2 & x_3 = 3 & \\ y_1 = 1 & y_2 = 2 & y_3 = 3 & N = 3 \end{array}$$

To find values for the slope and intercept, we must evaluate four sums.

$$\begin{aligned} \Sigma x_i &= 1 + 2 + 3 = 6 \\ \Sigma y_i &= 1 + 2 + 3 = 6 \\ \Sigma x_i y_i &= (1)(1) + (2)(2) + (3)(3) = 14 \\ \Sigma x_i^2 &= 1 + 4 + 9 = 14 \end{aligned}$$

Then we can find a and b by direct substitution.

$$a = \frac{(3)(14) - (6)(6)}{(3)(14) - 6^2} = 1, \qquad b = \frac{6 - (1)(6)}{3} = 0$$

The straight line that these two parameters describe ($y = 1x + 0$) is a precise fit to the three points. Thus it is "best" in that it has the least-square error.

Exercises

7–20 Calculate the values of the least-square errors of the slope and the intercept for the following data.

x	0.0	0.5	1.0	1.5	2.0
y	1.2	1.575	1.95	2.325	2.7

Calculate the total error.

7–21 Increase by 10% the value of the slope in Exercise 7–20 and recalculate the total error. Decrease the slope by 10% and recalculate the total error.

Exercises 7–20 and 7–21 should indicate that the least-square value of slope a results in the minimum total error.

We commonly refer to the quantities a and b as the *least-square coefficients.* This terminology is appropriate, since these two choices for slope and intercept result in the least-squared error between the straight line and the data points.

In this section we discussed only linear models, that is, those in which a straight line expresses the relationship between two experimental variables. Section 7–8 introduces techniques that will enable us to apply least-square concepts to other basic models.

7–8 EXTENSIONS OF THE LEAST-SQUARES METHOD

To begin this section, we rewrite the four basic deterministic models introduced earlier, Equations (7–9) through (7–12).

$$y = ax + b$$
$$y = c/x + d$$
$$y = Ae^{Bx}$$
$$y = Cx^D$$

The least-squares method discussed in Section 7–7 appears to be applicable only to the first equation (the linear model). After all, it is identical to the form that we analyzed in detail when we derived the least-squares coefficients. Like so many apparently true conclusions, this one is superficial. The key concept to recognize is that we can write any of these equations in linear form as long as we properly define the variables. For example, the exponential model will produce a straight-line relationship when we plot ln (y) as a function of x. And we can make similar statements for the other models, as Table 7–1 illustrates.

To amplify this point, let's rewrite Equation (7–11) with a minor change of variable.

$$z = Ae^{Bx}$$

Applying the least-squares method to data that fit this equation will produce useless results because a plot of z versus x is nowhere near linear. However, if we rewrite the equation as

$$y = \ln(z) = \ln(A) + Bx$$

a linear relationship now exists between the new variable y and x.

TABLE 7–1. CHOICES OF VARIABLES THAT GIVE LINEAR RELATIONSHIPS

Model	*Name*	*Linearly related variables*	*Slope*	*Intercept*
$y = ax + b$	Linear	y versus x	a	b
$y = c/x + d$	Hyperbolic	y versus $1/x$	c	d
$y = Ae^{Bx}$	Exponential	ln (y) versus x	B	ln (A)
$y = Cx^D$	Power law	ln (y) versus ln (x)	D	ln (C)

In short, if an exponential model fits an array of data, the proper procedure is to take the natural logarithm of the experimental variable analogous to z in our example and apply the least-squares algorithm to find the best values of slope and intercept.

The following data and analysis will clarify this procedure.

z	x	$y = \ln(z)$
0.5	0.0	−0.693
0.824	1.0	−0.194
1.359	2.0	0.307
2.241	3.0	0.807
3.694	4.0	1.307
6.091	5.0	1.807

To calculate the least-square coefficients, we have to evaluate four summations.

$$\Sigma x_i = 15.0$$
$$\Sigma y_i = 3.341$$
$$\Sigma x_i y_i = 17.104 \qquad N = 6$$
$$\Sigma x_i^2 = 55.0$$

With these values, we can find the slope and intercept directly.

$$\text{Slope} = \frac{6(17.104) - (15.0)(3.341)}{6(55.0) - (15.0)^2} = 0.5$$

$$\text{Intercept} = \frac{3.341 - (0.5)(15.0)}{6} = -0.693$$

For the exponential model, we may make the following identifications.

$$\text{Slope} = B = 0.5$$
$$\text{Intercept} = \ln(A) = -0.693$$

Since we need the quantity A for the model, we find it by applying the definition of a logarithm to the equation for the intercept.

$$A = e^{\text{intercept}} = e^{-0.693} = 0.5$$

Thus the final model is

$$z = Ae^{Bx} = 0.5e^{0.5x}$$

You can analyze any data that fit one of the four deterministic models by using the least-squares method if you make the proper transformation of variables. Table 7–1 illustrates the necessary changes of variable.

In the following example, we begin with a set of raw data. We then work through the technique for selecting one of the deterministic models as appropriate. We conclude by applying the least-squares method to determine the best values of the model's two parameters.

y	3.1	1.9	1.8	1.6	1.4	1.15	1.04
z	1.0	2.0	2.5	3.0	4.0	7.39	12.0

When you are confronted with a new set of experimental results, you should try to fit the available data to one of the four suggested deterministic models. You can use either graphical or computer-based techniques. In this example we shall use a graphical method. The following three rows of numbers are useful for preparing the plots.

ln (y)	1.13	0.642	0.588	0.470	0.336	0.140	0.039
ln (z)	0.000	0.693	0.916	1.099	1.396	1.000	2.485
$1/z$	1.000	0.500	0.400	0.333	0.250	0.135	0.083

In Figures 7–19 through 7–22 we plot the data in four different ways.

These figures indicate that the hyperbolic model is appropriate in this case. We need to perform additional analysis to determine accurate values of the two parameters (c,d). A brief aside is in order here. None of the four models may fit the data with one variable along the vertical axis and the other along the horizontal. In that case, we would interchange the variables in the hyperbolic and exponential models and retest them. Of course, we don't do it for the linear and power-law models, since interchanging variables in these models makes no difference.

For convenience, rewrite the data in two rows, y and $x = 1/z$.

y	3.1	1.9	1.8	1.6	1.4	1.15	1.04
x	1.0	0.5	0.4	0.33	0.25	0.135	0.083

The formulas for the least-square slope (c) and intercept (d) involve four summations.

FIGURE 7–19 A linear plot (y versus z)

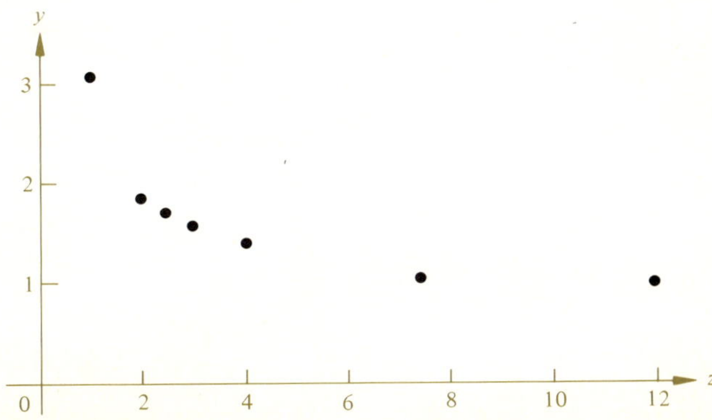

FIGURE 7–20 A hyperbolic plot (y versus $1/z$)

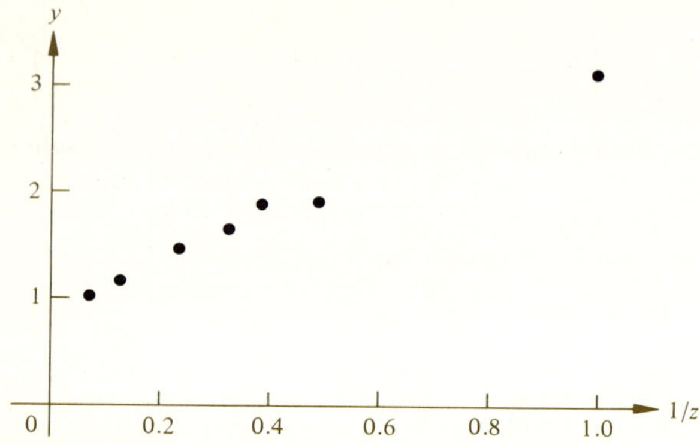

FIGURE 7–21 An exponential plot (ln (y) versus z) on semilog paper

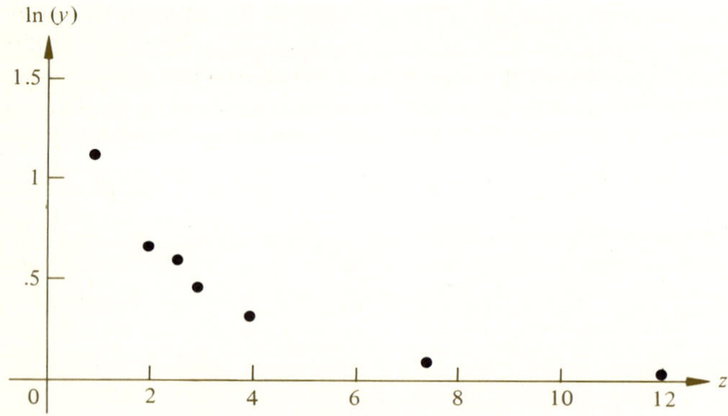

FIGURE 7–22 A power law plot (ln (y) versus ln (z)) on log-log paper

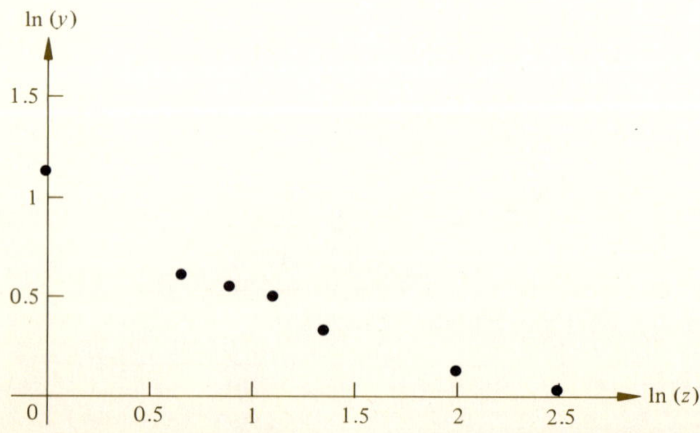

$$\Sigma x_i = 2.698 \qquad \Sigma y_i = 11.99 \qquad \Sigma x_i y_i = 5.89 \qquad \Sigma x_i^2 = 1.61 \qquad N = 7$$

Substituting these into the defining equations yields

$$\text{Slope} = \frac{(7)(5.89) - (2.698)(11.99)}{(7)(1.61) - (2.698)^2} = 2.23$$

$$\text{Intercept} = \frac{11.99 - (2.23)(2.698)}{7} = 0.853$$

The hyperbolic model is

$$y = 2.23x + 0.853 = \frac{2.23}{z} + 0.853$$

This final equation constitutes a deterministic model of the relationship between the quantities y and z for

$$1 \leq z \leq 12$$

The most common error normally made in interpreting such results is extending the model beyond the experimental range over which it is valid. There is no justification for extending the hyperbolic model in this example to values of z less than 1 or greater than 12. You must keep this limitation firmly in mind when you use such a model.

As you can see from these examples, a significant amount of computational effort is involved in calculating least-square coefficients. Again this is a task for which the digital computer and FORTRAN are well suited. The flow chart in Figure 7–23 illustrates the basic procedures for computing these coefficients. The figure shows four variations of one of the boxes, one for each of the four different models.

Exercises

7–22 Write and execute a FORTRAN program to calculate the least-square slope and intercept for a linear model that fits the following data.

y	−1.14	−0.52	0.11	0.76	1.31	1.95	2.45
z	−6.00	−4.00	0.02	0.00	2.00	4.00	6.00

7–23 Write and execute a FORTRAN program to calculate the least-square slope and intercept for a hyperbolic model that fits the following data.

y	11.5	−5.2	−8.8	−10.1	−11.0	−11.6	−12
x	1.1	3.2	5.4	7.2	9.3	11.7	13.9

FIGURE 7–23 A flow chart for the calculation of least-square coefficients. A different path is provided for each of the four models. The dots indicate missing statements to be supplied by the reader.

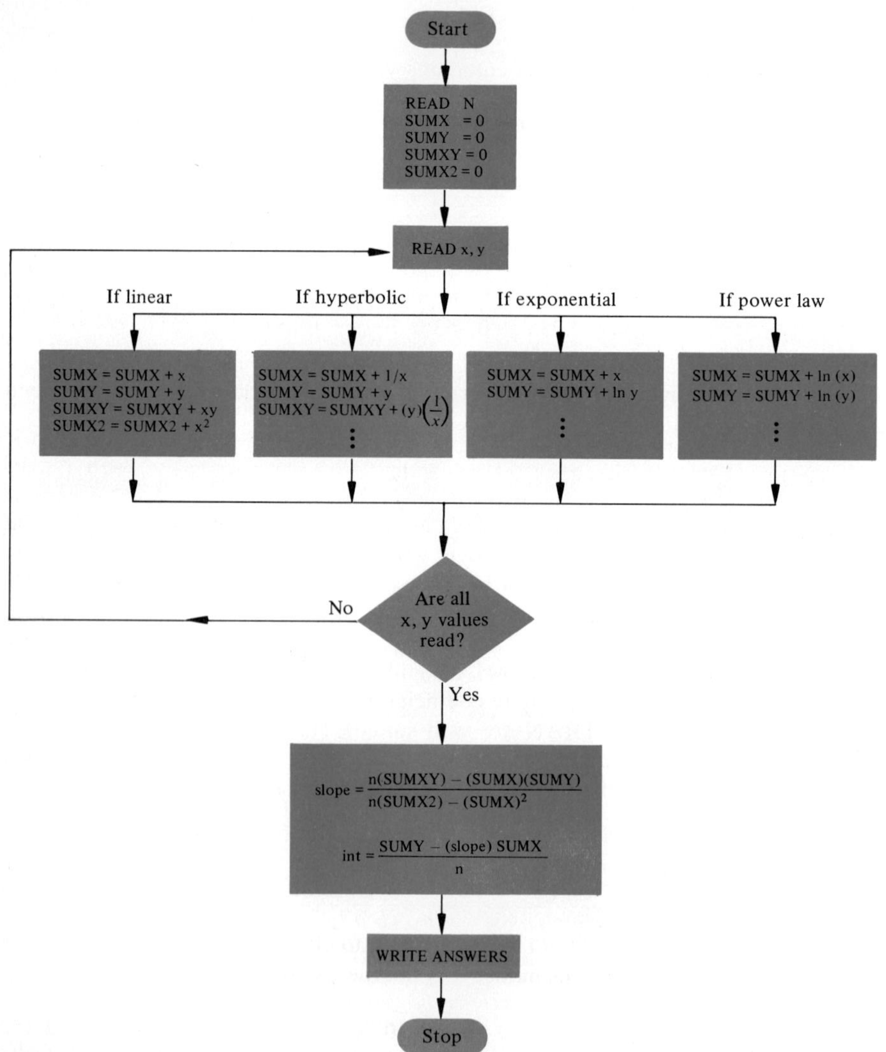

7–24 Write and execute a FORTRAN program to calculate the least-square slope and intercept for an exponential model that fits the following data.

y	7.1	5.03	3.62	2.52	1.76	1.27	0.4	0.078	0.017
x	0	0.11	0.22	0.34	0.46	0.57	0.95	1.5	2

7–25 Write and execute a FORTRAN program to calculate the least-square slope and intercept for a power-law model that fits the following data.

y	0	10.5	31.8	61	96.5	138	185	236
x	0	0.8	1.6	2.4	3.2	4.0	4.8	5.6

In the first seven sections of this chapter, we have covered a lot of material. We have placed the concept of a deterministic model on a much firmer footing than it was in the introductory chapters. We have mathematically defined four different classes of models, and we have introduced a simple test for selecting the most appropriate one. Further, we have rigorously derived the least-squares method of determining the "best" parameters for a given set of data, and we later applied them in examples.

Once we have obtained a model, the next logical step is to use it to obtain information to design a system and predict its behavior. After all, a deterministic model is not an end in itself. Earlier we discussed some methods of extracting information. In Section 7–9, we shall develop an additional technique for obtaining information from a model.

7–9 NUMERICAL INTEGRATION

We shall now use some basic engineering mechanics to introduce some of the basic concepts of integral calculus in a physical context. We can calculate distance from a starting point (for simplicity, we assume a one-dimensional universe) by using our knowledge of velocity as a function of time. For example, suppose you begin a journey at

$$x = 0 \qquad \text{when } t = 0.$$

You then proceed to walk at a constant speed of 0.5 meter per second (m/sec) for 1000 sec, then at a uniform 1.5 m/sec for 1000 sec, and finally at −1 m/sec for 2000 seconds. (The minus sign means that you are walking in a direction opposite to that of the first 2000 sec.) Your final position, x(4000 sec), is then

$$\begin{aligned} x(4000\text{ sec}) &= (0.5\text{ m/sec})(1000\text{ sec}) + (1.5\text{ m/sec})(1000\text{ sec}) \\ &\quad + (-1\text{ m/sec})(2000\text{ sec}) \\ &= 0\text{ meters} \end{aligned}$$

How did we calculate this quantity? We divided total travel time into three intervals. Over each, the velocity was constant. To find the total distance traveled in each of these intervals, we used the fundamental principle of mechanics:

$$\text{Distance covered} = (\text{constant velocity})\,(\text{elapsed time})$$

Adding the distances covered in each of these three intervals gave us the total distance.

Figure 7-24 shows, in graphical form, a plot of velocity versus time for this example. What is the total distance covered in the first 1000 seconds? It is simply the product of the velocity in that interval and the elapsed time. The figure shows this distance as a cross-hatched area labeled a. In an analogous way, the total distances covered during the second and third time intervals (also cross-hatched) are labeled b and c, respectively.

The total distance covered (zero) is simply the total shaded area. The area below the t axis is counted negatively. Take a moment to calculate the area of these three rectangles.

We reach a somewhat tentative conclusion based on this analysis. The total distance covered is the area under the curve of velocity versus time. Although this example involves only constant velocities, this principle is valid in general. One warning is in order. The following discussion is in no way mathematically rigorous. We intend it simply to be a demonstration of the calculus concepts involved.

Now consider a more complicated motion. You start from

$$x = 0 \qquad \text{at } t = 0$$

You run rapidly, tire, begin to slow down, and finally stop. Figure 7–25 shows a possible velocity–time profile. How can we calculate the distance covered? This problem is more difficult, since velocity continuously changes. We cannot calculate the area under the curve in this example as simply as we did in the first example.

For clarity, we have magnified the first 20% of the velocity–time curve in Figure 7–26. In addition, we have divided the time axis into segments Δt long. (As the discussion proceeds, the proper size of Δt will become clear.)

FIGURE 7–24 Curve of piecewise constant velocity v versus time t

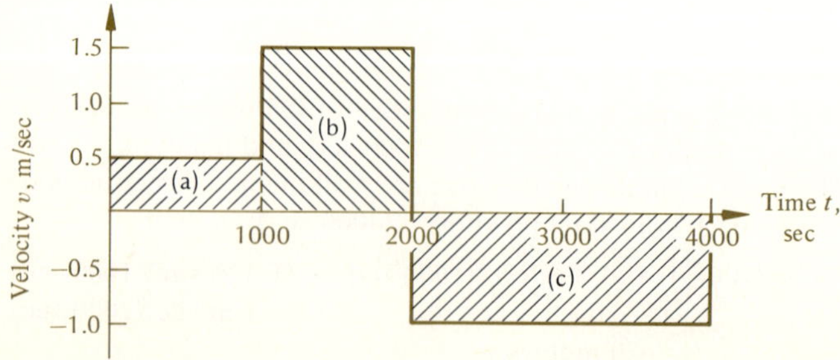

FIGURE 7–25 Curve of variable velocity v versus time t

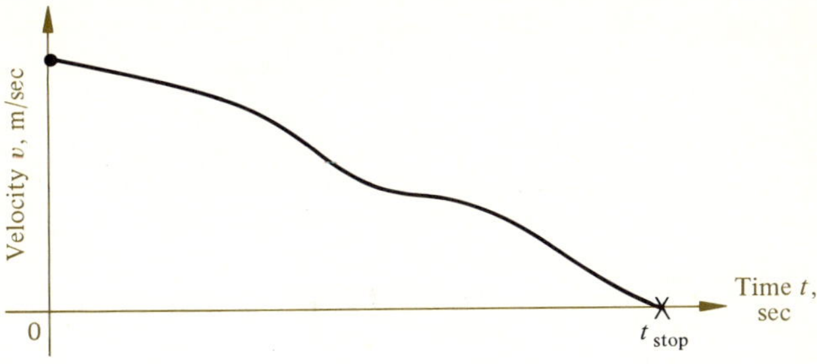

Examine Figure 7–26 carefully. You will see that it is possible to approximate the area under the velocity–time curve from 0 to Δt seconds by determining the area of the cross-hatched rectangle. The area of the rectangle is

$$\left[\frac{v(0) + v(\Delta t)}{2}\right] \Delta t$$

By extending this expression, we can approximate the total area under the part of the curve shown in Figure 7–26. The total area is

$$\left\{\left[\frac{v(0) + v(\Delta t)}{2}\right] + \left[\frac{v(\Delta t) + v(2\,\Delta t)}{2}\right] + \cdots + \left[\frac{v(4\,\Delta t) + v(5\,\Delta t)}{2}\right]\right\} \Delta t$$

FIGURE 7–26 Division of the time axis into "chunks" of width Δt for numerical integration

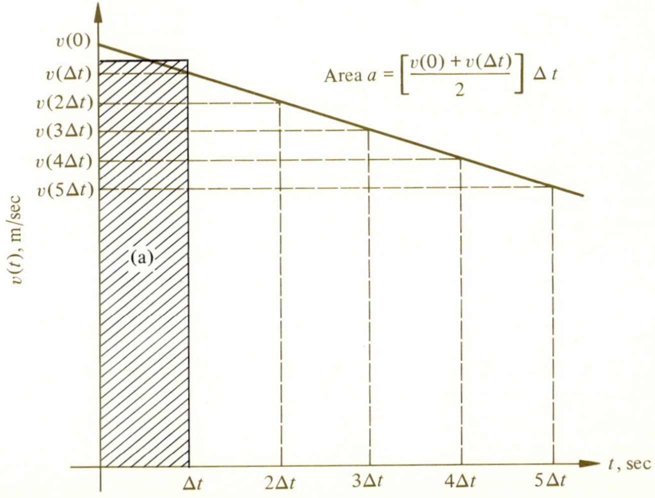

We can extend this result along the entire time interval of interest. To do so we define a quantity N by the equation

$$t_{\text{stop}} = N\,\Delta t$$

Then the total area is

$$\left\{\left[\frac{v(0) + v(\Delta t)}{2}\right] + \left[\frac{v(\Delta t) + v(2\,\Delta t)}{2}\right] + \cdots + \left[\frac{v[(N-1)\,\Delta t] + v(N\,\Delta t)}{2}\right]\right\}\Delta t$$

Combining terms then gives

$$\frac{\Delta t}{2}\left\{v(0) + 2v(\Delta t) + 2v(2\,\Delta t) + \cdots + 2v[(N-1)\,\Delta t] + v(N\,\Delta t)\right\}$$

As a general rule, the approximation becomes better as Δt becomes smaller. You may well ask: How small must Δt be before the approximation becomes acceptable? To provide some guidance in answering this question, let us consider Figure 7–27. Suppose we calculated the area under the curve using the rectangular approximation and the Δt shown in the figure. The result would be a wildly inaccurate estimate. Even cutting Δt in half would not improve the estimate very much. To get an accurate estimate, we have to slice the time axis into increments that are small compared to the times over which the velocity changes significantly. But this is not that useful because it is not quantitative. To overcome this limitation, we shall introduce a technique by which we can use the computer to successively reduce the quantity Δt until we get acceptable accuracy.

FIGURE 7–27 Effects of large Δt. If Δt is "large," the approximate area will be significantly different from the true area.

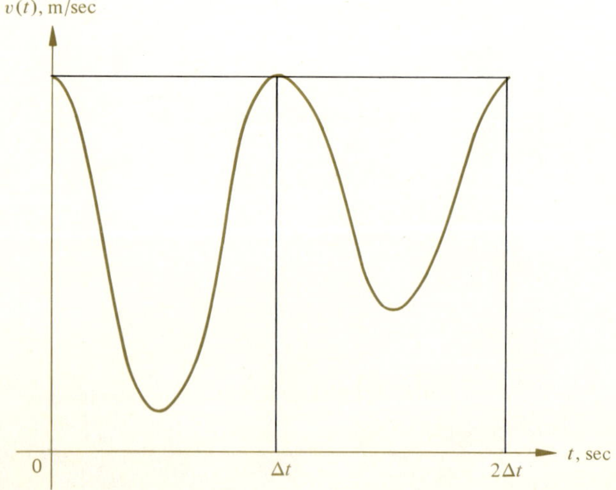

Now let's go one step further. If we make the quantity Δt so small that it almost vanishes, we can define the area under the curve in terms of a mathematical operation, the *integral.* The common notation is

$$x(t) = \int_0^t v(t)\,dt, \qquad \text{when } x(0) = 0$$

A more general equation is

$$x(t) = \int_0^t v(t)\,dt + x_0, \qquad \text{when } x(0) = x_0$$

Computers cannot find analytical expressions for integrals, but they can evaluate approximations by adding areas of geometric shapes.

To further develop this concept, let's consider a mechanical system. We know only the acceleration $a(t)$ of a body and its initial velocity and position. The following two equations give its subsequent motion.

$$v(t) = v_0 + \int_0^t a(t)\,dt,$$

$$x(t) = x_0 + \int_0^t v(t)\,dt,$$

where v_0 = velocity at $t = 0$ and x_0 = position at $t = 0$.

Engineers calculate the trajectories of spacecraft and rockets using precisely these formulas. Each vehicle contains a set of inertial guidance instruments that measure only the acceleration as a function of time. Integrating these data (and the model that describes them) twice then gives the vehicle's velocity and position at any time after its launch. (We shall consider this problem in Chapter 10.)

There are many other situations in which integrals are useful for extracting information from deterministic models. For example, if the power p of an electric motor is a function of the time t, the total energy e consumed by the motor is the integral of the power.

When the motor runs for T seconds, the total energy consumed is

$$e(T) = \int_0^T p(t)\,dt$$

Let's consider another example in which integrals are useful. A chemical engineer often keeps track of volumes of chemicals added to or removed from storage tanks by integrating the measured flow rate. If $f(t)$ is the flow in liters per second, the volume added or removed from a storage tank in T seconds is

$$V(T) = \int_0^T f(t)\,dt$$

We close this section by reconsidering how small the time increment Δt must be to give an accurate estimate of velocity. If you have access to a digital computer, you can determine the minimum size of Δt by using the looping algorithm shown in the flow chart of Figure 7–28.

FIGURE 7–28 Flow chart for numerical integration

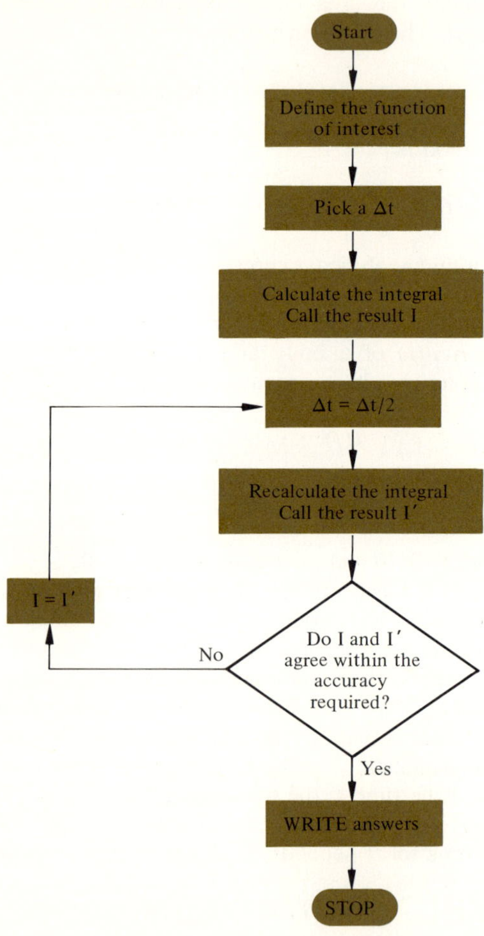

Start by choosing the initial time increment Δt. (Avoid taking such a small Δt that you will need excessive amounts of computer time to carry out the desired integration.) Next, calculate the integral and attach a label I to this numerical result. Then cut the time increment in half and recalculate the integral (with a label I' attached to the second result). Then compare the two approximations of the integral (I and I'). Do they agree with acceptable accuracy? It is in this step that you must apply some degree of engineering judgment. For example, must the integration be accurate to three significant figures or to five significant figures? There is no way to formulate a general answer to this question. The particular problem at hand will dictate the degree of accuracy needed.

If the two approximations do agree to the extent desired, you have finished. If they do not, cut Δt in half and repeat the calculation. On this second pass through the algorithm, you will be comparing numerical approximations to the integral that have been calculated using time increments of one-half and one-quarter of the original Δt. If these are not accurate enough, halve the increment again. In this way, you are making the increment smaller on each pass through the loop, but only small enough to ensure the required accuracy. Because computer time costs money, this algorithm is an efficient way, in terms of the dollars spent, to obtain a small enough Δt.

Exercises

7–26 An electric motor consumes power according to

$$p(t) = 1000 \text{ watts for } 0 \leqslant t \leqslant 1 \text{ hour},$$
$$p(t) = 1500 \text{ watts for } 1 \leqslant t \leqslant 1.5 \text{ hours},$$
$$p(t) = 1200 \text{ watts for } 1.5 \leqslant t \leqslant 2 \text{ hours}$$

Suppose that the local utility charges 0.005 cent/watt–hour. How much did it cost to run the motor for two hours?

7–27 A storage tank is empty at $t = 0$. Opening a valve allows a chemical flow f to occur.

$$f(t) = 150t \text{ liters/hour}$$

At what time will the volume of chemical in the tank be 7500 liters? [*Hint:* Consider the geometric shape of $f(t)$ versus t.]

7–28 Using an electronic calculator, approximate the integral

$$\int_1^5 y\,dx, \qquad \text{where } y = x \ln(x)$$

(a) First evaluate it using a value of $\Delta x = 4$.
(b) Then use a value of $\Delta x = 2$.
(c) And finally, repeat the calculation for $\Delta x = 0.5$.

7–29 Write and execute a FORTRAN program to evaluate

$$\int_1^5 x \ln(x)\,dx$$

to an accuracy of 0.01. Use the flow chart of Figure 7–28 as a starting point.

7–10 SUMMARY

Let us now round off this chapter with concise summaries of the material covered. We shall give two perspectives: The first presents a general view of the role of deterministic models in engineering. The other shows, in a more detailed way, the techniques introduced in this chapter.

The flow chart of Figure 7–29 illustrates the scientific method that engineers and scientists use to develop accurate models so that they can predict the behavior of systems and facilitate the design of new related systems. The flow begins at the top, on the basis of available experimental data. They develop a model which consists of an abstraction of these data, but which retains the essential characteristics of the experimental results. They refine the model by manipulating it theoretically as well as by precisely adjusting the parameters so that the model fits, as well as possible, all the available evidence. Once they have formulated an accurate model, they use it to predict and design, by using the techniques for extracting information.

The major concepts in this discussion correspond to the specific topics of this chapter, as the flow chart of Figure 7–30 shows. No further explanation of this figure is necessary, since we have now examined the major elements in detail.

In the next chapter, we shall consider probabilistic models. These are necessary for describing physical systems in which experimental outcomes cannot be predicted with certainty.

FIGURE 7–29 Flow chart of the design process

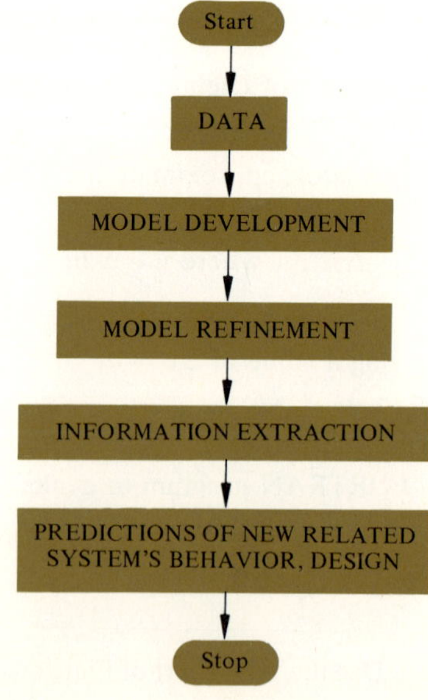

FIGURE 7–30 Flow chart of model formulation

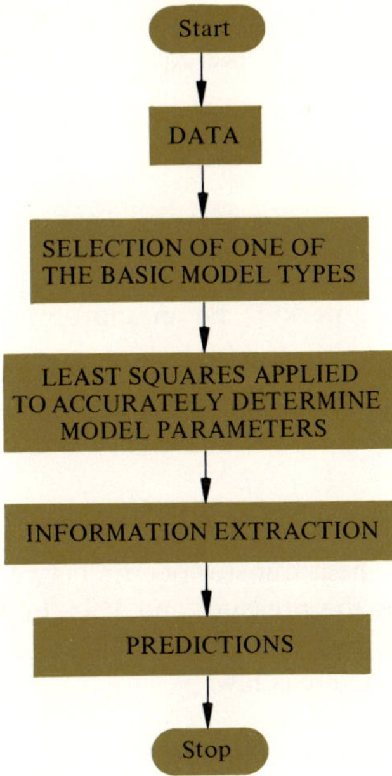

PROBLEMS

P7–1 Assume that automobile manufacturers had followed a different course over the past decade. The result is that the model for "average" fuel economy c (mpg) versus speed v (mph) for the automobile population is

$$c = 2000\left(\frac{v}{v^2 + 3025}\right)$$

Write and execute a FORTRAN program using a DØ loop that prints out the values of c and v for $v = 0, 5, 10, 15, 20, \ldots, 100$.

P7–2 Using the dichotomous search method written in Chapter 5, evaluate the maximum value of c and the corresponding value of v to within an uncertainty of $\Delta v = 0.5$ for the model of Problem 7–1.

P7–3 An initial amount of money M_0 is invested at an interest rate i (expressed as a fraction). After N years at compound interest, the amount to which it has grown is

$$M = M_0(1 + i)^N$$

The initial investment is \$12,000. Calculate the amount resulting after 15 years at 8% interest.

9 P7–4 Find the number B such that the equation

$$M = M_0(1 + i)^N$$

can be written in the form

$$M = M_0(2)^{N/B}$$

Why would "doubling period" be an appropriate name for B?

P7–5 We said in Section 7–4 that air may be compressed by decreasing the total volume of a fixed mass of air in a piston-cylinder apparatus. The symbolic model given earlier,

$$PV^{1.4} = C$$

is valid as long as no heat transfer occurs between the air and its surroundings. Recall that P is the pressure and V is the volume. Experiments have shown that a good approximation to the behavior of air at low pressures and temperatures is the ideal-gas law,

$$PV = AT$$

In this equation, A is a constant.

A sample of air occupies a volume of 35 m^3 at atmospheric pressure (10^5 N/m^2) and a temperature of 300 K. Assume that the gas is confined in a cylinder of fixed volume. Calculate the pressure P when the temperature increases to 350 K.

P7–6 An automobile traveling in a circular path (around a curve) at constant speed has the following accelerating force in the direction of the center of curvature.

$$\frac{Mv^2}{R}$$

where M is the car mass, v is the speed, and R is the radius of the curve.

Friction between the tires of the car and the pavement provides the force necessary to maintain this acceleration and hold the car on its circular path. The friction force is given by CF_N where C is the coefficient of friction and F_N is the normal force of the wheels on the pavement.

Sloping the road surface toward the center of the curve increases the force of contact between the car wheels and the pavement. Thus sloping the surface increases the rate at which a car can travel around a given curve. This is the reason that racetracks have banked curves.

FIGURE P7–1 Automobile on a banked curve

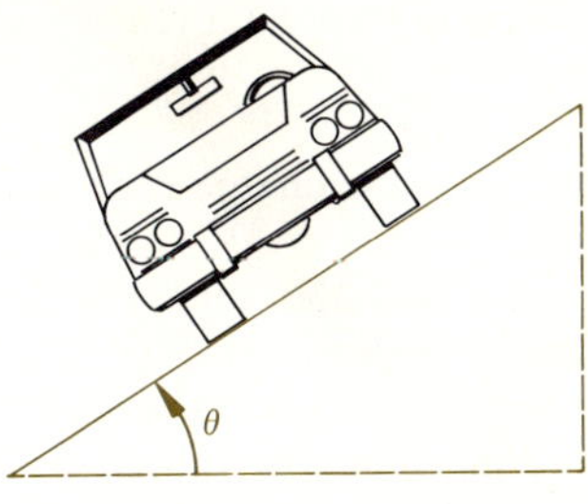

Slope $S = \tan\theta$

Solving the equations of engineering mechanics gives the formula for the necessary slope (Figure P7–1) of the road's surface. This formula has the form

$$S = \frac{(v^2/gR) - C}{C(v^2/gR) + 1}$$

where S is the slope and g is the acceleration of gravity.

Write and execute a FORTRAN program to evaluate the slope S and bank angle θ for a racetrack with a curve radius of 700 ft if cars are to travel at the following speeds (mph): 90, 100, 110, 120, 130, 140, 150, 160, 170.

Repeat these calculations for curves of radii 1000 ft and 1300 ft. Take $C = 0.70$. Be careful of the units involved in these equations.

P7–7 Scientists are often interested in the age of rocks or artifacts formed from organic material. One method to date these is to measure the amount of radioactive carbon-14 present. Extensive experimentation has revealed an equilibrium between the rate of decay and the rate of assimilation of radioactive carbon atoms for all living organisms. When the organism dies, the assimilation stops and the radioactive carbon-14 in the tissue decays. The *specific activity* (number of disintegrations per minute per gram of material) satisfies the following exponential model.

$$A(t) = 15.3e^{-\lambda t}$$

where $A(t)$ is the specific activity, t is the time since death in years, and λ [$= 0.0001244\ (\text{year})^{-1}$] is the decay constant for carbon-14.

Determine the time T (called the *half-life* of the radioactive material) at which the specific activity is reduced to one-half its original value.

A wooden slide rule recently discovered in ruins has a specific activity of two disintegrations per minute per gram. Estimate the age of the civilization that produced the slide rule.

You may use either calculator or computer.

P7–8 The following equation describes the decay of electrical charge on a capacitor of C farads discharging through a resistor of R ohms.

$$Q(t) = Q_0 \, e^{-t/RC}$$

where t is in seconds, Q_0 is the charge in coulombs at $t = 0$, and $Q(t)$ is the charge at time t.

The product RC is called the *time constant* of the discharge and is denoted by τ. In general, the time constant is the time τ at which the amount of the substance undergoing exponential decay has been reduced to $1/e$ of the original amount. Determine the time constant of a capacitor of $2(10)^{-5}$ farads discharging through a resistor of 10^6 ohms.

What fraction of the original charge is left on the capacitor after one time constant? after two time constants? after ten time constants?

P7–9 For the symbolic model in Problem 7–8, we define a half-life T as the time at which the charge remaining on the capacitor is one-half the original amount. Using this definition, find a relationship between the time constant τ and the half-life T.

P7–10 A 4000-liter tank is initially filled with a solution consisting of 200 kg of common salt (NaCl) dissolved in water. A valve is then opened, and fresh water flows into the tank at a constant rate of 40 liters per minute. A mixing impeller constantly stirs the solution in the tank, and the solution overflows from the tank at the inlet rate of 40 liters per minute. A chemical engineer who studied the problem found that the following deterministic model described the weight W of salt remaining in the tank t hours after the inlet valve was opened.

$$W = 200 \, e^{-0.6t} \text{ kg}$$

Find W for $t = 2$ hours, $t = 4$ hours, and $t = 1000$ hours.

What is the half-life of the weight of salt in the tank? What is the time constant of the decay of salt in the tank?

P7–11 A deterministic model for a nuclear reactor core predicts that in order for the reactor to obtain criticality—that is, for the fission process to be constant and self-sustaining—the following equation must be satisfied.

$$\frac{ke^{-\tau B^2}}{1 + L^2B^2} = 1$$

A nuclear engineer specifies the relative amounts of uranium, water, and other structural materials desired in the reactor and finds that

$$k = 1.745, \qquad \tau = 30 \text{ cm}^2, \qquad L^2 = 2.33 \text{ cm}^2$$

A design condition is that the height H of the cylindrical reactor must be equal to $R/0.55$. The quantity B is related to the reactor radius R through the equation

$$B^2 = \frac{8.77}{R^2}$$

Write and execute a computer program that uses the Newton–Raphson method to find the critical value of B^2. Use arithmetic-statement functions or function subprograms to define the function and the derivative needed. What are the critical values of R and H?

P7–12 An engineer studying the problem sketched in Figure P7–2 has gathered data concerning the flow rate f of liquid from the tank as a function of the liquid height h. The following table summarizes these data.

f (liters/min)	37.9	32.9	26.8	19.0	12.0	10.4	8.5	6.0	3.3	1.9
h (meters)	10	7.5	5	2.5	1	0.75	0.5	0.25	0.075	0.025

Using graphical techniques, find which of the four basic deterministic models fits these data.

P7–13 For the data given in Problem 7–12, carry out a least-squares calculation to determine the best values of the two parameters in the model that you selected in Problem 7–12. Use a suitable computer program to evaluate these parameters.

P7–14 An electrical engineer is testing power transistors that will be used under extremely heavy electrical loads. The evaluation involves applying a voltage to each transistor and measuring the maximum current that can flow through the device without an unacceptably large temperature increase. The following table of voltage V and maximum current I summarizes the experimental findings.

FIGURE P7–2 Storage tank emptying under the influence of gravity. Flow f is related to height h.

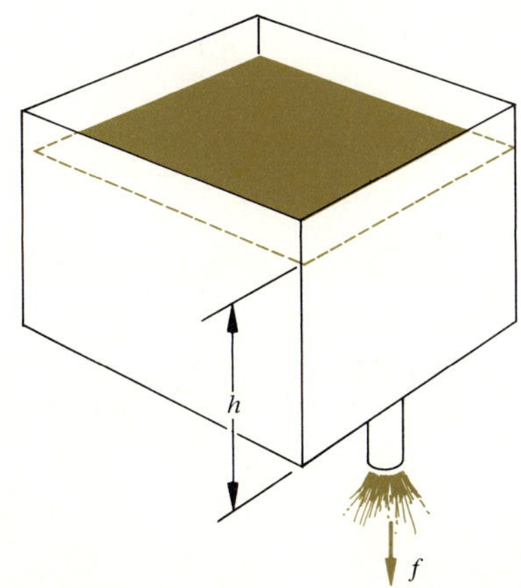

V (volts)	1500	300	150	75	38	25	15	7.5	3.8	1.9
I (amps)	0.01	0.05	0.1	0.2	0.4	0.6	1.0	2.0	4	8

Using graphical methods, find the basic deterministic model that best fits these data.

P7–15 For the data given in Problem 7–14, carry out a least-squares calculation to determine the best values of the two parameters in the model that you selected in Problem P7–14. Use a suitable computer program to evaluate these parameters.

P7–16 Evaluate the least-square coefficients (a and b) for a linear model fitting the following data.

y	7	10	13	16.2	19.2	23	26
x	0	1	2	3	4	5	6

Use a suitable FORTRAN program to calculate the total error resulting from these data and this model.

Increase the intercept b by 10%, and recalculate the total error. Keep the slope constant.

Decrease the original intercept b by 10%, and recalculate the total error. Keep the slope constant.

These numerical exercises indicate that the least-squares value of intercept b results in the minimum total error.

P7–17 Using graphical methods, find the model and associated parameters that best describe the relationship among the following data.

y	6	3.6	4.8	5.0	7.2	8.0	6.2	4.4	5.0	2.0	5.8
x	8	200	28	50	3	1	5	70	20	900	14

P7–18 Measuring the number N of radioactive nuclei present at four different times yields the following data.

N	969	880	880	741
t (minutes)	1	2	4	8

N follows the deterministic model

$$N(t) = N_0 e^{-\lambda t}$$

where λ is the decay constant and N_0 is the number of radioactive nuclei present at $t = 0$. Using least squares, find the "best" values of λ and N_0.

P7–19 Two cars begin traveling toward each other on a straight road from points 10 km apart. The following table lists the speeds of the cars as functions of time.

Time interval (hours)	0–0.25	0.25–0.50	0.50–0.75	0.75–1.00
Speed of car 1 (km/hour)	37.9	51.9	57.0	58.9
Speed of car 2 (km/hour)	42.6	50.0	52.9	54.5

How far will car 1 travel in one hour? Find the time at which the cars will pass each other.

P7–20 When a small particle settles in a fluid such as air or water, three forces act on it. The first (acting downward) is its weight (mg), where m is the particle mass and g is the acceleration of gravity. The second (acting upward) is the buoyant force given by $m_f g$, where m_f is the mass of fluid displaced by the particle. Finally, a viscous retarding force (acting upward) of magnitude cv is also present because the fluid acts to hold back the particle; c is a constant and v is the particle velocity.

(a) Write Newton's second law for this situation.

(b) What velocity v will the particle have when the sum of forces acting on it is zero?

(c) A particle at rest on the surface of a fluid is released. The following model gives its velocity-versus-time behavior.

$$v(t) = \frac{g(m - m_f)}{c}(1 - e^{-(ct/m)})$$

For

$$g = 9.80 \text{ m/sec}^2, \qquad m_f = 0.0009 \text{ kg}$$
$$m = 0.001 \text{ kg}, \qquad c = 0.050 \text{ kg/sec}$$

write a computer program that will print v as a function of time for $t = 0$, 0.01, 0.02, · · ·, 0.2 sec.

(d) Repeat part (c) for a planet on which $g = 0.920 \text{ m/sec}^2$.

(e) Solve the equation in part (c) for the time t_0 at which the particle velocity is one-half its ultimate velocity. Leave your answer in symbolic form.

(f) Using the numerical values given in part (c), numerical integration and a suitable FORTRAN program, find the total distance traveled by the particle from $t = 0$ until the time found in part (e).

P7–21 The following model gives the flow rate f (liters per minute) for a fluid draining from a storage tank.

$$f = 500\, e^{-0.01t}$$

where t is measured in seconds.

Using a suitable FORTRAN program, evaluate the total number of gallons that will drain from the tank between the times $t = 1$ sec and $t = 200$ sec.

P7–22 A projectile is fired vertically from the surface of the Earth. The following table gives its velocity each second after firing.

Velocity (meters/sec)	100	10	4	2	−4	−10
Time (seconds)	0	1	2	3	4	5

Estimate the height of the projectile at the times listed in the table. Estimate the time at which the projectile begins to fall back to the Earth.

P7–23 A railroad switching engine operates for five minutes. Its accelerations, recorded every minute, are as follows.

a (m/min^2)	33	21	9	−3	−15	−27
t (minutes)	0	1	2	3	4	5

(a) Estimate the speed of the engine at $t = 5$ minutes, given that the engine is initially at rest.
(b) Is the engine moving "forward" or "backward" at $t = 5$ minutes?
(c) Find the position of the engine at $t = 2$ minutes.

P7–24 A chemical reagent is being transferred to an initially empty tank at a rate of $40 + 2t$ liters/hour, where t is the time of transfer in hours. After the flow rate has reached 60 liters/hour, it remains constant. When the inlet flow first reaches 60 liters/hour, an outlet valve is opened and the reagent is allowed to flow out of the tank at a rate of $10 + 3T$ liters/hour, where T is the time in hours since the opening of the outlet valve.
(a) How much fluid is in the tank when the drain is opened?
(b) How long will it take to empty the tank after the drain is opened?
(c) After the tank is empty, what is the outlet flow rate?

P7–25 The equation $y = Ax^2 + B$ (A,B are constants) is a possible model relating two variables, y and x. Find the "best" values of A and B for the following data.

y	0.4	1.4	3.6	20.7	81
x	0.0	0.4	0.8	2	4

CHAPTER EIGHT
PROBABILISTIC MODELS

We stated in Chapter 2 that one means of classifying models is according to a deterministic versus probabilistic scheme. In deterministic models, specifying the values of a set of control variables yields a precisely predictable outcome. For example, specifying the current I and resistance R in a simple electrical circuit yields a unique voltage E according to Ohm's law. Thus Ohm's law is a deterministic model. Although most of the models with which engineers must deal are deterministic, they frequently encounter problems involving uncertainty. The electrical engineer in a power plant is unable to predict with certainty what the electric power demand will be on a given day. The civil engineer designing a roadway intersection cannot determine exactly what traffic patterns will develop there. The chemical engineer in a petroleum refinery cannot predict precisely the future demand for gasoline. The industrial engineer in a manufacturing plant cannot describe with certainty the arrival of parts at an assembly station. There are innumerable situations in which engineers must deal with uncertainty in designing, constructing, and controlling systems made up of people, materials, energy, and equipment.

Fortunately, there are laws governing uncertainty, and a well-developed science called *probability* that systematizes these laws. This chapter examines some of the basic concepts of probability and illustrates their application to some typical engineering problems. Hence this chapter enlarges on the notion of probabilistic models in engineering. For mathematical simplicity, we shall present only the concepts and models that we can develop without using integral calculus.

8–1 BASIC PROBABILITY

RANDOM EXPERIMENT

Fundamental to the basic concepts of probability is the notion of a random experiment or, as it is often called, a chance experiment. A random experiment is a process that, in a given trial, yields any one of several possible outcomes. Chance completely determines the outcome that actually occurs on any given trial; we cannot predict it with certainty before the trial. Common examples of random or chance experiments include tossing a coin, rolling a die, and pulling the starter rope on a gasoline-powered lawnmower. Examples of random experiments encountered in engineering include inspecting a manufactured assembly for defects, observing the status of several channels of communication at a specific instant in time, and operating an electric circuit for an extended period of time. In each of these instances, we cannot state with certainty what the outcome of a trial will be until we have observed it.

SAMPLE SPACE

Although we cannot predict with certainty the precise outcome of a single trial of a random experiment, we can list all possible outcomes. This set of n outcomes, called the sample space, is written as

$$S = \{a_i\} = \{a_1, a_2, \ldots, a_n\} \tag{8–1}$$

where a_i ($i = 1, 2, \ldots, n$) represents the ith possible outcome of the random experiment. For the toss of a coin, the outcomes are a head H or a tail T. The sample space for this simple experiment is

$$S = \{H,T\}$$

Rolling a single die produces an outcome in which the die shows 1,2,3,4,5, or 6 spots on the top face. The sample space for this experiment is

$$S = \{1,2,3,4,5,6\}$$

Pulling the starter rope on a lawnmower produces just two outcomes, so that the sample space is

$$S = \{\text{Start, No start}\}$$

We might also regard the inspection of a manufactured assembly as having just two possible outcomes, defective D or nondefective N. Thus the sample space for the inspection of a single assembly is

$$S = \{D,N\}$$

For the example involving channels of communication, consider the situation in which just three channels are available. If a channel is not in use, it is assigned the value 0, whereas a 1 indicates that the channel is in operation. The sample space for this random experiment is therefore

$$S = \{(0,0,0), (1,0,0), (0,1,0), (0,0,1), (1,1,0), (1,0,1), (0,1,1), (1,1,1)\}$$

This example points out that the outcomes in the sample space show all possible ways that the channels can be either idle or operational. For instance, the outcome (0,1,0) shows that only channel 2 is operational at a given time and that channels 1 and 3 are idle. This is physically different from the outcomes (1,0,0), and (0,0,1). This sample space has two important properties. The outcomes are *mutually exclusive,* and they are *exhaustive.* Mutually exclusive simply means that the outcomes are distinct; if one occurs, another cannot occur. Exhaustive simply means the only possible outcomes of the experiment are those delineated in the sample space S.

Consider the operation of an electrical circuit for a fixed period of time T. We can view this as a simple experiment. If the circuit survives the entire period T, we denote the outcome by a 1; if it fails at some time $t < T$, we represent the outcome by a 0. This sample space is simply

$$S = \{0,1\}$$

Now let's take a more complicated view. Suppose that the circuit operates for some random time $t \leq T$. Then the sample space contains an infinite number of possible values of t. The concept of a sample space as we have stated it is not very useful in this situation. Since the mathematical modeling of such random experiments requires the use of integral calculus, we shall not consider such models any further in this book. Instead, we shall concentrate on those random experiments whose outcomes we can enumerate. This branch of probability is called *discrete probability,* or finite probability.

PROBABILITY

For a given trial of a random experiment, one does not know precisely which outcome in the sample space will result. But repeating the experiment many times lets us estimate the frequency of a given outcome. That is, if we repeat the random experiment M times and the outcome a_i results m_i times, the outcome a_i's frequency of occurrence is m_i. If we divide the frequency m_i by the total number of trials M, we get the probability p_i of outcome a_i.

$$p_i = \frac{m_i}{M} \tag{8–2}$$

The sum of the frequencies of the n outcomes must equal the total number of trials of the experiment; that is,

$$m_1 + m_2 + \cdots + m_i + \cdots + m_n = M \tag{8–3}$$

Dividing both sides of (8–3) by the total number of trials M shows that

$$p_1 + p_2 + \cdots + p_i + \cdots + p_n = 1 \tag{8–4}$$

Associated with each outcome in the sample space is a probability, so that we can write the *set* of probabilities as

$$P = \{p_1, p_2, \ldots, p_n\} \tag{8–5}$$

If the outcome a_i never occurred in M trials, its probability would be

$$p_i = \frac{0}{M} = 0$$

Similarly, if the outcome a_i occurred with every trial, its probability would be

$$p_i = \frac{M}{M} = 1$$

These observations enable us to state two important laws of probability:

$$0 \leqslant p_i \leqslant 1, \qquad i = 1, \ldots, n \tag{8–6}$$

$$\sum_{i=1}^{n} p_i = 1 \tag{8–7}$$

Let's see what these laws mean. In the toss of a fair coin, for example, the head is just as likely to occur as the tail, so that the probabilities are

$$p(\text{H}) = 1/2, \qquad p(\text{T}) = 1/2, \qquad P = \{1/2, 1/2\}$$

In the roll of a fair die, each of the six outcomes is equally likely to occur, so that the set of probabilities is

$$P = \{1/6, 1/6, 1/6, 1/6, 1/6, 1/6\}$$

These two examples illustrate the notion of equally likely outcomes. This concept is simple: Each of the n outcomes a_i in the sample space S has just as good a chance of occurring as any other outcome. The probability of the outcome a_i is thus simply $1/n$. Unfortunately, not all random experiments produce such convenient results. If the starting of a lawnmower had ever yielded equally likely outcomes, solid-state ignition systems might never have been developed for power mowers. The "week-end groundskeeper" might have been perfectly content with having this random experiment produce the same results as tossing a coin.

Exercises

8–1 Show the outcomes and probabilities for the random experiment of tossing two fair coins simultaneously.

8–2 Show the sample space for drawing two screws from a set of six screws (numbered from 1 to 6). If any one screw is just as likely to be selected as any other screw in the set, show the probabilities for the sample space.

8–3 Suppose that a random experiment consists of receiving a signal from each of four satellites. The probability of receiving the signal from a given satellite is 3/4. Show the sample space and associated probabilities for this random experiment.

8–4 Write a computer program that will develop the sample space for the random experiment of tossing three dice. Assume that the dice are distinguishable. Have the program compute the probability of each outcome in this sample space.

EVENTS

Another fundamental concept in probability is that of an event. An event is a collection of those outcomes in the sample space that have an attribute or characteristic of interest. To illustrate this, let us consider the problem of rolling a pair of dice. Each die can show 1,2,3,4,5, or 6 spots on the top face. Hence the pair of dice produces the following outcomes:

$$S = \{(1,1),(1,2),(1,3),(1,4),(1,5),(1,6),(2,1),(2,2),(2,3),(2,4),(2,5),(2,6),(3,1),(3,2),(3,3),(3,4),(3,5),(3,6),(4,1),(4,2),(4,3),(4,4),(4,5),(4,6),(5,1),(5,2),(5,3),(5,4),(5,5),(5,6),(6,1),(6,2),(6,3),(6,4),(6,5),(6,6)\}$$

Structuring the sample space in this fashion implies that we can distinguish the outcome (1,3), say, from (3,1). It is simple to make this distinction if we have one red die and one green die, for example. Then these outcomes are (Red 1, Green 3) and (Red 3, Green 1). Thus there are 36 outcomes in this sample space, and each outcome is equally likely to occur. Hence the probability of each outcome is 1/36.

Suppose we are interested in the event R that exactly one 3 appears. Collecting the outcomes in the sample space that produce exactly one 3, we have the smaller space

$$R = \{(3,1),(3,2),(3,4),(3,5),(3,6),(1,3),(2,3),(4,3),(5,3),(6,3)\}$$

Ten outcomes yield the event R. How do we find the probability of the event R? Just as we aggregate the set of outcomes that define the event R, we must similarly aggregate the probabilities of these outcomes to determine the probability of the event, that is, $P(R)$. Since 10 of the 36 equally likely outcomes yield event R, we quickly establish that

$$\begin{aligned} P(R) &= p(3,1) + p(3,2) + p(3,4) + p(3,5) + p(3,6) + p(1,3) + p(2,3) \\ &\quad + p(4,3) + p(5,3) + p(6,3) \\ &= \frac{1}{36} + \frac{1}{36} + \frac{1}{36} + \frac{1}{36} + \frac{1}{36} + \frac{1}{36} + \frac{1}{36} + \frac{1}{36} + \frac{1}{36} + \frac{1}{36} \\ &= \frac{10}{36} \end{aligned}$$

Thus we sum the probabilities for those outcomes in the sample space S that yield the event R. That is, if r of the n outcomes in the sample space S yield the event R, then the probability of event R is

$$P(R) = \sum_{j=1}^{r} p_j \tag{8–8}$$

where we sum over all outcomes in the event R. This notation suggests that the first r outcomes in the sample space S produce the event R, but we can reorder the outcomes in the sample space without changing the nature of the space.

OPERATIONS WITH EVENTS

Consider two events B and C. Two *operations* with events are of fundamental importance in the study of probability: union and intersection. The symbol $B + C$ denotes the *union* operation, shown in Figure 8–1(a); that is, *at least one* of the two events B *or* C occurs. The symbol BC represents the *intersection* operation, depicted in Figure 8–1(b); that is, *both* events B and C occur. Suppose that b of the n outcomes in the sample space S form the event B, c outcomes form C, and d outcomes yield both events B and C. Then the probability of the *intersection* BC is

$$P(BC) = \sum_{k=1}^{d} p_k \tag{8–9}$$

The probability of the *union* $B + C$ is

$$P(B + C) = \sum_{i=1}^{b} p_i + \sum_{j=1}^{c} p_j - \sum_{k=1}^{d} p_k \tag{8–10}$$

FIGURE 8–1 Operations with events—a Venn-diagram approach: (a) union $(B + C)$, (b) intersection (BC), (c) mutually exclusive

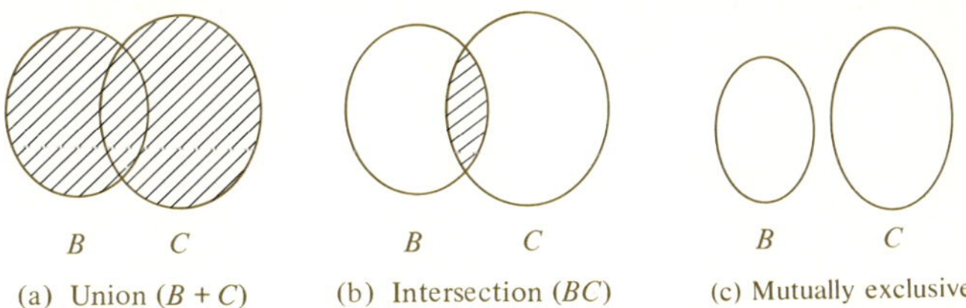

Since d of the b outcomes in event B are also in event C, the probabilities of these d outcomes appear in both the first and second terms in (8–10). Hence we subtract them out in the third term. This leads to the *additive law of probability;* that is,

$$P(B + C) = P(B) + P(C) - P(BC) \tag{8–11}$$

Now suppose that event C cannot occur if event B does, or vice versa. Then we say that events B and C are *mutually exclusive*. That is, the probability $P(BC)$ is zero, and Equation (8–11) reduces to

$$P(B + C) = P(B) + P(C) \tag{8–12}$$

As an example of the additive law of probability, consider a manufactured lot of 1000 parts. Of this lot, 84 parts have a type B defect, and 53 parts have a type C defect. Now 13 of these parts have both types of defects. Therefore the probabilities of these events are

$$P(B) = \frac{84}{1000}, \qquad P(C) = \frac{53}{1000}, \qquad P(BC) = \frac{13}{1000}$$

The probability of a defect is then the probability of a type B defect *or* a type C defect, that is, the probability of the union of B and C:

$$P(B + C) = \frac{84}{1000} + \frac{53}{1000} - \frac{13}{1000} = \frac{124}{1000} = 0.124$$

As an example of mutually exclusive events, consider a machine that is supposed to automatically fill boxes with 28 grams of breakfast cereal. According to manufacturing specifications, the range of acceptable net weights of cereal per box is from 27.0 to 29.0 grams. Suppose that the probability of an underweight box (less than 27.0 grams) of cereal is 1/100, and that of an overweight box (more than 29.0 grams) is 2/100. Representing these as events B and C, respectively, we see that

$$P(B + C) = \frac{1}{100} + \frac{2}{100} = \frac{3}{100}$$

It is clear that events B and C are mutually exclusive, because an underweight box cannot also be an overweight box. Observe that, in this example, $P(B + C)$ is the probability that a given box of cereal does not meet manufacturing specifications.

Frequently we are interested in the composite event BC. If the occurrence of event B in no way affects the occurrence of event C, and vice versa, we say that they are *independent*. For independent events B and C, the probability of the composite event is simply

$$P(BC) = P(B) \cdot P(C) \tag{8–13}$$

We can generalize this relationship to any number of events $B,C,D,E, \ldots$

$$P(BCDE \ldots) = P(B) \cdot P(C) \cdot P(D) \cdot P(E) \cdots \tag{8–14}$$

For example, consider the receipt of a transmitted signal from each of two satellites. The probability of receiving the signal from satellite B is 0.9, and from satellite C it is 0.8. These are independent events, so that the probability of receiving both signals is 0.72.

Exercises

8–5 Consider the toss of a pair of dice. Let B represent the event that a 5 occurs on a single die and C the event that a 6 occurs. Which outcomes yield the event B? event C? What are the probabilities of these events?

8–6 In Exercise 8–5, what is the probability of a 5 *or* a 6 occurring? What is the probability of a 5 *and* a 6 occurring?

8–7 In the example of receiving signals from satellites B and C with probabilities 0.9 and 0.8, respectively, what is the probability of receiving at least one of the two signals? neither of the two signals?

8–8 In the example of filling boxes of breakfast cereals, the probability of a box containing less than 27.0 grams of cereal is 1/100, while the probability of a box containing more than 29.0 grams is 2/100. If someone consumed two boxes of cereal for breakfast, what is the probability that both boxes were underweight? both overweight? both within specifications ($27 \leq x \leq 29$)?

CONDITIONAL PROBABILITY

The engineer sometimes faces the problem of determining the probability that an event C will occur, given that another event B occurs. This situation implies that the occurrence of event B somehow affects the occurrence of event C; that is, events B and C are *dependent*. The probability $P(BC)$ of the composite event in this case is

$$P(BC) = P(B) \cdot P(C|B) \tag{8–15}$$

The symbol $P(C|B)$ is read "the conditional probability of event C, given that event B occurs." By rearranging, we can determine this conditional probability $P(C|B)$ from the equation

$$P(C|B) = \frac{P(BC)}{P(B)} \tag{8–16}$$

We can extend the principle of conditional probability given in (8–16) to establish the probability that an event A_i, which we know has occurred, might have occurred in a particular manner. That is, suppose that a random experiment has n mutually exclusive outcomes $A_1, A_2, \ldots, A_n$, and that there is an event B such that $P(B) \neq 0$. Then the conditional probability of outcome A_i given that event B has occurred is

$$P(A_i|B) = \frac{P(B|A_i) \cdot P(A_i)}{\sum_{j=1}^{n} P(B|A_j) \cdot P(A_j)} \tag{8–17}$$

Equation (8–17), called *Bayes's theorem,* is a very useful tool in engineering analysis. For example, suppose that a part is produced on three machines: A, B, and C. Machine A yields 40% of the output, machine B 35%, and machine C 25%. The percentage of defective output from the three machines is 8%, 5%, and 3%, respectively. We select one part at random from the total output of the three machines and find that it is defective. What is the probability that it was produced on machine A? Let D denote a defective part; let A denote the event that the part came from machine A, B from machine B, and C from machine C. Then the probabilities of interest are

$$\begin{aligned} P(A) &= 0.40 \qquad & P(D|A) &= 0.08 \\ P(B) &= 0.35 \qquad & P(D|B) &= 0.05 \\ P(C) &= 0.25 \qquad & P(D|C) &= 0.03 \end{aligned}$$

The statement $P(A) = 0.40$ says that the probability that a given part is produced on machine A is 0.4, since machine A produces 40% of all parts. The statement $P(D|A) = 0.08$ says that the conditional probability that a part is defective, given that it was produced on machine A, is 0.08, since 8% of all parts produced on machine A are defective. From (8–17),

$$\begin{aligned} P(A|D) &= \frac{P(D|A) \cdot P(A)}{P(D|A) \cdot P(A) + P(D|B) \cdot P(B) + P(D|C) \cdot P(C)} \\ &= \frac{(0.08)\,(0.40)}{(0.08)\,(0.40) + (0.05)\,(0.35) + (0.03)\,(0.25)} \\ &= \frac{0.0320}{0.0320 + 0.0175 + 0.0075} \\ &= \frac{0.0320}{0.0570} \\ &= 0.561 \end{aligned}$$

This result says that, given a defective part, the probability that it was produced on machine A is 0.561. Similarly, the probabilities that the defective part was produced on machines B and C are found to be 0.307 and 0.132, respectively.

Exercise

8–9 Part of the automatic quality control is to test a brake drum for a certain defect. If the drum is acceptable, a green light flashes. If the drum is questionable, a yellow light flashes. If the drum is definitely defective, a red light flashes. We know that 95% of the manufactured brake-drum assemblies are not defective. Tests with the automatic quality-control equipment produce the following results: (1) When units known to be *non*defective were tested, the green light flashed 88% of the time, the yellow light 10% of the time, and the red light 2% of the time. (2) When units known to be defective were tested, the green light flashed 15% of the time, the yellow light 30% of the time, and the red light 55% of the time.

(a) Define the events in this experiment, and show their probabilities.

(b) Determine the probability that the brake drum is defective when the yellow light is observed.

(c) What should we conclude when the yellow light flashes?

RANDOM VARIABLES

An important concept associated with the notion of the sample space is the random variable. It is a real-valued function of the outcomes a_i in the sample space S. The n outcomes in the sample space produce k possible values of the random variable, where $k \leq n$. Denoting the random variable as X and its possible values as $x_1, x_2, \ldots, x_k$, we see that one or more outcomes a_i aggregated yield a possible value x_j. Similarly, the probability of value x_j, denoted $f(x_j)$, is the sum of the probabilities for those outcomes a_i that combine to produce the possible value x_j.

For example, consider a system of three channels of communication and a random experiment that consists of checking which channels are in use at randomly selected times. The sample space is

$$S = \{(0,0,0),(1,0,0),(0,1,0),(0,0,1),(1,1,0),(1,0,1),(0,1,1),(1,1,1)\}$$

Suppose the probabilities for this random experiment are

$$P = \{0.25, 0.15, 0.15, 0.15, 0.09, 0.09, 0.09, 0.03\}$$

Define the random variable X as the number of channels in use. Table 8–1 shows the possible values of X, the outcomes in S that produce these values, and the associated probabilities.

FREQUENCY DISTRIBUTION

The probabilities $f(x_j)$ for possible values of X form the frequency distribution of the random variable X. We denote the distribution by $f(X)$. Note that these probabilities satisfy the basic laws of probability stated in Equations (8–6) and (8–7); that is,

TABLE 8–1. PROBABILITIES OF VALUES x_j OF RANDOM VARIABLE X

Possible value x_j	*Outcomes* a_i *in S that yield value* x_j	$f(x_j) = \Sigma p_i$
0	(0,0,0)	0.25
1	(1,0,0), (0,1,0), (0,0,1)	0.45
2	(1,1,0), (1,0,1), (0,1,1)	0.27
3	(1,1,1)	0.03

$$0 \leq f(x_j) \leq 1, \qquad j = 1, \ldots, k \tag{8–18}$$

$$\sum_{j=1}^{k} f(x_j) = 1 \tag{8–19}$$

Thus the frequency distribution $f(X)$ is often called the *probability function* of the random variable X.

It is often helpful to visualize probability functions by drawing graphs like the one in Figure 8–2. This graph is called a *histogram*. The height of the rectangle for the value x_j is proportional to the probability $f(x_j)$, and the bases of the rectangles touch. We recognize the histogram as an application of the bar graph that we discussed in Chapter 2.

In many problems, we are interested not only in the probability $f(x_j)$ of the value x_j, but also in the probability that the random variable X has a value less than or equal to x_j, or $P(X \leq x_j)$. Denoting this probability by $F(x_j)$, we see that

$$P(X \leq x_j) = F(x_j) = \sum_{l=1}^{j} f(x_l) = f(x_1) + f(x_2) + \cdots + f(x_j) \tag{8–20}$$

FIGURE 8–2 Probability function of a random variable x

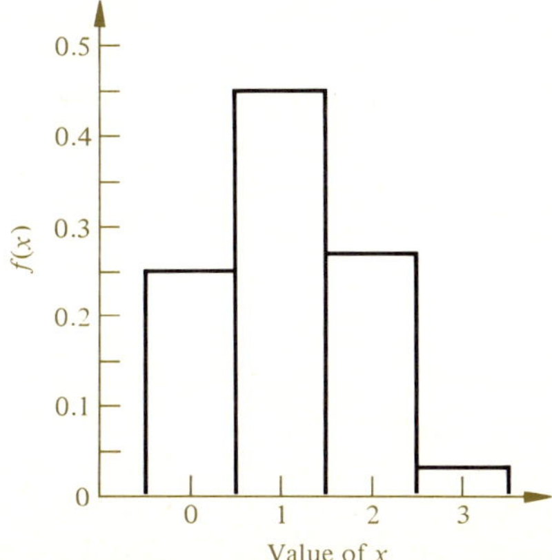

In our example about channels of communication, the probability that two or fewer channels are in use is

$$P(X \leq 2) = F(2) = f(0) + f(1) + f(2)$$
$$= 0.25 + 0.45 + 0.27 = 0.97$$

Often we want to know the probability that the random variable X has a value greater than x_j; that is, we want to know $P(X > x_j)$. This is given by

$$P(X > x_j) = 1 - P(X \leq x_j) = 1 - F(x_j) = 1 - \sum_{l=1}^{j} f(x_l) \qquad (8\text{–}21)$$

In our example, the probability that more than one channel is in use is

$$P(X > 1) = 1 - P(X \leq 1) = 1 - F(1)$$
$$= 1 - [f(0) + f(1)] = 1 - [0.25 + 0.45]$$
$$= 0.30$$

A probability $f(x_j)$ is associated with each possible value x_j of the random variable X; likewise a cumulative probability $F(x_j)$ is associated with each x_j. Thus the *cumulative probability function* gives the probabilities $F(x_j)$ expressed by Equation (8–20). Table 8–2 shows the cumulative probabilities for the problem about the channels of communication. Figure 8–3 shows the cumulative probability function $F(x_j)$ for the random variable X, the number of communications channels in operation.

EXPECTED VALUE

One might ask: What is the average value of the random variable X? The answer to this question lies in the concept of expectation. The *expectation* of a function $g(X)$ of the random variable X is

$$E\,[g(X)] = \sum_{j=1}^{k} g(x_j) \cdot f(x_j) \qquad (8\text{–}22)$$

The notation $E[\ \ ]$ means "the expectation of [ ]." Observe that the summation is over all k possible values of X. The average or mean value of X, denoted μ, is simply the expectation of X,

TABLE 8–2. CUMULATIVE PROBABILITIES $F(x_j)$

Possible value x_j	*Probability* $f(x_j)$	*Cumulative probability* $F(x_j)$
0	0.25	0.25
1	0.45	0.70
2	0.27	0.97
3	0.03	1.00

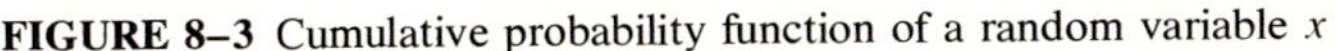

FIGURE 8–3 Cumulative probability function of a random variable x

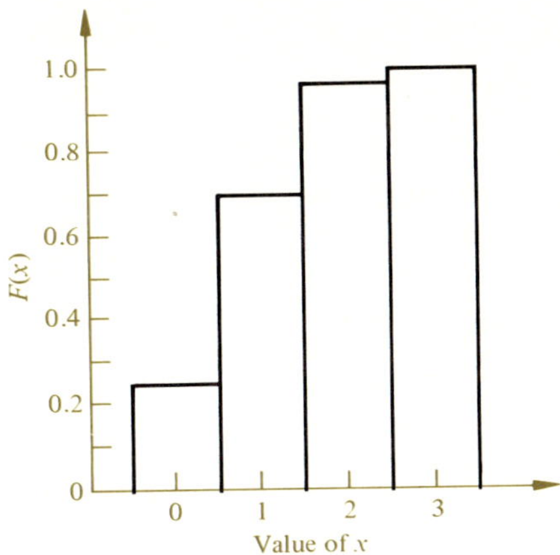

$$\mu = E[X] = \sum_{j=1}^{k} x_j \cdot f(x_j) \tag{8–23}$$

In the communications-channels problem, the expected number of channels in use is

$$\begin{aligned}\mu &= \sum_{j=1}^{4} x_j \cdot f(x_j) \\ &= (0)(0.25) + (1)(0.45) + (2)(0.27) + (3)(0.03) = 1.08\end{aligned}$$

That is, the *mean* number of channels in use is 1.08. Although the random variable X is a *discrete* quantity (0, 1, 2 or 3), its expected value is not. This concept might at first be troublesome. How can one "expect" to obtain a value that the random variable X cannot even assume? This is the case in this example, where the *mean* μ is 1.08. It is important to understand exactly what the *expected value* or *mean* really is. In this example, if one were to observe the status of this communications system a very large number of times, the *average* number of channels in use would be 1.08.

VARIANCE

Another important parameter that characterizes a frequency distribution of a random variable X is its *variance*. This parameter is a measure of the *spread* or *dispersion* of the values that the random variable assumes. The *variance,* denoted σ_x^2, also depends on the concept of expectation. That is,

$$\sigma_x^2 = E\{[X - E(X)]^2\} = E(X^2) - [E(X)]^2 \tag{8–24}$$

The definition of expectation gives

$$E(X^2) = \sum_{j=1}^{k} x_j^2 \cdot f(x_j) \tag{8–25}$$

And we have the following equation for computing the variance of X:

$$\sigma_x^2 = E[X^2] - \mu^2 \tag{8–26}$$

In our example, the variance of the random variable X is

$$\begin{aligned}\sigma_x^2 &= [(0)^2(0.25) + (1)^2(0.45) + (2)^2(0.27) + (3)^2(0.03)] - (1.08)^2\\ &= (0.0 + 0.45 + 1.08 + 0.27) - (1.1664)\\ &= 0.6336\end{aligned}$$

Exercises

8–10 In the problem of tossing a pair of dice, determine the probability function and cumulative probability function of the random variable X, the sum of the spots on the two dice.

8–11 Find the mean and the variance of the random variable X in Exercise 8–10.

8–12 Prove that $E\{[X - \mu]^2\}$ is $E(X^2) - \mu^2$.

8–2 PROBABILITY DISTRIBUTIONS

So far we have based the concept of a probability function on outcomes. We defined a random variable X and enumerated those *outcomes* in the sample space S that gave a particular value x_j of the random variable X. Then we summed the probabilities for those outcomes to find the probability of the value x_j. This development is straightforward, but it depends entirely on enumerating the outcomes in a sample space. We can often approach random experiments by developing a *probability model* of a random process. In the following sections, we shall describe several such models.

BERNOULLI TRIAL

A Bernoulli trial is a random experiment that has exactly two outcomes. These outcomes are generally termed success and failure; 1 denotes success and 0 denotes failure. Tossing a coin, pulling the starter rope on a lawnmower, and inspecting a manufactured part are familiar examples of a Bernoulli trial. We can represent the probability model for such a random experiment by

$$f(x) = p^x q^{1-x}, \qquad x = 0,1 \tag{8–27}$$

where p = probability of success and $q = 1 - p$ = probability of failure.

For example, suppose that 10% of the parts produced in a manufacturing operation are defective. If discovery of a defective part denotes success, the probability of success is $p = 0.1$. Then the probabilities of the values 0 and 1 are

$$f(0) = (0.1)^0(0.9)^1 = 0.9, \qquad f(1) = (0.1)^1(0.9)^0 = 0.1$$

The set of values $f(x)$ is commonly called the *probability distribution* of X. Figure 8–4 shows a histogram for this probability distribution.

BINOMIAL DISTRIBUTION

The binomial random variable X is the number of successes in n independent Bernoulli trials. By independent we mean that the outcome of a given trial of the random experiment in no way affects the outcome of any other trial. The probability model is represented by

$$f(x) = \binom{n}{x} p^x q^{n-x}, \qquad x = 0, 1, \ldots, n \tag{8–28}$$

The expression $\binom{n}{x}$ denotes the combination of n things taken x at a time. It is given by

$$\binom{n}{x} = \frac{n!}{x!\,(n-x)!} \tag{8–29}$$

where $n!$ is read "n factorial," and we define it as

$$n! = n(n-1)(n-2)\cdots(2)(1)$$

Hence we can rewrite Equation (8–28) as

$$f(x) = \frac{n!}{x!\,(n-x)!} p^x q^{n-x}, \qquad x = 0, 1, \ldots, n \tag{8–30}$$

FIGURE 8–4 A Bernoulli trial: $p = 0.1$

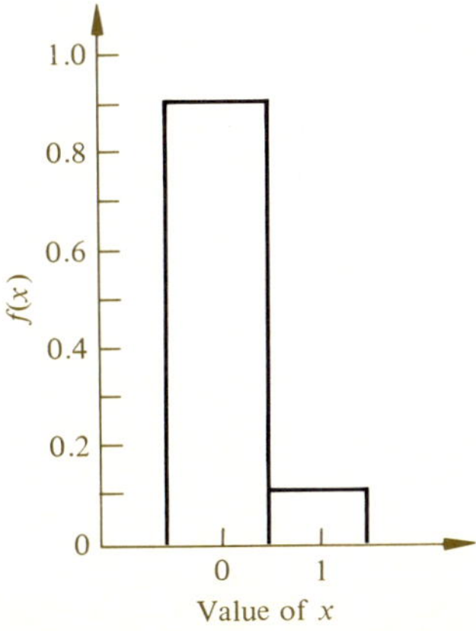

Equation (8–30) is called the binomial distribution. When we say "the combination of n things taken x at a time," we mean simply that we have a sample space containing n outcomes and that exactly x of these outcomes yield the event that we are considering. For example, suppose that in the previous example, we inspect exactly four manufactured parts. If N denotes a nondefective part and D denotes a defective part, the sample space is

$$S = \{(N,N,N,N), (D,N,N,N), (N,D,N,N), (N,N,D,N), (N,N,N,D), (D,D,N,N), (D,N,D,N), (D,N,N,D), (N,D,D,N), (N,D,N,D), (N,N,D,D), (D,D,D,N), (D,D,N,D), (D,N,D,D)\ (N,D,D,D), (D,D,D,D)\}$$

There are 16 outcomes in this sample space. The number of outcomes x in S in which exactly one of the four parts ($n = 4$) is defective is

$$\frac{n!}{x!(n-x)!} = \frac{4!}{1!3!} = 4$$

Examining the sample space, we see that these four outcomes are (D,N,N,N), (N,D,N,N), (N,N,D,N), (N,N,N,D). Thus the term $[n!/x!(n-x)!]$ in the binomial-distribution model gives the number of outcomes in S that produce the value x. The term $p^x q^{n-x}$ yields the probability of each of these $[n!/x!(n-x)!]$ outcomes. Thus, in computing the probability $f(x)$ of the value x, we are essentially summing $[n!/x!(n-x)!]$ separate probabilities. The binomial-probability-distribution model for our example is thus

$$f(x) = \frac{4!}{x!(4-x)!}(0.1)^x\,(0.9)^{4-x}, \qquad x = 0,1,2,3,4$$

where x is the number of defective parts. Evaluating this model at each of the five possible values of x gives the probabilities and cumulative probabilities in Table 8–3. Figure 8–5 shows the histogram for this binomial process.

The mean of the binomial distribution is

$$\mu = np \tag{8–31}$$

TABLE 8–3. BINOMIAL PROBABILITY DISTRIBUTION OF INSPECTION OF MANUFACTURED PARTS, $n = 4$, $p = 0.1$

Possible value x_j	*Probability* $f(x_j)$	*Cumulative probability* $F(x_j)$
0	0.6561	0.6561
1	0.2916	0.9477
2	0.0486	0.9963
3	0.0036	0.9999
4	0.0001	1.0000

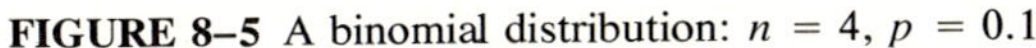

FIGURE 8–5 A binomial distribution: $n = 4$, $p = 0.1$

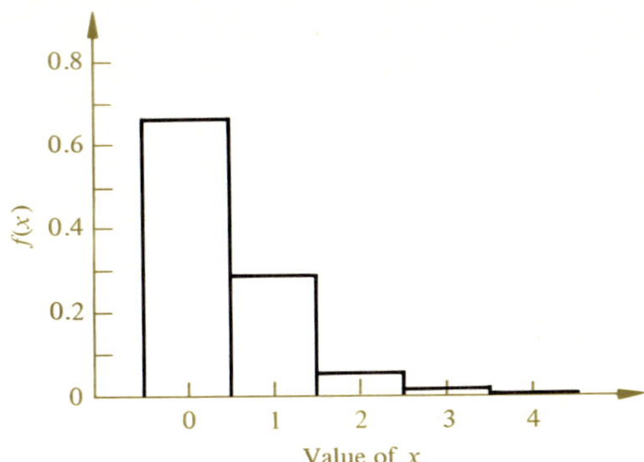

and the variance is

$$\sigma_x^2 = npq \qquad (8\text{–}32)$$

For our example, the mean and variance are

$$\mu = (4)(0.1) = 0.4, \qquad \sigma_x^2 = (4)(0.1)(0.9) = 0.36$$

From the binomial-distribution model given by Equation (8–30), as well as from its mean (Equation 8–31) and variance (Equation 8–32), we can see that this distribution is completely specified if we know the number of independent trials n and the probability of success p in each trial. These quantities are thus known as the *parameters* of the binomial distribution. Each probability distribution can be characterized by one or more parameters.

POISSON DISTRIBUTION

An important probability distribution in engineering is the Poisson distribution. The model for this distribution is

$$f(x) = \frac{e^{-\lambda}\lambda^x}{x!}, \qquad x = 0,1,2, \ldots \qquad (8\text{–}33)$$

This distribution models random processes such as the rate at which vehicles arrive at a highway toll booth, the number of defects per manufactured assembly, the number of telephone calls per minute at a telephone exchange, and the number of defects per 1000 feet of electrical cable. That is, we typically use the Poisson probability distribution to describe the rate at which an event occurs in a given time, length, area, volume, assembly, and so forth. We use it when the opportunity for occurrence of the event is large, but the probability of its occurrence in a given interval is relatively small. The mean of the Poisson distribution is

$$\mu = \lambda \qquad (8\text{–}34)$$

TABLE 8–4. PROBABILITY DISTRIBUTION OF RADIOISOTOPE DISINTEGRATION

Possible value x_j	*Probability* $f(x_j)$	$F(x_j)$
0	0.061	0.061
1	0.170	0.231
2	0.238	0.469
3	0.222	0.691
4	0.156	0.847
5	0.087	0.934
6	0.041	0.975
7	0.016	0.991
8	0.006	0.997
9	0.002	0.999

and its variance is

$$\sigma_x^2 = \lambda \tag{8–35}$$

As an example of a Poisson process, consider measuring radioactive disintegrations of a radioisotope with a Geiger counter. The Geiger counter measures the number of disintegrations per minute. Suppose the mean rate of disintegration is 2.8 counts per minute. Thus the Poisson model for this process is

$$f(x) = \frac{e^{-2.8}(2.8)^x}{x!}, \qquad x = 0,1,2, \ldots$$

The mean and variance are both 2.8. Table 8–4 gives the probability distribution and the cumulative probabilities. Figure 8–6 shows the histogram for this Poisson

FIGURE 8–6 A Poisson distribution: $\lambda = 2.8$

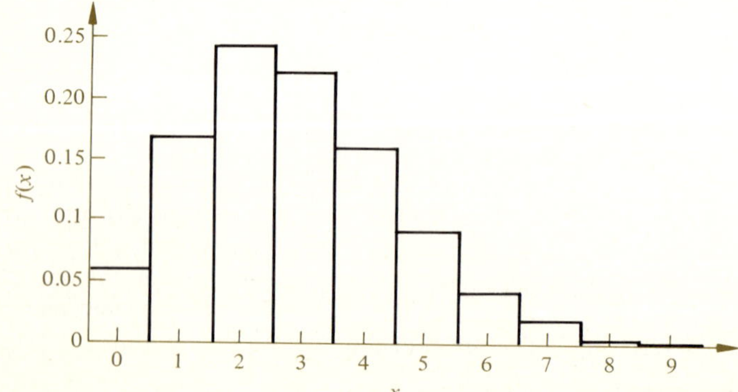

process. Note that values of x greater than 9 counts per minute have an extremely low probability of occurrence.

Exercises

8–13 The probability is 0.03 that a part manufactured by a certain machine is defective. We have five items produced by this machine.
(a) What is the probability that none is defective?
(b) What is the probability that at most one is defective?
(c) What is the average number of defective items in a lot of these items?

8–14 Defects occur randomly in electrical cable at a mean rate of one defect per 2000 feet of cable. Using the Poisson model, compute the probability that no more than one defect occurs per 800 feet of cable.

8–3 ANALYSIS OF STATISTICAL DATA

Almost all engineers must plan, conduct, and evaluate experiments to obtain information about the systems with which they are involved. An industrial engineer might be concerned with measuring an attribute of some raw material or manufactured product. A chemical engineer might be interested in measuring the yield from a chemical process. A mechanical engineer might need to measure the wear on the tread of an experimental tire. In each case, the engineer observes and records one or more measurements of some quantity. The measured quantity often varies from one observation to the next, because of numerous factors whose influence cannot be predicted. Hence this variation is random.

In most cases, measuring every item in a great number of items is prohibitively expensive or completely impractical. It may even involve the destruction of the item, as in a strength test. Hence the engineer measures just a few of the items. The set of items measured is called a *sample*. All the items from which the sample is taken are called a *population*. The objective of measuring items in the sample is to infer information about the population on the basis of information obtained from the sample.

For example, suppose that we draw 100 screws from a production lot of 10,000 screws and that we find that 5 of the 100 screws are defective. Then we can conclude that about 5% of the production lot are defective. We are willing to draw this conclusion because we believe that the sample is random. We would naturally be suspicious if the set of 100 screws happened to be the first 100 produced on the screw machine, or the last 100, for that matter. We would also be suspicious if the sample consisted of precisely every hundredth screw produced on the machine. Our willingness to base a conclusion about the population on the results of the sample also depends on the size of the sample. We are more confident that 5% are defective than we would have been had we found 1 defective screw in a set of 20 screws, but we would be even more confident of our conclusion if we had observed 25 defective screws in a sample of 500 screws.

The science that enables us to extract information about populations by analyzing samples is called *statistics*. The study of statistics is closely related to the study of probability. Engineers usually approach the study of probability and statistics together. Typically they take two quarters or semesters of probability and statistics in their junior or senior years. Since these studies depend on integral calculus, our treatment is necessarily brief and greatly simplified. Nevertheless the basic concepts established here should help you collect, organize, analyze, and interpret the experimental data in science and engineering courses that you encounter early in your academic training.

Problems in different engineering settings often require different methods and techniques, but the basic steps in approaching these problems are the same. The important steps in an experimental investigation and the activities associated with each of these steps are as follows.

Step 1. *Formulate the problem.* Delimit the investigation, so that you can obtain useful results with the time and funds available. Identify the key variables affecting the problem, and hypothesize a mathematical model relating these variables.

Step 2. *Design the experiment.* Select the statistical method to be employed in the last step of the investigation, determine the sample size n (the number of items to be drawn from the population and measured), decide what physical procedures you will use in the experiment, and determine the order in which you will take the observations if randomization is required.

Step 3. *Experiment or collect data.* Perform the physical process the prescribed number of times (n) strictly according to the rules established in the second step. Record the data. For a given variable X, these data consist of values $x_1 \dots, x_n$, called *observations*.

Step 4. *Tabulate and analyze the data.* Arrange the observations in a clear and simple tabular form, and represent the data in histograms and graphs. Compute the descriptive measures, called *sample statistics,* that is, functions $g(x_1, \dots, x_n)$ of the n observations. (We shall expand on the concept of a statistic later in this section.)

Step 5. Draw conclusions from the data. Attempt to reach conclusions about the nature of the population based on the properties discovered about the sample of n observations. This is called *statistical inference,* the procedure for obtaining the solution to the initial problem.

TABULAR AND GRAPHICAL REPRESENTATION OF SAMPLES

Consider an investigation intended to examine the tensile strength of structural concrete. The experiment consists of preparing test concrete cylinders, 0.2 meter in diameter and 0.3 meter in length, and performing tests of tensile strength 30 days later. Table 8–5 gives the results of a sample of 100 observations of tensile strength (in kilonewtons/m²) from such an investigation.

To develop a frequency histogram of the values shown in Table 8–5, we have to examine each observation and place a tally next to the appropriate value in the frequency histogram, as Table 8–6 shows. After we have completed the frequency histogram, we divide the absolute frequency by the sample size, in this case 100, to obtain the *relative frequency*. Figure 8–7 shows the plot of the relative frequency. These follow exactly the same principle as the probability function and cumulative probability function seen earlier.

SAMPLE STATISTICS

After we have tabulated and graphically displayed the set of n observations, we compute various descriptive measures called *sample statistics*. These descriptive measures are classified according to the type of information they reveal about the sample, and hence about the population. Measures that reveal the sample values' tendency to occur toward the middle value of the group are called measures of *central tendency*. Those that show the tendency of the values to spread out over a range are called measures of *dispersion*.

TABLE 8–5. SAMPLE OF 100 OBSERVATIONS OF TENSILE STRENGTH (kilonewtons/m²) OF CONCRETE CYLINDERS

60	63	56	68	61
56	57	58	60	63
62	59	65	61	61
59	60	60	54	59
61	58	57	58	64
57	61	56	60	63
62	59	64	63	58
59	58	61	57	61
64	62	59	60	61
59	54	58	59	60
58	63	62	58	58
66	59	61	64	62
63	60	64	61	64
57	63	57	66	59
58	61	63	56	60
61	59	60	62	61
60	66	59	61	63
61	61	60	65	62
62	60	60	59	63
59	60	64	60	58

TABLE 8–6. FREQUENCY HISTOGRAM OF A SAMPLE OF 100 VALUES OF TENSILE STRENGTH (kilonewtons/m²) OF CONCRETE CYLINDERS

x	*Absolute frequency*		*Relative frequency*	*Cumulative frequency*	*Cumulative relative frequency*
54	11	2	0.02	2	0.02
55		0	0.00	2	0.02
56	1111	4	0.04	6	0.06
57	HHT 1	6	0.06	12	0.12
58	HHT HHT 1	11	0.11	23	0.23
59	HHT HHT 1111	14	0.14	37	0.37
60	HHT HHT HHT 1	16	0.16	53	0.53
61	HHT HHT HHT	15	0.15	68	0.68
62	HHT 111	8	0.08	76	0.76
63	HHT HHT	10	0.10	86	0.86
64	HHT 111	8	0.08	94	0.94
65	11	2	0.02	96	0.96
66	111	3	0.03	99	0.99
67		0	0.00	99	0.99
68	1	1	0.01	100	1.00

The most important measures of central tendency are the arithmetic mean, the median, and the mode. The *arithmetic mean* $\bar{x}$ of n observations $x_1, x_1, \ldots, x_n$ is

$$\bar{x} = \frac{1}{n} \sum_{i=1}^{n} x_i \tag{8–36}$$

For *grouped data,* like those in Table 8–6, a more convenient form of Equation (8–36) is

$$\bar{x} = \frac{1}{n} \sum_{j=1}^{k} f_j x_j \tag{8–36a}$$

where f_j is the absolute frequency of the value x_j and k is the number of groups or values of x. For the data in Table 8–6, the arithmetic mean is 60.47 kN/m², as you can verify. The *median* is the middle value in a set of observations ordered from highest to lowest. If there is an even number of observations, the median is the average of the two middle values. Table 8–6 shows that the median value is 60 kN/m². The *mode* is the value that occurs most frequently in the set of observations. You can determine it from a histogram or bar chart like the one in Figure 8–7. For example, the mode of the distribution of tensile strength values is 60 kN/m².

The most important measures of dispersion are the variance, the standard deviation, and the range. The *variance* of a sample of n observations is

FIGURE 8–7 Relative frequency and cumulative relative frequency of a sample of 100 observations: (a) relative frequency, (b) cumulative relative frequency

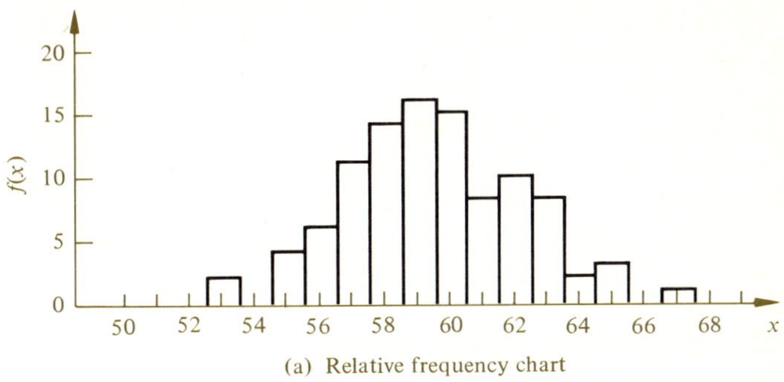

(a) Relative frequency chart

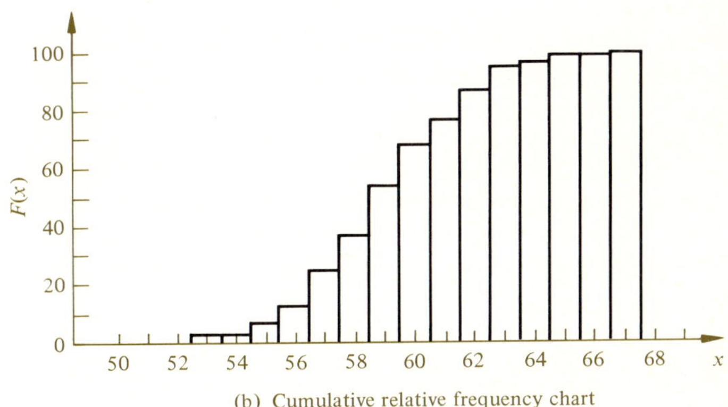

(b) Cumulative relative frequency chart

$$s_x^2 = \frac{1}{n} \sum_{i=1}^{n} (x_i - \bar{x})^2 \tag{8–37}$$

For grouped data, the variance is

$$s_x^2 = \frac{1}{n} \sum_{j=1}^{k} f_j (x_j - \bar{x})^2 \tag{8–37a}$$

By expanding the right side of Equation (8–37), we get

$$s_x^2 = \frac{1}{n} \sum_{i=1}^{n} x_i^2 - \frac{2}{n} \left(\sum_{i=1}^{n} x_i \right) \bar{x} + \bar{x}^2$$

But in the middle term, we see that

$$\frac{1}{n}\sum_{i=1}^{n} x_i = \bar{x}$$

So, combining the middle and last terms, we can obtain a form of Equation (8–37) that is easier to compute.

$$s_x^2 = \frac{1}{n}\sum_{i=1}^{n} x_i^2 - \bar{x}^2 \tag{8–38}$$

For grouped data, we have the equation

$$s_x^2 = \frac{1}{n}\sum_{j=1}^{k} f_j x_j^2 - \bar{x}^2 \tag{8–38a}$$

For the data in Table 8–6, we get a variance $s_x^2 = 7.13$ (kN²/m⁴). The *standard deviation* is simply the square root of the variance,

$$s_x = \sqrt{s_x^2} \tag{8–39}$$

The standard deviation of the data in Table 8–6 is $s_x = 2.67$ kN/m². The *range* of a set of n observations is the difference between the highest and lowest values in the sample,

$$r = x_{\max} - x_{\min} \tag{8–40}$$

For the tensile-strength data in Table 8–6, the range is $r = 68 - 54 = 14$ kN/m². For large samples ($n \geqslant 100$), the range r is about seven or eight times as large as the standard deviation s_x.

Exercises

8–15 The following data give the weight loss (in grams) of 20 balls used to grind granulated material into fine powder in a mill after 10 hours of use.

0.094	0.108	0.114	0.132	0.128
0.117	0.099	0.093	0.105	0.119
0.126	0.122	0.125	0.115	0.097
0.113	0.109	0.111	0.130	0.120

Find (a) the mean of the sample, (b) the variance of the sample, (c) the median, and (d) the range.

8–16 The following 16 readings (in degrees Celsius) were taken at various locations in a kiln.

225, 245, 310, 260, 275, 250, 270, 280,
275, 270, 265, 280, 290, 240, 255, 270

Find (a) the mean of the sample, (b) the variance of the sample, (c) the median, and (d) the range.

8–4 QUEUING MODELS

One of the familiar settings in which engineers encounter probabilistic models is the so-called queue or waiting line. In this situation, items arrive at a station and undergo a service operation. Examples of such arrival processes are the arrival of vehicles at a highway toll station, the departure of aircraft on a runway, and the arrival of completed assemblies at an inspection station in a production plant. The associated service operations for these settings are collecting the toll at the toll station, aircraft takeoff, and inspecting the manufactured assembly, respectively.

Both the arrival process and the service process are usually random or probabilistic in nature. If one item is in service when another item arrives, the new item must wait in line until the service operation is completed. Meanwhile, other items might have joined the queue.

Some of the important variables associated with a queue are the number of items waiting and time spent waiting. Two variables important in the analysis of such systems are the *arrival rate* (in items per period) and *service time* (in periods).

Figure 8–8 depicts two of the most common classes of queuing models. Figure 8–8(a) shows a *single-channel queue;* it has only one service channel and hence only one waiting line. Figure 8–8(b) illustrates a queuing model with two service channels, but only one waiting line. The first item in the waiting line takes the first available service channel.

The most common queuing models are single-channel queues. They have Poisson arrival and service rates. Figure 8–8(a) shows a single-channel queue. The model for the Poisson process is Equation (8–33). The mean number of arrivals per period is λ, and the mean service rate is μ items per period. Thus the mean service time is $1/\mu$ periods. The population of items demanding service is assumed to be infinite.

The probability of n units in the system at a given time is

FIGURE 8–8 Queuing systems: (a) single-channel queue, (b) two-channel queue

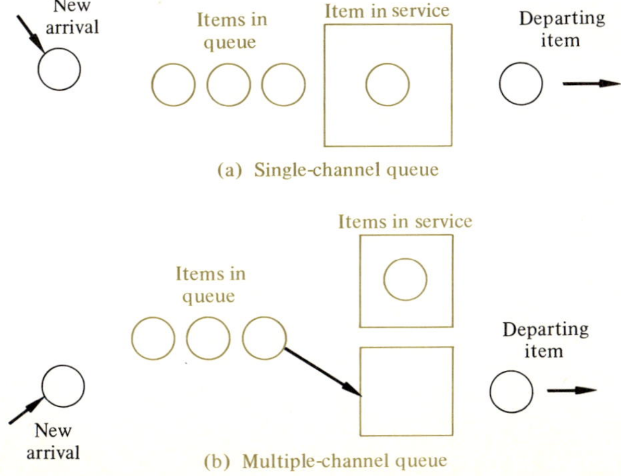

$$P_n = \left(1 - \frac{\lambda}{\mu}\right)\left(\frac{\lambda}{\mu}\right)^n \tag{8–41}$$

The mean number of units in the system is

$$\bar{n} = \frac{\lambda}{\mu - \lambda} \tag{8–42}$$

Thus the mean queue length is $\bar{n}$ minus the mean number of units in service, or

$$\begin{aligned} \bar{m} &= \bar{n} - \frac{\lambda}{\mu} = \frac{\lambda}{\mu - \lambda} - \frac{\lambda}{\mu} \\ \bar{m} &= \frac{\lambda^2}{\mu(\mu - \lambda)} \end{aligned} \tag{8–43}$$

The mean waiting time is

$$\bar{w} = \frac{\lambda}{\mu(\mu - \lambda)} \tag{8–44}$$

while the mean time spent in the system is the mean waiting time $\bar{w}$ plus the mean service time, or

$$\bar{t} = \bar{w} + \frac{1}{\mu} = \frac{\lambda}{\mu(\mu - \lambda)} + \frac{1}{\mu}$$

Thus

$$\bar{t} = \frac{1}{\mu - \lambda} \tag{8–45}$$

To see how to apply these models, consider the departure of aircraft at a busy airport with a single runway. The mean departure rate λ is 0.5 unit per minute. The mean time on the runway is 0.8 minute, so the mean service rate μ is 1/0.8 or 1.25 units per minute. Now we can calculate the probability distribution of n aircraft in the system, that is, on the taxi strip and runway.

$$\begin{aligned} P_0 &= \left(1 - \frac{\lambda}{\mu}\right)\left(\frac{\lambda}{\mu}\right)^0 = (0.6)(0.4)^0 = 0.600 \\ P_1 &= (0.6)(0.4)^1 = 0.240 \\ P_2 &= (0.6)(0.4)^2 = 0.0960 \\ P_3 &= (0.6)(0.4)^3 = 0.0384 \\ P_4 &= (0.6)(0.4)^4 = 0.0154 \\ P_5 &= (0.6)(0.4)^5 = 0.0061 \\ P_6 &= (0.6)(0.4)^6 = 0.0025 \\ P_7 &= (0.6)(0.4)^7 = 0.0010 \end{aligned}$$

Figure 8–9 exhibits the probability of n units in the system. We can extract certain important characteristics of the system from this distribution. For example, the

FIGURE 8-9 Probability of n units in the system

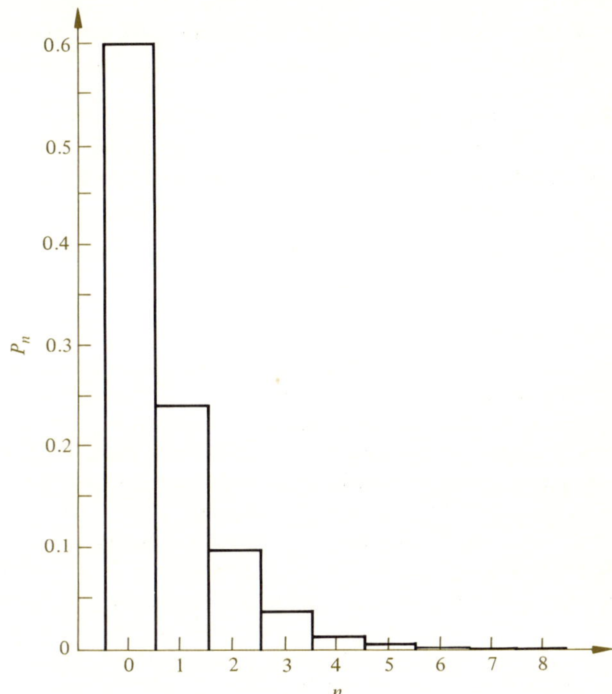

probability of one or more units in the system is 0.4, the probability of an idle system is 0.6, and so forth. The mean number of units in the system is

$$\bar{n} = \frac{\lambda}{\mu - \lambda} = \frac{0.5}{1.25 - 0.5} = 0.667 \text{ aircraft}$$

The mean queue length, that is, the mean number of aircraft waiting for takeoff, is

$$\bar{m} = \frac{\lambda^2}{\mu(\mu - \lambda)} = \frac{(0.5)^2}{(1.25)(0.75)} = 0.267 \text{ aircraft}$$

The mean waiting time is

$$\bar{w} = \frac{\lambda}{\mu(\mu - \lambda)} = \frac{0.5}{(1.25)(0.75)} = 0.533 \text{ minute}$$

while the mean time in the system is

$$\bar{t} = \frac{\lambda}{\mu(\mu - \lambda)} + \frac{1}{\mu} = 0.533 + 0.8 = 1.333 \text{ minutes}$$

Exercise

8–17 The arrival rate for a certain queuing system obeys a Poisson distribution with a mean of 0.6 unit per hour. The service rate is also Poisson distributed, and the mean service rate is 0.9 unit per hour.
(a) Develop the distribution of the number of units in the system.
(b) Determine the mean number of units in the system, mean queue length, mean waiting time, and the mean time in the system.

8–5 RELIABILITY MODELS

The severity of environmental conditions under which a product must perform increases the difficulty of designing and supervising the manufacture of that product. For instance, the problems of designing a tire that will perform well for at least 20,000 miles on a standard passenger automobile are quite different from the problems of designing a tire that will withstand the rigorous conditions of the Indianapolis 500. Such problems involve the element of uncertainty.

The engineering term describing this situation is reliability. The *reliability* of a product is the *probability that the product will perform within design specifications for a specified period of time under specified environmental conditions*. Thus the *reliability* of a standard automobile tire is nearly 1 for 20,000 miles of normal road use, but almost 0 for 500 miles at high speeds on a racetrack.

Since the reliability R of a product is a probability, it obeys the laws of probability stated in relations (8–6) and (8–7). That is,

$$0 \leqslant R \leqslant 1 \tag{8–46}$$

and

$$R + (1 - R) = 1 \tag{8–47}$$

Equation (8–47) states that the sum of the probability of reliable performance R and the probability of failure $(1 - R)$ is unity. These and other probability relationships prove useful in analyzing the reliability of systems made up of several components, each of which has a reliability R_i. Such systems can be considered series, parallel systems, or combinations of these. In a *series* system, the entire system fails if a single component fails. A *parallel* system fails only if all its components fail.

Let us first examine a series system with n components, as shown in Figure 8–10.

FIGURE 8–10 A series reliability system

Suppose that the components are independent, so that the reliability of one component does not affect the reliability of another. Under these conditions, the probability that the system will function is given by

$$R_s = \prod_{i=1}^{n} R_i \tag{8–48}$$

where R_i is the reliability of the ith component and R_s is the reliability of the series system. The symbol $\prod_{i=1}^{n}$ is read "the product of n terms."

For example, consider the reliability of a series system made up of five components, each with reliability 0.97. The reliability of the system is

$$R_s = (0.97)^5 = 0.859$$

One way to increase the reliability of a system is to replace certain "high-risk" components by several similar components connected in parallel. If a system consists of n independent components connected in parallel, it will fail to perform only if all n components fail. Thus if the probability of failure of the ith component is

$$F_i = 1 - R_i \tag{8–49}$$

then the probability that the system of n components in parallel will fail is

$$F_p = \prod_{i=1}^{n} F_i \tag{8–50}$$

Thus the reliability of the parallel system is

$$R_p = 1 - F_p = 1 - \prod_{i=1}^{n} (1 - R_i) \tag{8–51}$$

Figure 8–11 shows a parallel system. Consider such a parallel system with three components, each with reliability 0.7. The reliability of the system is then

$$R_p = 1 - (1 - 0.7)^3 = 1 - 0.027 = 0.973$$

FIGURE 8–11 A parallel reliability system

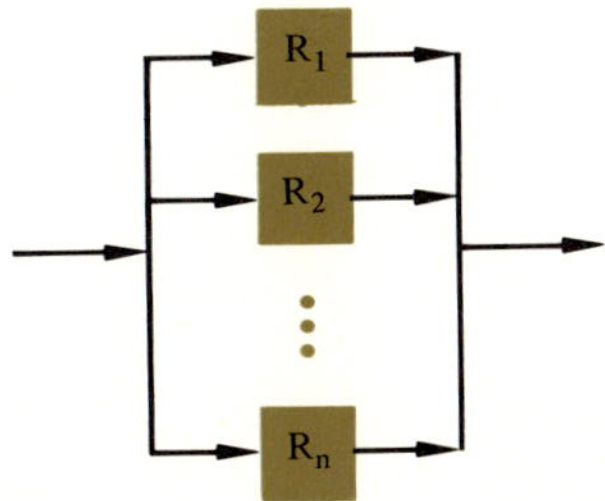

FIGURE 8–12 A mixed series–parallel reliability system

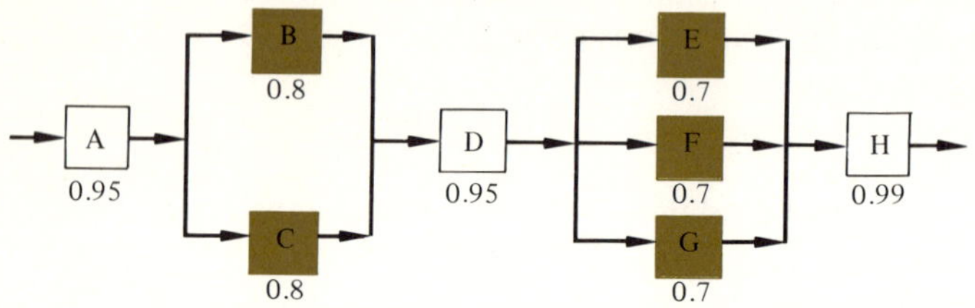

We can combine these two basic formulas for the system's reliability to calculate the reliability of a system consisting of both series and parallel parts.

Consider the problem depicted in Figure 8–12. We can resolve this problem by first finding the equivalent reliability of each of the parallel sectors of the system and then solving a series system of five components. The parallel system consisting of components B and C has the equivalent reliability

$$R_{\mathrm{BC}} = 1 - (1 - 0.8)^2 = 0.96$$

while the EFG sector has reliability

$$R_{\mathrm{EFG}} = 1 - (1 - 0.7)^3 = 0.973$$

The remaining series system has reliability

$$R_s = (0.95)(0.96)(0.95)(0.973)(0.99) = 0.8346$$

Exercises

8–18 Find the reliability of a series system consisting of six components with reliabilities 0.98, 0.995, 0.95, 0.97, 0.96, and 0.99.

8–19 Find the reliability of a parallel system with four components, each with reliability 0.5.

8–20 Find the reliability of the system in the following figure.

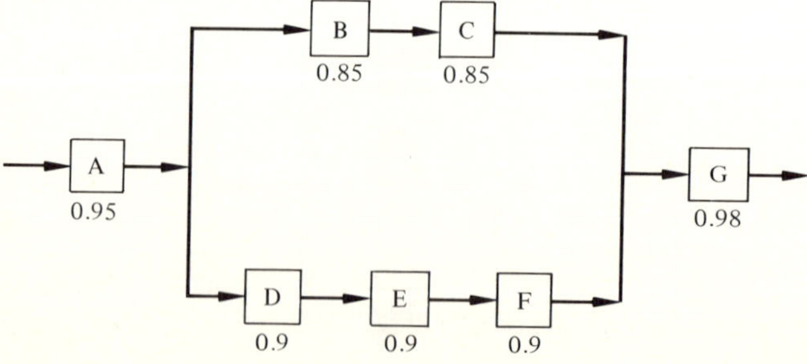

8–6 MONTE CARLO ANALYSIS

Probabilistic models of operational systems are often so complex that manipulating them mathematically is extremely difficult, if not impossible. Under these conditions, it may be necessary to use a simulation technique known as *Monte Carlo analysis* to obtain a solution from the model. Moreover, Monte Carlo simulation can often be used instead of expensive physical experiments to determine optimum conditions for operating a real system.

The step-by-step procedure for performing a Monte Carlo simulation consists of (1) formalizing the system logic, (2) determining the probability functions for the random processes controlling the system, and (3) generating outcomes for these random processes. This section describes this technique in detail.

In formalizing the logic of the system, we usually assume that the system under study operates according to a logical pattern. We have to formalize this pattern by constructing a model. If we use a computer program to represent this logic, we have to prepare an accurate flow chart that patterns the process under study before we code the logic in FORTRAN or some other programming language.

To illustrate this technique, let us consider the problem of an inspector randomly checking roller-bearing assemblies after they have completed the manufacturing process. Smooth-Roll Bearing Company has a quality-control policy of randomly inspecting 40% of the roller-bearing assemblies produced on a certain production line. Of those assemblies selected for inspection, 25% receive a class A inspection, which takes 0.5 hour per assembly. The remaining 75% receive a class B inspection, requiring 0.1 hour per assembly. The rejection rate from the class A inspection is 20%, while that from the class B inspection is 10%. Figure 8–13 shows a flow chart describing the basic logic of the system.

The second step of a Monte Carlo simulation is to develop probability models for the random procedures embedded in the logical model. The first random process in the logic of inspecting roller bearings is the type of inspection performed, if any. If we denote this random experiment as X, the sample space is

$$X = \{x_1, x_2, x_3\} = \{\text{None, A, B}\}$$

The probabilities for the outcomes in the sample space for this random experiment are

$$P_x = \{p(x_1), p(x_2), p(x_3)\} = \{0.6, 0.1, 0.3\}$$

Now let Y denote the random process corresponding to the outcome of the class A inspection. Then the sample space is

$$Y = \{y_1, y_2\} = \{\text{Reject, Accept}\}$$

For this sample space the probabilities are

$$P_y = \{p(y_1), p(y_2)\}, = \{0.2, 0.8\}$$

Likewise, the sample space for the outcome Z of the class B inspection is

$$Z = \{z_1, z_2\} = \{\text{Reject, Accept}\}$$

FIGURE 8–13 Flow chart of the Monte Carlo simulation of a random inspection procedure for roller-bearing assemblies

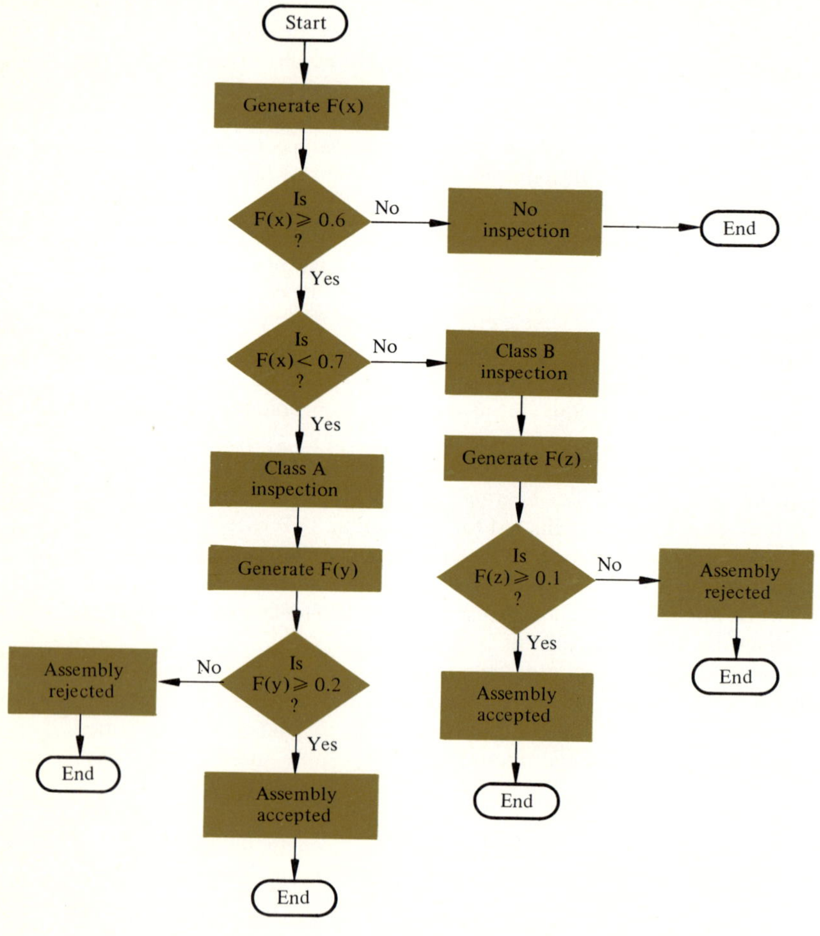

and the probabilities are

$$P_z = \{p(z_1), p(z_2)\} = \{0.1, 0.9\}$$

The third step in a Monte Carlo simulation is to generate outcomes for the random processes modeled in Step 2. To accomplish this, we first construct a cumulative probability distribution for each random process. We generate uniform random numbers in the interval [0,1] and use these random numbers with the associated cumulative probability distribution to determine the appropriate values for the given random process. For instance, in our example, consider the random process of rejecting or accepting a bearing assembly in a class B inspection. The probabilities are

$$P_z = \{p(z_1), p(z_2)\} = (0.1, 0.9)$$

The cumulative probability distribution is

$$F_z = \{F(z_1), F(z_2)\} = \{0.1, 1.0\}$$

Therefore, for those random numbers corresponding to a cumulative probability $0 \leq F(z_1) < 0.1$, the outcome is "reject the roller-bearing assembly." Similarly, if the random number corresponds to the cumulative probability $0.1 \leq F(z_2) \leq 1.0$, the outcome is "accept the roller-bearing assembly." We repeat this procedure N times, where N is the sample size. Any Monte Carlo simulation consists of a predetermined sample size N.

In Monte Carlo simulation, each of the uniform random numbers in the interval [0,1] has an equal chance of occurring. One method for obtaining such random numbers is to use a table such as those given in Appendix A. Table A–1 presents uniformly distributed two-digit random numbers; Table A–2 gives four-digit random numbers. To use these tables, one randomly selects a column and row in the table and proceeds down the column for N numbers, moving to another column if necessary. Another technique is to employ a computer program to generate random numbers. Almost any computer has a random number generator.

For our problem, we shall use the 100 two-digit random numbers in the interval [00,99] for each of the three random processes. For the random process of choosing the type of inspection procedure, the random numbers corresponding to the three outcomes are as follows.

Random numbers	$F(x_i)$	*Type of inspection*
00–59	$0 \leq F(x_1) < 0.6$	None
60–69	$0.6 \leq F(x_2) < 0.7$	A
70–99	$0.7 \leq F(x_3) \leq 1.0$	B

For the outcome of the class A inspection, the associated random numbers are as follows.

Random numbers	$F(y_i)$	*Outcome of inspection*
00–19	$0 \leq F(y_1) < 0.2$	Reject
20–99	$0.2 \leq F(y_2) \leq 1.0$	Accept

Likewise, the random numbers corresponding to the outcomes of the class B inspection are as follows.

Random numbers	$F(z_i)$	*Outcome of inspection*
00–09	$0 \leq F(z_1) < 0.1$	Reject
10–99	$0.1 \leq F(z_2) \leq 1.0$	Accept

Thus, when divided by 100, these random numbers produce the cumulative probabilities corresponding to the outcomes of our random experiment.

Table 8–7 presents the results of the simulated inspection of 20 roller-bearing assemblies. These results compare favorably with the expected behavior of the inspection process for roller bearing assemblies. That is, 12 assemblies received no inspection at all; this is exactly what we expect, since (0.6)(20) = 12. One assembly received a class A inspection, compared with the expected (0.1)(20) = 2.0. Of the 7 assemblies inspected by the class B procedure, 6 were accepted, compared to an expected (0.90)(7) = 6.3.

One should not expect a Monte Carlo simulation with so few trials to always behave as well as this simulation. In general, a Monte Carlo simulation more closely approaches the system's expected behavior as the sample size N increases. We can illustrate this by using a larger sample size to simulate the inspection of the roller-bearing assembly on a computer. Figure 8–14 lists a FORTRAN program representing the simulation model of the inspection process. Figure 8–15 shows the simulated inspection of 50 roller-bearing assemblies.

The random numbers employed in this Monte Carlo model are generated by

TABLE 8–7. MONTE CARLO SIMULATION OF A ROLLER-BEARING ASSEMBLY INSPECTION

Assembly	*Random number*	*Type of inspection*	*Random number*	*Outcome of inspection*
1	84	B	81	Accept
2	43	None	—	—
3	46	None	—	—
4	73	B	56	Accept
5	12	None	—	—
6	87	B	95	Accept
7	91	B	41	Accept
8	82	B	31	Accept
9	99	B	02	Reject
10	37	None	—	—
11	02	None	—	—
12	15	None	—	—
13	68	A	73	Accept
14	24	None	—	—
15	27	None	—	—
16	57	None	—	—
17	10	None	—	—
18	12	None	—	—
19	73	B	72	Accept
20	59	None	—	—

FIGURE 8–14 FORTRAN program for a Monte Carlo simulation of a random inspection procedure for roller-bearing assemblies

```
C*****  PROGRAM FOR MONTE CARLO SIMULATION OF A RANDOM
C*****  INSPECTION PROCEDURE FOR ROLLER BEARING ASSEMBLIES
C
        WRITE (6, 1)
      1 FORMAT (1H1)
        IPART = 0
        IX = 2759
      5 CALL RANDU (IX,IY,PX)
        IX = IY
        IPART = IPART + 1
        IF (IPART.EQ.51) GO TO 100
        IF (PX.LE.0.4) GO TO 15
        WRITE (6, 10) IPART
     10 FORMAT (5X,I5,5X, 'NO INSPECTION')
        GO TO 5
     15 IF (PX.LE.0.1) GO TO 20
        CALL RANDU (IX,IY,PZ)
        IX = IY
        IF (PZ.LE.0.90) GO TO 16
        WRITE (6, 18) IPART
     18 FORMAT (5X,I5,5X, 'CLASS B INSPECTION—ASSEMBLY FAILS')
        GO TO 5
     16 WRITE (6, 19) IPART
     19 FORMAT (5X,I5,5X, 'CLASS B INSPECTION—ASSEMBLY PASSES')
        GO TO 5
     20 CALL RANDU (IX,IY,PY)
        IX = IY
        IF (PY.LE.0.8) GO TO 21
        WRITE (6, 22) IPART
     22 FORMAT (5X,I5,5X, 'CLASS A INSPECTION—ASSEMBLY FAILS')
        GO TO 5
     21 WRITE (6, 23) IPART
     23 FORMAT (5X,I5,5X, 'CLASS A INSPECTION—ASSEMBLY PASSES')
        GO TO 5
    100 STØP
        END
```

FIGURE 8–15 Results from a Monte Carlo simulation of a random inspection of 50 roller bearing assemblies

ASSEMBLY	*RESULT*
1	NO INSPECTION
2	NO INSPECTION
3	NO INSPECTION
4	NO INSPECTION
5	CLASS B INSPECTION—ASSEMBLY PASSES
6	CLASS B INSPECTION—ASSEMBLY PASSES
7	NO INSPECTION
8	CLASS B INSPECTION—ASSEMBLY PASSES
9	NO INSPECTION
10	NO INSPECTION
11	NO INSPECTION
12	NO INSPECTION
13	NO INSPECTION
14	NO INSPECTION
15	NO INSPECTION
16	NO INSPECTION
17	NO INSPECTION
18	NO INSPECTION
19	NO INSPECTION
20	CLASS B INSPECTION—ASSEMBLY PASSES
21	NO INSPECTION
22	CLASS B INSPECTION—ASSEMBLY PASSES
23	NO INSPECTION
24	NO INSPECTION
25	CLASS B INSPECTION—ASSEMBLY PASSES
26	NO INSPECTION
27	NO INSPECTION
28	NO INSPECTION
29	NO INSPECTION
30	NO INSPECTION
31	NO INSPECTION
32	CLASS B INSPECTION—ASSEMBLY PASSES
33	NO INSPECTION
34	NO INSPECTION
35	CLASS A INSPECTION—ASSEMBLY PASSES
36	NO INSPECTION
37	NO INSPECTION
38	NO INSPECTION
39	NO INSPECTION
40	NO INSPECTION
41	CLASS B INSPECTION—ASSEMBLY PASSES
42	NO INSPECTION
43	CLASS B INSPECTION—ASSEMBLY PASSES
44	NO INSPECTION
45	NO INSPECTION
46	NO INSPECTION
47	NO INSPECTION
48	NO INSPECTION
49	NO INSPECTION
50	CLASS B INSPECTION—ASSEMBLY PASSES

FIGURE 8–16. Random number generator RANDU

```
      SUBROUTINE RANDU (IX,IY, RNUM)
      IY = IX * 899
      IF (IY) 5, 6, 6
5     IY=IY+32767+1
6     RNUM = IY
      RNUM = RNUM/32767.
      RETURN
      END
```

calling the subroutine RANDU, a random number generator available on the IBM 1130 computer. The calling statement in the main program shown in Figure 8–15 is structured as follows.

CALL RANDU (IX, IY, RNUM)

That is, a random number seed IX is the argument passed into subroutine RANDU. The subroutine then uses this seed to produce the next seed IY and the random number RNUM, a value between 0 and 1. A random number seed is any integer less than or equal to the largest integer the given computer can store. After RANDU is called, IX must be immediately set equal to IY in order to maintain the string of random number seeds. Likewise the appropriate quantity PX, PY, or PZ must be set equal to the uniformly distributed random number RNUM. However, a more direct way to accomplish this is simply to show PX, PY, or PZ as the third argument in the call list. For example, the statement

CALL RANDU (IX, IY, PX)

randomly generates a value for $F(x_i)$. The subroutine used to generate random numbers on a given computer may vary somewhat from this one.

Exercises

8–21 Using the four-digit random numbers in Table A–2 in Appendix A, simulate 360 tosses of a pair of dice, where the random variable is the number of spots on the top faces of the dice. Develop a histogram of the outcomes. How does this distribution compare with the expected frequency distribution for this random experiment?

8–22 Develop a computer program for a Monte Carlo simulation of the reliability problem in Exercise 8–19. Have the program simulate 100 operations of the system. How does the simulated system compare with the expected behavior of the system?

8–7 SUMMARY

In this chapter, we have tried to show you that engineers must be prepared to cope with elements of uncertainty in designing, developing, and controlling systems composed of people, materials, energy and equipment. Probability is the science that systematically treats these elements of uncertainty. Every engineer should have a basic understanding of probability and its many applications.

Engineers also need to be acquainted with statistical data analysis, another topic that evolves from probability. This chapter has provided a brief glimpse of statistical data analysis; you will probably see more of this topic after you have studied second-year calculus. Many engineering schools offer courses called engineering statistics. Lack of familiarity with these basic tools can result in incomplete or erroneous interpretation of engineering data, or in excessively costly engineering experiments. You would be well advised to prepare yourself for coping with uncertainty and randomness in engineering problems.

The computer is an essential tool for solving problems that require the statistical analysis of data. Numerous FORTRAN program packages for analyzing statistical engineering data are available. Thus knowing how to prepare involved computer programs is less important than understanding how to gain access to and use available programs.

PROBLEMS

P8–1 Consider a random experiment consisting of rolling three 6-sided dice. Develop and execute a FORTRAN program that accomplishes the following.
- (a) Enumerates the points in the sample space.
- (b) Computes values for the random variable X, the total number of spots showing on the top faces of the three dice for each outcome in the sample space.
- (c) Develops the frequency distribution for the random variable X.
- (d) Develops the cumulative distribution for the random variable X.
- (e) Computes the mean and variance of the random variable X.

P8–2 Parts produced on a numerically controlled lathe are subject to three types of defects: X, Y, and Z. A sample of 10,000 parts produced over the last six months gave the following results: 2.4% had only type X defect, 2.1% had only type Y defect, 2.5% had only type Z defect; 0.2% had both X and Y defects, 0.3% had both X and Z defects, 0.2% had both Y and Z defects, 0.05% had all three types of defects.

(a) What percentage of the parts were free of defects?
(b) What percentage had at least one type of defect?
(c) What percentage were free of type Y and type Z defects?
(d) What percentage had no more than one type of defect?
(e) What percentage of the parts had type Z defects?

P8-3 An air-to-ground missile has a device that tells the aircraft whether the missile hit its intended target. The missile has probability 0.8 of hitting the target. The device returns a correct signal with probability 0.9.

(a) The pilot receives a signal that the target was hit. What is the probability that it was actually hit?
(b) The pilot receives a signal that the target was missed. What is the probability that it was actually hit?

P8-4 An aircraft is armed with six air-to-ground missiles. The probability that a missile hits its intended target is 0.8.

(a) What kind of random process describes the firing of one of the six missiles?
(b) What kind of random process describes the firing of all six missiles if the random variable X is the number of missiles that hit their intended targets?
(c) What are the mean and the variance of the random variable X in part (b)?
(d) What is the probability that none of the six missiles hits its target? that all six hit their targets?
(e) What is the probability that two or more missiles hit their targets?

P8-5 Defects in plate glass average one defect per 100 m^2 of glass.

(a) What random process best describes the occurrence of defects per 100 m^2 of glass produced? Why?
(b) Consider a production lot of 800 m^2 of glass. What is the probability of fewer than five defects in the lot? What is the probability of zero defects in the lot?
(c) Compute the mean and the variance of the number of defects in a day's production of 3200 m^2 of plate glass.

P8-6 Develop and execute a FORTRAN program that will read N observations of a variable X from cards and compute the sample mean and the standard deviation. Also have the program construct a relative-frequency chart having M intervals, where M is read from the same data card as N. Use this program to compute the mean and the standard deviation of the yield (weight fraction of the coal fed into the coke oven) from a certain coke oven over 50 days of operation. Develop a relative frequency chart with intervals of 0.005 weight fraction.

0.687	0.711	0.716	0.677	0.718
0.707	0.711	0.675	0.702	0.693
0.715	0.727	0.721	0.681	0.697
0.695	0.691	0.693	0.687	0.713
0.708	0.684	0.702	0.693	0.703
0.716	0.722	0.712	0.699	0.712
0.694	0.709	0.692	0.694	0.709
0.723	0.712	0.699	9.698	0.696
0.671	0.699	0.711	0.713	0.682
0.730	0.684	0.706	0.701	0.694

P8–7 A Poisson distribution describes the arrival of assembled CB radios at an inspection station. The mean rate of arrival is 12 units per hour. The mean time of the inspection operation is 4.2 minutes per unit.

(a) Compute the distribution of n units in the system at the inspection station for $0 \leqslant n \leqslant 10$.
(b) What is the mean number of units in the system?
(c) What is the mean number of radios awaiting inspection?
(d) What is the mean time in the system at the inspection station?
(e) What is the mean time spent waiting at the inspection station?

P8–8 A certain four-engine aircraft is capable of flying as long as at least one of the two engines mounted on each wing (left and right) is operating. The probability that an engine operates for the duration of a flight is 0.98. Find the probability that an aircraft is forced to crash-land on any given flight.

P8–9 Find the reliability of the system shown in Figure P8–1.

P8–10 Develop and execute a FORTRAN program for simulating 1000 flights of the four-engine aircraft described in Problem P8–8.

P8–11 Develop and execute a FORTRAN program for simulating the inspection of 1000 parts produced on the numerically controlled lathe described in Problem P8–2.

FIGURE P8–1

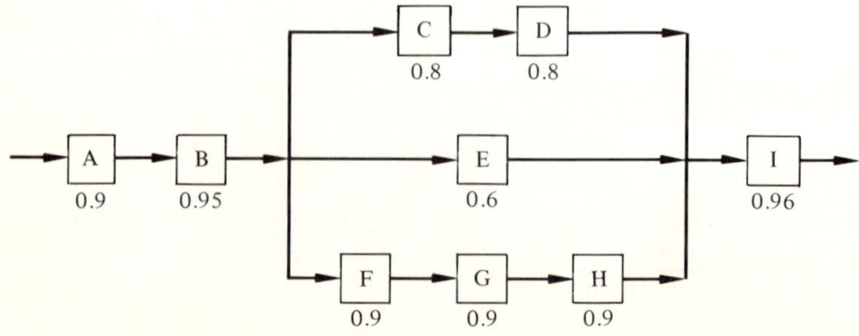

P8–12 A mission consists of eight aircraft. Each has six air-to-ground missiles like those described in Problem P8–3. Using the table of two-digit random numbers in Appendix A, simulate an air attack on an armored unit with 64 tanks.

P8–13 Develop and execute a FORTRAN program for Problem P8–12. Simulate 50 such missions. Compute the mean and the standard deviation of the number of tanks destroyed per mission.

CHAPTER NINE
ECONOMIC MODELS IN ENGINEERING

The engineer seeks to transform the materials and forces found in the natural environment into products and services that are even more useful to humanity. To achieve this objective, the engineer designs, constructs, and controls systems consisting of people, materials, energy, and equipment. To serve the needs of humanity, such systems must function economically. That is, these systems must provide goods and services that are within the economic reach of a sufficiently large segment of society, or they must satisfy some special need of the community, the nation, or the world. Engineers must heed the economic aspects of the systems with which they are involved. This chapter introduces the basic concepts of economic models as they apply to engineering.

9–1 COST CONCEPTS

Engineers are constantly seeking ways to produce new and improved goods and services, or to produce goods and services more efficiently and economically. Stated simply, a new product or service must either perform better than an existing one at the same price or provide the same benefit at a lower price. This competitive economic environment is one of the founding principles of our capitalist society. The objective of an enterprise in our society is to produce profits. Profits induce investment, which in turn expands production and creates jobs. Many practicing engineers in capitalist countries are employed either in private industry or in private engineering firms that market specialized professional services to industry and government.

A growing number of engineers are employed in government, at federal, state, county, and municipal levels. These engineers are engaged in activities that generally fall into the domain of the general welfare: defense, transportation, flood control, harbors and canals, power development, fire and police protection, health services, sanitation and environmental services. Government agencies do not seek to produce profits in their activities; in many instances, there is not even a cash return on the investment. Nevertheless they seek to provide the greatest benefit possible for a given expenditure of public funds. Hence engineers employed in government agencies must also consider economic factors.

The engineer in a private enterprise seeks to increase the profit gained by that enterprise. A simple statement of the profit equation is

$$\text{Profit} = \text{Income} - \text{Costs} \tag{9–1}$$

The engineer in a government agency seeks to maximize the ratio of benefits—measured in terms of money—to costs. That is, the benefit–cost ratio is

$$B/C = \frac{\text{Benefits}}{\text{Costs}} \tag{9–2}$$

In each of these different economic environments, the engineer must come to grips with costs. Indeed, the engineer's major concern may be to produce goods and services at minimum cost, leaving it to managers and accountants to concern themselves with the other financial elements in these equations. Before we develop economic models in engineering, we must discuss several classifications of costs: first cost, fixed cost, variable cost, incremental cost, and sunk cost.

FIRST COST

First cost is the expense involved in acquiring an asset, initiating a project, or getting an activity underway. First cost is ordinarily limited to expenses that occur only at the outset of the economic life of a venture: the cost of buying a crane, the cost of constructing a section of highway, and the cost of buying land for a public park. The first cost of an asset is also called the acquisition cost.

FIXED COST

Fixed cost is the constant portion of costs that arise over the duration of a project or activity. For instance, the cost of owning, operating, and maintaining a machine over the life of a production activity is relatively constant. In contrast, the costs of materials and labor are not fixed because they are proportional to the number of units produced. Typical items contributing to the fixed costs of an engineering activity are depreciation, maintenance, taxes, insurance, interest on invested capital, sales programs, and research.

VARIABLE COST

Variable cost is proportional to the number of units, the size of the entity, or the level of operational activity. The costs of labor and material are generally functions of the number of units produced. The cost of materials used in a construction project is usually proportional to the size of the structure.

Variable costs are not always linearly related to the quantity of output or the size of a project. For instance, the cost of material may increase at a decreasing rate because of the price reductions often available with high-volume purchases. Figure 9-1 illustrates the contributions of fixed cost and variable cost to the total cost of an activity.

INCREMENTAL COST

Incremental cost is the increase or decrease in the cost of output due to a given increment in the quantity of output. If the change in cost is denoted ΔC and the change in output ΔO, the incremental cost is

$$IC = \frac{\Delta C}{\Delta O} \tag{9–3}$$

Figure 9–1 illustrates the concept of incremental cost.

SUNK COST

Sunk cost is a past expense that does not affect the present or future feasibility of a venture. Thus if you are comparing the economic effectiveness of several present or future investment alternatives, you would not include sunk costs. For example,

FIGURE 9-1 Fixed, variable, and incremental costs

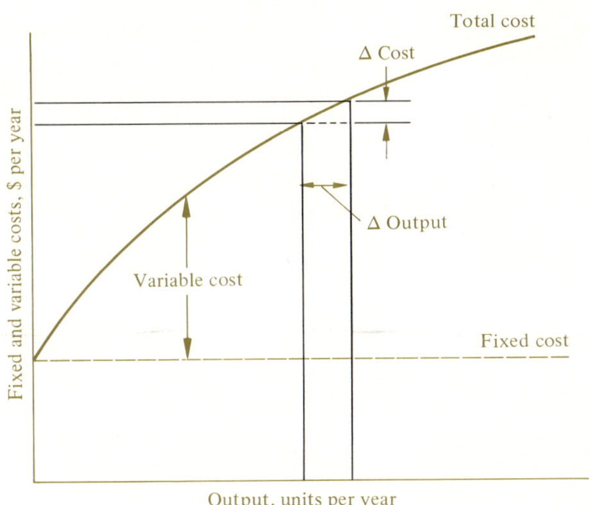

suppose the owner of an automobile has spent $600 on maintenance in the past year and is now thinking about buying a new automobile. The total annual cost of operating the new model, including amortized purchase cost, maintenance, and fuel, is less than that of the old car. Should the rather sizable expense of maintenance affect the owner's decision about buying a new car? Definitely not. The $600 maintenance expense is a sunk cost. Since it does not affect the future, the owner must ignore it.

There is a natural reluctance to disregard sunk cost in an economic analysis of a proposed engineering activity. But the economic consequences of failing to treat sunk cost properly can mean the difference between a profitable venture and a financial loss.

9-2 INTEREST AND ECONOMIC EQUIVALENCE

Initiating an engineering operation or project typically requires an investment of capital, the first cost. There are two sources of capital for such an investment. One is the firm's *equity capital.* Equity is the amount by which the firm's marketable assets exceeds its liabilities. Equity provides an available source of capital for new economic ventures. The second is *debt capital.* If equity capital is unavailable for a proposed investment, the firm may be able to borrow from a lending agency, if it is willing to pay interest on the borrowed sum. Funds thus obtained are called debt capital.

THE TIME VALUE OF MONEY

A fundamental precept of the free-enterprise marketplace is that someone who wishes to borrow capital must pay "rent" for the use of that capital. This rent is called interest. For example, someone who borrows $10,000 at 10% interest for one year must repay the $10,000 borrowed plus $1000 in interest at the end of the year. Although philosophers in past centuries have debated the morality of such usury, it has become an underlying mechanism of the modern marketplace. In fact, it is so prevalent that firms considering a new venture usually compare the economic gain from the venture with the gain they could obtain simply by lending the equivalent capital. The basic mechanism of this process is the time value of money.

SIMPLE INTEREST

We express the interest paid on a sum of borrowed money as a proportion or rate i of the original sum P for a period of one year. We typically state the interest rate i as a percentage; for example, 8%. When we compute the interest charge, we treat i as a decimal fraction; for example, 0.08. The total interest I charged when the loan is repaid is proportional to the interest rate i and the loan's duration n. For simple interest, the interest charge is

$$I = Pni \tag{9–4}$$

The total sum repaid after n periods under simple interest is

$$F = P(1 + ni) \tag{9–5}$$

For example, suppose that you have borrowed $100,000 at an annual interest rate of 9%, and that you have agreed to repay it in a single payment after five years. The total interest for this loan under the simple interest concept would be

$$I = (\$100{,}000)(5)(0.09) = \$45{,}000$$

The sum that you would repay is

$$F = (\$100{,}000)[1 + (5)(0.09)] = \$145{,}000$$

COMPOUND INTEREST

A more common practice in charging "rent" on borrowed capital is to compute the interest charges at the end of each period, say one year, and to charge interest on the accumulated interest charges at the end of each period. Thus interest is paid on interest. This concept is called compound interest. To illustrate this concept, let us consider our previous example. You have borrowed $100,000 at an annual interest rate of 9% for 5 years. Table 9–1 shows the schedule of interest charged and the total amount you will owe at the end of each year for the 5-year period. You will have to pay a total of $53,862 to "rent" $100,000 for 5 years at 9% interest compounded annually. This is significantly greater than the $45,000 you would pay with simple interest. The difference, $8862, is "interest paid on interest."

Compound interest is more common; thus this scheme more realistically represents the situation one encounters in seeking debt capital on the money market. The

TABLE 9–1. CALCULATION OF COMPOUND INTEREST

Year	*Amount owed at beginning of year*	*Interest added at end of year*	*Total owed at end of year*	*Amount repaid*
1	\$100,000	\$ 9,000	\$109,000	—
2	109,000	9,810	118,810	—
3	118,810	10,693	129,503	—
4	129,503	11,655	141,158	—
5	141,158	12,704	153,862	\$153,862

actual compounding period might be one year, six months, three months, one month, or an even smaller period. The formulas used in evaluating economic alternatives are based on this concept of compound interest.

INTEREST FORMULAS AND EQUIVALENCE

With compound interest, either the borrower repays the interest at the end of an interest period or the interest itself begins to earn interest. Table 9–1 demonstrates the compounding effect of leaving the interest charges unpaid. This section examines several important relationships in engineering economic analysis. These relationships are based on the notion of compounding interest at the end of each period over the life of an economic venture. We shall adopt the following terms and notation to develop these economic models.

i = nominal annual interest rate, written as a decimal fraction
n = number of interest periods
P = principal sum or present amount
F = future sum after n periods
A = amount of each payment in a series of uniform payments

In the example shown in Table 9–1, the loan is repaid in a single payment after five periods. By substituting the general terms for the numerical values in Table 9–1, we have the quantities shown in Table 9–2. The sum F repaid at the end of n periods is equal to P, the sum borrowed, multiplied by a factor $(1 + i)^n$. That is,

$$F = P(1 + i)^n \tag{9–6}$$

The factor $(1 + i)^n$ is termed the single-payment compound-amount factor; the abbreviation for it is $(F|P,i,n)$, read "F given P at interest rate i for n periods." Verifying this with the previous example gives

$$F = \$100{,}000(1.09)^5 = \$100{,}000(F|P,9,5) = \$100{,}000(1.5386) = \$153{,}860$$

Figure 9–2(a), called a cash-flow diagram, illustrates this concept. From a borrower's viewpoint, the upward arrow ↑ represents the amount P received at time zero. The downward arrow ↓ represents the amount F paid out at time n. From the lender's viewpoint, exactly the opposite is true. The amount P represents a disbursement of money, while the sum F is a receipt. The cash-flow diagram in

TABLE 9–2. DEVELOPMENT OF THE SINGLE-PAYMENT COMPOUND-AMOUNT FACTOR

Year	*Amount owed at beginning of year*	*Interest added at end of year*	*Total owed at end of year*	*Amount repaid*
1	P	Pi	$P(1 + i)$	—
2	$P(1 + i)^1$	$P(1 + i)i$	$P(1 + i)^2$	—
3	$P(1 + i)^2$	$P(1 + i)^2 i$	$P(1 + i)^3$	—
4	$P(1 + i)^3$	$P(1 + i)^3 i$	$P(1 + i)^4$	—
.	.			.
.	.			.
.	.			.
n	$P(1 + i)^{n-1}$	$P(1 + i)^{n-1} i$	$P(1 + i)^n$	F

Figure 9–2(b) shows this view. This convention of representing a receipt with an upward arrow ↑ and a disbursement with a downward arrow ↓ is quite common in engineering economic analysis. You will find it helpful to draw such graphs before proceeding far into an economic analysis.

Suppose that someone is willing to lend \$20,000 with a stipulation that the borrower repay \$29,280 in a single payment at the end of the fourth year. What is the annual interest rate i? Applying Equation (9–6), we have

$$\$29{,}280 = \$20{,}000(1 + i)^4$$

Therefore

$$(1 + i)^4 = \frac{\$29{,}280}{20{,}000} = 1.464$$
$$(1 + i) = (1.464)^{1/4}$$
$$i = 0.1$$

Thus the annual interest rate is 10%. We could also have obtained this result directly by examining the interest tables in Appendix B.

FIGURE 9–2 Cash flow diagrams of single-payment equivalence relationships: (a) borrower's viewpoint, (b) lender's viewpoint

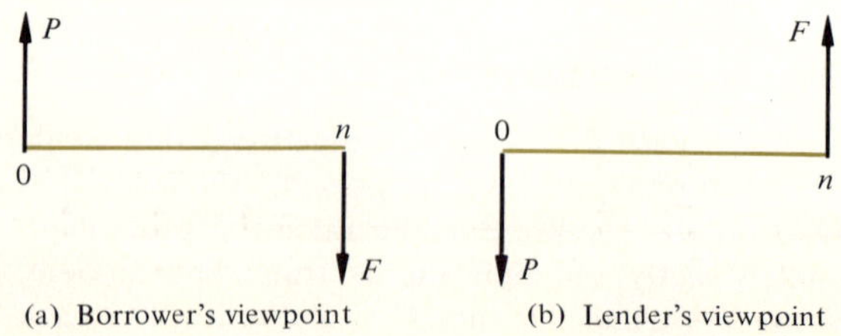

(a) Borrower's viewpoint (b) Lender's viewpoint

Now consider depositing an amount P into an account earning compound interest at a rate i. The depositor wants to have an amount F after n periods. Rearranging Equation (9–6), we obtain the formula

$$P = F\left[\frac{1}{(1 + i)^n}\right] = F(P|F,i,n) \tag{9–7}$$

The term $[1/(1 + i)^n]$ is known as the single-payment present-worth factor. It gives the present worth P of a future amount F. For example, assume that you want to accumulate \$500,000 in six years. At 8% interest, the amount of money that you would have to deposit is

$$\begin{aligned} P &= F\left[\frac{1}{(1 + 0.08)^6}\right] = \$500{,}000(P|F,8,6) \\ &= \$500{,}000(0.6302) = \$315{,}085 \end{aligned}$$

Observe that the single-payment present-worth factor is the *reciprocal* or *inverse* of the single-payment compound-amount factor, for the same P,F,i and n. That is,

$$(P|F,i,n) = 1/(F|P,i,n)$$

In many instances, it is either necessary or more convenient to make a series of uniform payments rather than a single payment. To account for these payments, we have to adopt a "year-end" convention, as shown in the cash flow diagram in Figure 9–3. That is, we make a payment A at the end of each period for n periods. At the end of the nth period, we "exchange" a payment A for the final sum F. This last payment earns no interest, but the payment A at the end of period $n - 1$ earns an amount Ai, so that $A(1 + i)$ accumulates toward the final sum F. Likewise, the payment at the end of period $n - 2$ is twice compounded, adding an amount $A(1 + i)^2$ to the final sum F. If we write the payments in reverse chronological order, we get

$$F = A(1 + i)^0 + A(1 + i)^1 + \cdots + A(1 + i)^{n-2} + A(1 + i)^{n-1}$$

Multiplying both sides of this equation by $(1 + i)$ gives

FIGURE 9–3 Cash flow diagram of a sinking fund: an equal-payment series equivalent to a future amount F

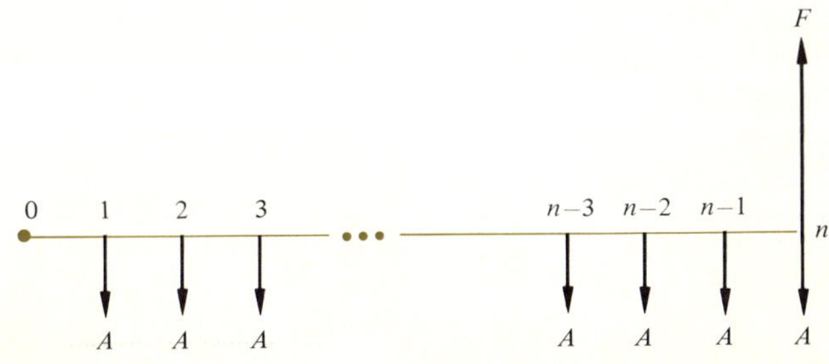

$$F(1 + i) = A(1 + i)^1 + A(1 + i)^2 + \cdots + A(1 + i)^{n-1} + A(1 + i)^n$$

Subtracting the first equation from the second yields

$$Fi = A[(1 + i)^n - 1]$$

Thus we can determine the future amount F accumulated by a series of equal payments A by dividing both sides of this equation by i,

$$F = A\left[\frac{(1 + i)^n - 1}{i}\right] = A(F|A,i,n) \qquad (9\text{–}8)$$

The term $\{[(1 + i)^n - 1]/i\}$ is known as the equal-payment series compound-amount factor. We use it to determine the future sum F that is equivalent to a series of uniform payments of amount A at interest rate i for n periods. For example, consider payments of \$5000 deposited at the end of each year for five years at 8% interest. The sum accumulated is

$$\begin{aligned} F &= \$5000\left[\frac{(1 + 0.08)^5 - 1}{0.08}\right] \\ &= \$5000(F|A,8,5) = \$5000(5.8666) \\ &= \$29,333 \end{aligned}$$

We can rearrange the equal-payment series compound-amount relationship expressed in (9–8) to find the uniform payment A necessary to accumulate a future sum F after n periods at interest rate i. That is,

$$A = F\left[\frac{i}{(1 + i)^n - 1}\right] = F(A|F,i,n) \qquad (9\text{–}9)$$

The term $\{i/[(1 + i)^n - 1]\}$ is known as the equal-payment series sinking-fund factor. It is the inverse of the equal-payment series compound-amount factor. For example, the annual deposit needed to produce an amount \$10,000 after six years at 10% interest is

$$\begin{aligned} A &= \$10,000\left[\frac{0.10}{(1.10)^6 - 1}\right] \\ &= \$10,000\,(A|F,10,6) = \$10,000\,(0.1296) \\ &= \$1296 \end{aligned}$$

Suppose we are interested in determining A, the amount that we can receive in each of a series of equal payments for n periods if we place an amount P into an account at interest rate i. Figure 9–4 shows the cash-flow diagram for this process. Substituting $P(1 + i)^n$ for F in Equation (9–9) gives

$$A = P\left[\frac{i(1 + i)^n}{(1 + i)^n - 1}\right] = P(A|P,i,n) \qquad (9\text{–}10)$$

The term $\{i(1 + i)^n/[(1 + i)^n - 1]\}$ is called the equal-payment series capital-recovery factor. That is, if an amount of capital P is invested at interest rate i for n periods, it is "recovered" in a series of n equal payments of amount A. For

FIGURE 9–4 Cash flow diagram of capital recovery: an equal-payment series equivalent to a present amount P

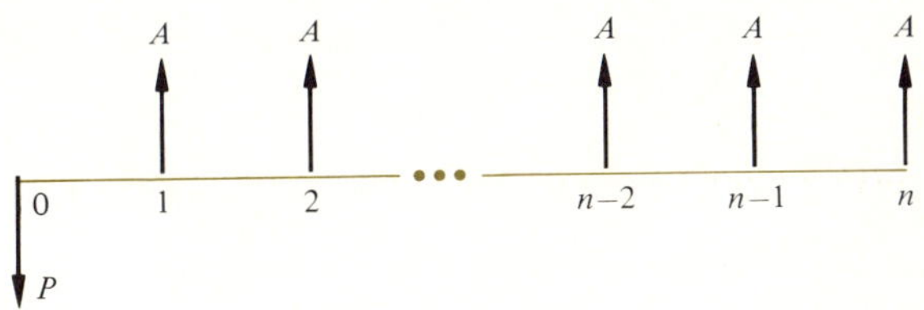

example, assume that a company invests \$100,000 at 8% interest for 8 years. This capital is recovered in annual payments equal to

$$
\begin{aligned}
A &= \$100{,}000 \left[\frac{0.08\,(1 + 0.08)^8}{(1 + 0.08)^8 - 1}\right] \\
&= \$100{,}000(A|P,8,8) = \$100{,}000\,(0.1740) \\
&= \$17{,}400
\end{aligned}
$$

Solving Equation (9–10) for P produces

$$
P = A\left[\frac{(1 + i)^n - 1}{i(1 + i)^n}\right] = A(P|A,i,n) \tag{9–11}
$$

The term $\{[1 + i)^n - 1]/i(1 + i)^n\}$ is known as the equal-payment series present-worth factor. This equation is useful for determining the present sum P equivalent to a series of uniform payments A at interest rate i for n periods. For example, suppose we want to know the present sum that is equivalent to a series of uniform disbursements of \$1000 per year for 12 years at an interest rate of 8%. This present worth is

$$
\begin{aligned}
P &= \$1000 \left[\frac{(1 + 0.08)^{12} - 1}{0.08(1 + 0.08)^{12}}\right] \\
&= \$1000\,(P|A,8,12) \\
&= \$1000\,(7.536) \\
&= \$7536
\end{aligned}
$$

This result means that depositing a present sum of \$7536 at 8% interest enables us to withdraw \$1000 at the end of each year for 12 years.

The interest formulas presented in this section provide six useful relationships among the quantities P, F, A, i, and n. Each of these relationships establishes the equivalence between any two of the quantities P, F, and A. This concept of equivalence is of fundamental importance to an engineer who is evaluating the economic characteristics of systems that will operate for a finite period of time. Whether the particular application is the installation of an electric power substation, the construction of a bridge along a new highway, or the purchase of ventilation equipment for an iron foundry, the correct application of economic analysis could very well be the crucial factor in the success or failure of the venture. Table 9–3

TABLE 9–3. SUMMARY OF INTEREST FORMULAS

Given	*Find*	*Factor*	*Complete formula*	*Abbreviated formula*
P	F	Single payment compound amount	$F = P[(1+i)^n]$	$F = P(F\|P,i,n)$
F	P	Single payment present worth	$P = F\left[\frac{1}{(1+i)^n}\right]$	$P = F(P\|F,i,n)$
A	F	Equal-payment series compound amount	$F = A\left[\frac{(1+i)^n - 1}{i}\right]$	$F = A(F\|A,i,n)$
F	A	Equal-payment series sinking fund	$A = F\left[\frac{i}{(1+i)^n - 1}\right]$	$A = F(A\|F,i,n)$
P	A	Equal-payment series capital recovery	$A = P\left[\frac{i(1+i)^n}{(1+i)^n - 1}\right]$	$A = P(A\|P,i,n)$
A	P	Equal-payment series present worth	$P = A\left[\frac{(1+i)^n - 1}{i(1+i)^n}\right]$	$P = A(P\|A,i,n)$

summarizes the six interest formulas discussed in this section. These formulas are at the core of most economic analyses of engineering designs.

Appendix B provides values for each of the factors enclosed in brackets [ ] in Table 9–3, for various combinations of i and n. For example, the tabulated value for the single-payment compound-amount factor at 6% interest and 7 years is 1.5036 (Table B–5). The value of the equal-payment series capital-recovery factor at 10% interest and 15 years, found in Table B–7, is 0.1315.

EFFECTIVE INTEREST VERSUS NOMINAL INTEREST

An interest rate is typically stated as an annual rate regardless of the compounding period actually used (a year, six months, three months, one month, and so forth). This rate is called the *nominal* rate of interest. One can quickly surmise that if interest is compounded several times each year, the actual or effective interest rate is somewhat higher because of the effect of interest paid on interest. For example, consider a nominal interest rate of 12% compounded quarterly. Thus the interest rate each quarter is 12/4% or 3%. The value of \$100 after one year under this interest scheme is

$$F = \$100(1.03)^4 = \$100(1.126) = \$112.60$$

This is equivalent to an effective interest rate of 12.6% compounded once at the end of the year.

We can derive an expression for the effective annual interest rate as a function of the nominal rate as follows: Let

r = nominal annual interest rate
i = effective annual interest rate
c = number of compounding periods per year

Then

$$i = \left(1 + \frac{r}{c}\right)^c - 1 \qquad (9\text{–}12)$$

In our example, $r = 12\%$, $c = 4$, and substituting these values gives

$$i = \left(1 + \frac{0.12}{4}\right)^4 - 1 = (1.03)^4 - 1$$
$$= 0.126$$

Note that we apply the formulas for compound interest without modifying them even when there are multiple compounding periods per year. The interest rate i is then the rate per period, and n is the total number of compounding periods in the life of the venture. For example, if 12% interest is compounded monthly for five years, i is 1% and n is 60 periods.

Exercises

9–1 An engineering firm borrows $75,000 from a bank to buy a piece of test equipment. It agrees to repay this loan in a single sum after three years at 8% interest compounded annually. What amount must the firm repay?

9–2 A construction firm pays $120,000 for a piece of heavy equipment. The equipment has a life of 10 years. After 5 years of use, an overhaul costing $12,000 will be needed. How much must the firm deposit in a 6% fund at the time it purchases this equipment in order to meet this $12,000 obligation after 5 years?

9–3 The equipment described in Exercise 9–2 will yield a return of $25,000 each year over its 10-year life. At an interest rate of 10%, what is the present worth of this annual income?

9–4 Suppose the construction firm pays for the equipment described in Exercise 9–2 in eight annual installments. The interest rate is 8%. What is the amount of each of these annual payments?

9–5 Suppose that in Exercise 9–2 the firm decides to make annual payments into a 6% fund in order to meet the $12,000 overhaul after five years. What is the amount of each of these annual payments?

9–6 A couple wish to invest in a fund to enable their newborn child to attend college at age 18. They estimate that by the time their child reaches college age, the expense each year will be $6000. They want to withdraw $6000 from the fund at the beginning of each of the four years that their child will be in college and exhaust the fund after the final withdrawal. How much must they invest each month for 18 years in a fund paying 6% interest compounded monthly (that is, ½% each month) to do this?

9–7 In Exercise 9–6, what is the effective annual interest rate of the fund?

9–3 DEPRECIATION MODELS

Engineered systems typically use items of capital equipment. For instance, a construction project might require a crane, a concrete mixer, a diesel generator, and several trucks. A firm acquires each item at an initial purchase cost, operates it over a period of months or years, and sells it for whatever value it might have at the end of its useful life. The accounting process by which an item of capital equipment declines in value over its operating life is known as *depreciation*. Physical depreciation results from wear and tear or from deterioration, from physical damage due to accident or corrosion, or from obsolescence due to the availability of functionally superior equipment.

When a firm buys equipment, it expects that the equipment will earn more than it cost. The firm uses the capital equipment over its operational life to produce goods and services. The price of the product or service includes the recovered capital, operating costs, and profit. At the end of the equipment's useful life, the firm can often sell it for an amount called its *salvage value,* thus recovering another part of the original investment.

There are a number of schedules for depreciating capital assets over their operational lives. A depreciation schedule enables the firm's accounting system to assign to the asset a monetary value that is commensurate with its physical worth at any given time. This worth, called the *book value,* contributes to the overall worth of the firm.

STRAIGHT-LINE DEPRECIATION

The simplest scheme of depreciation is the straight-line method, which assumes that the value of an asset declines at a constant rate. Suppose that P is the first cost of the asset, and F is its salvage value at the end of n years. Then, by the straight-line method, the amount by which the asset depreciates during the kth year is

$$D_k = (P - F)/n \qquad (9\text{–}13)$$

Since D_k is constant for straight-line depreciation, we shall denote it simply as D. The book value of the asset at the end of the kth year is then

$$B_k = P - k\left[\frac{(P - F)}{n}\right] = P - kD \qquad (9\text{–}14)$$

After n years, the book value B_n is exactly equal to the salvage value F.

FIGURE 9–5 Straight-line depreciation

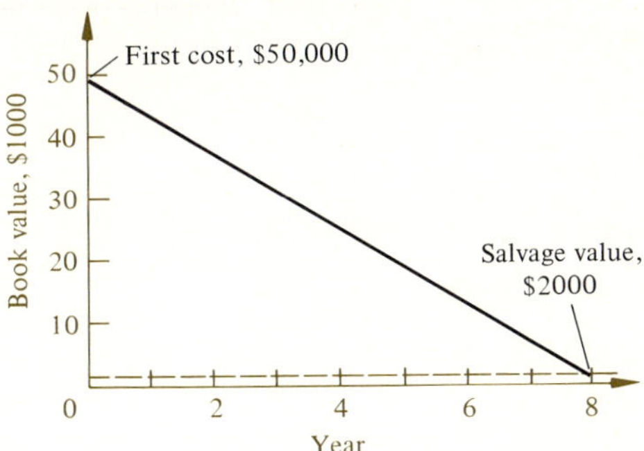

Figure 9–5 illustrates the value of the asset as a function of time for a hypothetical piece of equipment having a first cost of $50,000, a salvage value of $2000, and an operational life of 8 years. The annual depreciation amount is

$$D = \frac{\$50{,}000 - \$2000}{8} = \$6000$$

The book value after 6 years is

$$B_6 = \$50{,}000 - (6)(\$6000) = \$14{,}000$$

FIXED-PERCENTAGE DEPRECIATION

Sometimes a company will want to depreciate an asset by larger amounts early in its operational life and smaller amounts later. Thus the asset decreases in value at a decreasing rate. One depreciation model that accomplishes this is the *fixed-percentage method,* which reduces the book value of an asset by a fixed rate R in each year of the operational life. Thus the amount of depreciation in the kth year is

$$D_k = RB_{k-1}$$

where B_0 is the first cost P. The book value after the first year is

$$B_1 = (1 - R)B_0 = (1 - R)P$$

After the second year, it is

$$B_2 = (1 - R)B_1 = (1 - R)^2 B_0 = (1 - R)^2 P$$

Thus the book value after year k is

$$B_k = (1 - R)^k P \tag{9–15}$$

and the depreciation during the kth year is

$$D_k = RB_{k-1} = R(1 - R)^{k-1} P \tag{9–16}$$

In general, D_n need not be equal to the salvage value F.

To illustrate the fixed-percentage method of depreciation, let us suppose that we are depreciating an asset worth \$50,000 at 20% per year over its 8-year life. Therefore R is 0.20. The depreciation for the first year is then

$$D_1 = (0.2)(\$50{,}000) = \$10{,}000$$

which yields a book value of

$$B_1 = (0.8)(\$50{,}000) = \$40{,}000$$

For the second year, the depreciation is

$$D_2 = (0.2)(0.8)^1(\$50{,}000) = \$8000$$

and the book value is

$$B_2 = (0.8)^2(\$50{,}000) = \$32{,}000$$

After 8 years, the book value is

$$B_8 = (0.8)^8\,(\$50{,}000) = \$8387$$

So the book value of this asset is not the same as its \$2000 salvage value.

However, we can select a fixed-percentage rate R so that the book value of the asset after n years will be exactly equal to its salvage value. To find this R, we manipulate Equation (9–15) as follows:

$$B_n = (1 - R)^n P$$

Since we want B_n to equal F, we write

$$F = (1 - R)^n P$$

Rearranging gives

$$(1 - R)^n = \frac{F}{P} \qquad \text{and} \qquad (1 - R) = \left(\frac{F}{P}\right)^{1/n}$$

Finally

$$R = 1 - \left(\frac{F}{P}\right)^{1/n} \tag{9–16}$$

Now, substituting the specific values for our example, we obtain

$$R = 1 - \left(\frac{2{,}000}{50{,}000}\right)^{1/8} = 1 - (0.04)^{1/8}$$
$$= 0.33126$$

There is just one difficulty here. Since the depreciation charge is tax deductible, the U.S. Internal Revenue Service will not allow rates of depreciation greater than the ratio $2/n$. In this example, the largest permissible fixed-percentage rate is 2/8 or 0.25, so it is not possible to have B_8 equal to the salvage value F. This $2/n$ rate is called *double-declining-balance depreciation*. It is widely used because it greatly reduces income taxes.

SUM-OF-THE-YEARS'-DIGITS DEPRECIATION

Another depreciation method that rapidly reduces book value early in an asset's life is the sum-of-the-years'-digits method, so called because the depreciation in the kth year of an n-year life is a function of the year's digit k and the sum of the n digits 1, 2, . . . , n. This sum is

$$\sum_{j=1}^{n} j = 1 + 2 + \cdots + (n - 1) + n = \frac{n(n + 1)}{2} \qquad (9\text{–}17)$$

Thus, for an asset with an 8-year life, the sum of the years' digits is (8)(9)/2 or 36.

The asset is depreciated from its first cost P to its salvage value F; that is, the total depreciation in n years is $(P - F)$. Each year for n years, some fraction of this amount is depreciated. In the kth year, the proportion depreciated is the ratio of the number of years remaining at the beginning of the kth year, $(n - k + 1)$ to the sum of the years' digits, $n(n + 1)/2$. Thus the depreciation charge is

$$D_k = \frac{n - k + 1}{n(n + 1)/2}(P - F) \qquad (9\text{–}18)$$

This leaves the book value

$$B_k = P - \sum_{j=1}^{k} D_k \qquad (9\text{–}19)$$

Table 9–4 shows the depreciation schedule for a \$50,000 asset having a \$2000 salvage value after a life of 8 years. Observe that the numerator of the ratio, which determines the portion of $(P - F)$ depreciated each year, starts at n in year 1 and ends at 1 in year n.

COMPARING METHODS OF DEPRECIATION

By examining the book value of an asset over its operational life for these three methods of depreciation, one can easily see which methods produce larger amounts of depreciation in the early years of the asset's life. Figure 9–6 clearly demonstrates

TABLE 9–4. SUM-OF-THE-YEARS'-DIGITS DEPRECIATION

Year	*Depreciation charge*	*Book value*
0	—	\$50,000
1	$\frac{8}{36}$ (\$50,000 – \$2000) = \$10,667	39,333
2	$\frac{7}{36}$ (48,000) = 9333	30,000
3	$\frac{6}{36}$ (48,000) = 8000	22,000
4	$\frac{5}{36}$ (48,000) = 6667	15,333
5	$\frac{4}{36}$ (48,000) = 5333	10,000
6	$\frac{3}{36}$ (48,000) = 4000	6,000
7	$\frac{2}{36}$ (48,000) = 2667	3,333
8	$\frac{1}{36}$ (48,000) = 1333	2,000

FIGURE 9–6 A comparison of depreciation methods

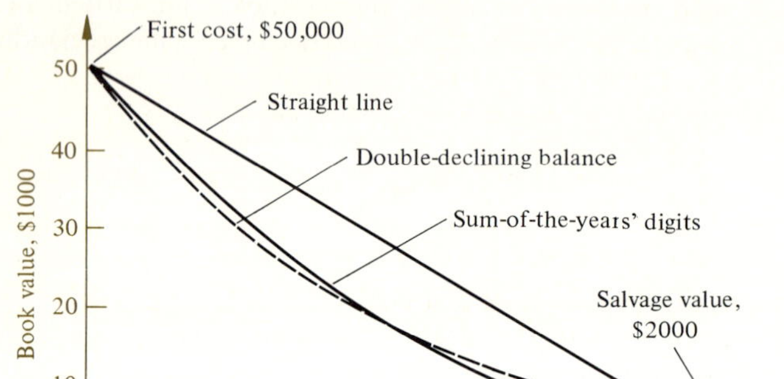

this comparison for the fixed-percentage method in the specific case of double-declining balance ($R = 2/n$). Since the depreciation charge in a given year represents a tax deduction, many companies favor the double-declining-balance and sum-of-the-years'-digits depreciation methods because they can thus postpone paying some taxes.

Exercises

9–8 A construction firm bought a piece of earth-moving equipment for $130,000. This equipment has a life of 12 years and a salvage value of $10,000. Compute the annual book value and depreciation charge for this equipment.
(a) Use the straight-line depreciation method.
(b) Use the fixed-percentage method such that the final book value exactly equals the salvage value.
(c) Use the sum-of-the-years'-digits method.

9–9 An engineering professor who buys a pocket calculator for $170 can sell it after five years for $10. Under the double-declining-balance method of depreciation, how much can the professor deduct for depreciation in each of the five years?

9–4 EVALUATING ECONOMIC ALTERNATIVES IN ENGINEERING DESIGN

To evaluate alternative engineering designs, we have to consider the projected receipts and disbursements for each design over its operational life. By *receipts* we mean the cash income that we receive when we sell goods and services over the life of the asset and any income we obtain by selling the asset at the end of its life. By *disbursements* we mean cash outlays to acquire and operate the asset, including the purchase cost, operating cost, maintenance, and labor.

The engineer has two basic ways of evaluating economic alternatives. The first is to compare a single alternative to the desired return from the venture. The second is to compare two or more alternatives to each other in order to select the best one. The use of multiple alternatives assumes that each of the ventures is acceptable, but to be safe the engineer should make sure that the best alternative is itself acceptable. Thus the second method implicitly involves the first.

EVALUATING A SINGLE ALTERNATIVE

To evaluate a single engineering design economically, we can use several procedures. Three of these methods employ one of the following.

1. Present worth
2. Equivalent annual value
3. Rate of return

To illustrate and compare these three procedures, we shall consider the purchase of a piece of earth-moving equipment used in construction projects. The initial cost of this equipment is $52,000, and its operational life is 6 years. An overhaul costing $8000 is needed after 3 years. The equipment yields receipts of $12,000 each year over its life. Its salvage value at the end of the sixth year is $10,000. Table 9–5 presents the cash flows for this venture.

PRESENT-WORTH METHOD

To use the present-worth method of evaluating a single engineering design or economic venture, we compare the present worth of the receipts P_R with the present worth of the disbursements P_D at some interest rate $i*$. The present worth is

TABLE 9–5. RECEIPTS AND DISBURSEMENTS FOR A SINGLE ALTERNATIVE

Item	*End of year*	*Receipts*	*Disbursements*
Purchase cost	0	—	$52,000
Income	1	$12,000	—
Income	2	12,000	—
Income	3	12,000	—
Overhaul	3	—	8,000
Income	4	12,000	—
Income	5	12,000	—
Income	6	12,000	—
Salvage	6	10,000	—

the sum that is the present equivalent of all future sums involved in the venture. The venture is economically acceptable if P_R is greater than P_D.

In the example of the earth-moving equipment, there are two kinds of receipts: (1) a series of equal amounts of income A equal to \$12,000 and (2) a single sum F of \$10,000 received at the end of the sixth year, the salvage value. The interest rate is 8%. The present worth of the *receipts* is then

$$\begin{aligned} P_R &= \$12{,}000\left[\frac{(1+0.08)^6-1}{0.08\,(1+0.08)^6}\right] + \$10{,}000\left[\frac{1}{(1+0.08)^6}\right] \\ &= \$12{,}000(P|A,8,6) + \$10{,}000(P|F,8,6) \\ &= \$12{,}000(4.6229) + \$10{,}000(0.6302) \\ &= \$55{,}474 + \$6302 = \$61{,}776 \end{aligned}$$

(The values of the compound-interest factors are from Table B–6 in Appendix B. We have not tried in these and later calculations to achieve accuracy exactly to the penny, as a bank would.)

The present worth of the *disbursements* is

$$\begin{aligned} P_D &= \$52{,}000 + \$8000\left[\frac{1}{(1+0.08)^3}\right] \\ &= \$52{,}000 + \$8000\,(P|F,8,3) \\ &= \$52{,}000 + \$8000\,(0.7938) \\ &= \$52{,}000 + \$6351 \\ &= \$58{,}351 \end{aligned}$$

The investment in the earth mover is justified, since the present worth of the receipts derived from its operation exceeds the present worth of the disbursements incurred over its life.

EQUIVALENT-ANNUAL-VALUE METHOD

The equivalent-annual-value method of analysis compares the equivalent annual value of the receipts A_R to that of the disbursements A_D. The venture is acceptable if A_R exceeds A_D. In our example, the equivalent annual value of receipts is the annual income plus the equal-payment series that is equivalent to selling the equipment at the end of the sixth year; that is,

$$\begin{aligned} A_R &= \$12{,}000 + \$10{,}000\left[\frac{0.08}{(1+0.08)^6-1}\right] \\ &= \$12{,}000 + \$10{,}000\,(A|F,8,6) \\ &= \$12{,}000 + \$10{,}000\,(0.1363) \\ &= \$13{,}363 \end{aligned}$$

The equivalent annual value of the disbursements is the equal-payment series equivalent to the cost of purchasing the earth mover plus the equal-payment series equivalent to spreading the present worth of the \$8000 overhaul over the 6-year operational life of the mover. That is,

$$A_D = \$52{,}000\left[\frac{0.08(1 + 0.08)^6}{(1 + 0.08)^6 - 1}\right] + \$8000\left[\frac{1}{(1 + 0.08)^3}\right]\left[\frac{0.08(1 + 0.08)^6}{(1 + 0.08)^6 - 1}\right]$$

$$= \$52{,}000\,(A|P,8,6) + \$8000(P|F,8,3)(A|P,8,6)$$

$$= \$52{,}000(0.2163) + \$8000(0.7938)(0.2163)$$

$$= \$11{,}248 + \$1374 = \$12{,}622$$

Observe that to get the second term in this calculation, we first have to determine the present worth of the overhaul at time zero, and then spread this amount uniformly over the 6-year life. The venture is acceptable because the equivalent annual value of the receipts exceeds that of the disbursements.

RATE-OF-RETURN METHOD

The rate of return is a universal measure of the economic acceptability of an engineering design. In effect, we compute the interest rate $i*$ that will make the present worth of the receipts exactly equal to that of the disbursements. For values of i less than $i*$, $P_R > P_D$. For values of i greater than $i*$, $P_R < P_D$. At $i*$, $P_R = P_D$. If this interest rate $i*$, called the *rate of return,* exceeds the firm's minimum attractive rate of return (MARR), the venture is deemed acceptable. We often interpolate graphs or tables of numbers to determine $i*$. In our example, suppose that MARR is 8%. Let us first consider the present worth of receipts and disbursements at an interest rate of 10%.

$$P_R = \$12{,}000(P|A,10,6) + \$10{,}000(P|F,10,6)$$

$$= \$12{,}000(4.3552) + \$10{,}000(0.5645)$$

$$= \$52{,}262 + \$5645 = \$57{,}907$$

$$P_D = \$52{,}000 + \$8000(P|F,10,3)$$

$$= \$52{,}000 + \$8000(0.7513)$$

$$= \$52{,}000 + \$6010 = \$58{,}010$$

$$P_R - P_D = \$57{,}907 - \$58{,}010 = -\$103$$

Thus 10% is too high an interest rate, since $P_R < P_D$. Trying an i of 8%, we get

$$P_R = \$12{,}000(P|A,8,6) + \$10{,}000(P|F,8,6)$$

$$= \$12{,}000(4.6229) + \$10{,}000(0.6302)$$

$$= \$55{,}474 + \$6302 = \$61{,}776$$

$$P_R = \$52{,}000 + \$8000(P|F,8,3)$$

$$= \$52{,}000 + \$8000(0.7938)$$

$$= \$52{,}000 + \$6350 = \$58{,}350$$

$$P_R - P_D = \$61{,}776 - \$58{,}350 = \$2426$$

Hence 8% is too low an interest rate, since $P_R > P_D$. Interpolating between 8 and 10%, we get

$$i* = 8\% + (10\% - 8\%)\left[\frac{0 - \$2425}{-\$103 - \$2425}\right]$$

$$= 8\% + (2\%)(0.9593)$$

$$= 8\% + 1.92\% = 9.92\%$$

Since $i*$ exceeds the minimum attractive rate of return of 8% this economic venture is deemed acceptable.

Since $P_R > P_D$ at 8% implies that $i*$ is indeed greater than MARR, you may wonder whether it is necessary to perform the final interpolation. We highly recommend that you do so, since most decision-makers want to know exactly what the rate of return is.

Note that each method of comparing alternatives yields the same decision. This should always be the case.

Exercise

9–10 A 500-ton press can be purchased for \$75,000. Its operational life is 12 years. During its life, it produces net annual revenues of \$9500 per year. It requires one overhaul after 4 years and another after 8 years; each costs \$6800. The salvage value of the press at the end of its 12-year life is \$7500. Determine its economic feasibility at an interest rate of 6% by using (a) the present-worth method, (b) the equivalent-annual-value method, and (c) the rate-of-return method.

EVALUATING MULTIPLE ALTERNATIVES

To evaluate multiple alternatives, we use the same procedures as with a single alternative. The only difference is that we select the best alternative and assume that it is economically feasible. The objective is to select the alternative that provides the desired service at least cost.

Consider the example of constructing a facility for automatically storing and handling material. Alternative A is a high-rise warehouse (40 feet high) served by two stacker cranes and two electric fork trucks. It would employ 20 warehouse workers at \$14,000 per year each. Alternative B is a standard-height structure (25 feet high) with eight electric fork trucks; it would employ 32 warehouse workers at \$14,000 per year each. The stacker cranes cost \$80,000 each, and the electric fork trucks cost \$16,000 each. Their salvage values are \$8000 and \$1600, respectively, after 12 years. The cost of constructing the high-rise warehouse is \$2.3 million, compared to \$1.6 million for the standard-height facility. We expect each facility to be worth \$0.5 million after 12 years. The minimum attractive rate of return is 8%. Table 9–6 summarizes the data for these two alternatives. Each will produce the same annual revenues. Hence we must compare them on the basis of their cost over their 12-year lives.

Present-worth method We compute the present worth of the disbursements for the high-rise facility as follows:

TABLE 9–6. COMPARISON OF COSTS FOR TWO ALTERNATIVE DESIGNS FOR AN AUTOMATED WAREHOUSE

Year	*Item*	*High-rise warehouse*	*Standard warehouse*
0	Construction	\$2,300,000	\$1,600,000
0	Equipment	192,000	128,000
0	Total installation	2,492,000	1,728,000
1 to 12	Labor	280,000	448,000
12	Salvage (receipt)	(519,200)	(512,800)

$$\begin{aligned} P_D &= \$2{,}492{,}000 + \$280{,}000(P|A,8,12) \\ &= \$2{,}492{,}000 + \$280{,}000(7.5360) \\ &= \$2{,}492{,}000 + \$2{,}110{,}108 = \$4{,}602{,}108 \end{aligned}$$

The present worth of receipts, neglecting the revenues, is simply the present value of the \$519,200 salvage price.

$$\begin{aligned} P_R &= \$519{,}200\ (P|F,8,12) \\ &= \$519{,}200\ (0.3971) = \$206{,}174 \end{aligned}$$

Thus the difference between receipts and disbursements is the present worth of the cost of operating the high-rise warehouse; that is,

$$P_H = \$4{,}602{,}108 - \$206{,}174 = \$4{,}395{,}934$$

We evaluate the standard facility in the same way.

$$\begin{aligned} P_D &= \$1{,}728{,}000 + \$448{,}000(P|A,8,12) \\ &= \$1{,}728{,}000 + \$448{,}000(7.5361) \\ &= \$1{,}728{,}000 + \$3{,}376{,}173 = \$5{,}104{,}173 \\ P_R &= \$512{,}800\ (P|F,8,12) \\ &= \$512{,}800\ (0.3971) = \$203{,}633 \end{aligned}$$

The present worth of the total cost of operating the standard facility for 12 years is

$$P_S = \$5{,}104{,}173 - \$203{,}633 = \$4{,}900{,}540$$

Since it costs less to operate the high-rise warehouse than the standard warehouse, we select the high-rise warehouse. That is, the high-rise warehouse produces the same revenue at less cost than the standard facility.

Equivalent-annual-cost method The equivalent annual cost of an asset is an equal-payment series over the operational life of the asset. The equivalent annual cost of each warehouse design is equal to the sum of the annual operating costs, plus the equal series that results from spreading the initial cost over the 12-year life, less the equal-payment series that is equivalent to the salvage value. For the high-rise warehouse, the equivalent annual cost is

$$\begin{aligned} A_D &= \$2{,}492{,}000(A|P,8,12) + \$280{,}000 \\ &= \$2{,}492{,}000(0.1327) + \$280{,}000 \\ &= \$330{,}688 + \$280{,}000 = \$610{,}688 \end{aligned}$$

The equivalent annual value of the \$519,200 salvage value is

$$\begin{aligned} A_R &= \$519{,}200(A|F,8,12) \\ &= \$519{,}200(0.0527) = \$27{,}362 \end{aligned}$$

Thus the net equivalent annual cost for the high-rise warehouse is

$$A_H = \$610{,}688 - \$27{,}362 = \$583{,}326$$

The comparable equivalent annual cost for the standard warehouse is

$$\begin{aligned} A_S &= A_D - A_R \\ &= \$1{,}728{,}000(A|P,8,12) + \$448{,}000 - \$512{,}800(A|F,8,12) \\ &= \$1{,}728{,}000(0.1327) + \$448{,}000 - \$512{,}800(0.0527) \\ &= \$229{,}306 + \$448{,}000 - \$27{,}025 = \$650{,}281 \end{aligned}$$

Hence we would select the high-rise facility because its equivalent annual cost is lower. It is worthwhile to note that this lower cost results from savings in labor.

Exercise

9–11 The Environmental Protection Agency has cited a manufacturing company for permitting too much smoke to issue from its chimneys. The company is considering the installation of an electrostatic precipitator to remove solid particles from the hot gases leaving the chimney. The plant engineer has compiled the following estimates.

	Precipitators		
Costs	*A*	*B*	*C*
First cost installed	\$6000	\$7500	\$9000
Life (years)	10	10	10
Salvage value	0	0	0
Annual operating costs:			
Water and power	850	800	750
Cleaning	500	450	400
Maintenance	700	600	500
Taxes and insurance	120	150	180
Total annual costs	\$2170	\$2000	\$1830

The company's minimum attractive rate of return is 10%. Which precipitator should the company install? (Each has the same capacity.)

9–5 REPLACING ENGINEERING EQUIPMENT

One of the important economic decisions confronting engineers is when to replace equipment that still works, but which, with the passage of time, either declines in productivity or incurs increased maintenance costs. This is a decision that each of us eventually faces when we consider whether to keep an old car and pay the maintenance costs or buy a new car and pay the high initial cost.

In the replacement decision, there are two alternatives: the existing equipment and its potential replacement. One should choose the alternative that offers the greater economic advantage. The procedure most often used for evaluating these two alternatives is the equivalent-annual-cost method because it ignores any differences in the lengths of their operational lives. This feature is important, since the newer equipment will usually still be working after the old equipment is defunct.

To illustrate this problem, let's consider the following example. Machine A, which is presently used to perform a certain task, has a present market value of $1000 and an estimated remaining life of 2 years, with no salvage value. Operating costs with machine A are $1000 per year. After 2 years, we can replace machine A with machine B. This machine costs $25,000 to purchase, has an operational life of 10 years, and a salvage value of $100. Since the present market value of machine A is $1000, retaining machine A deprives the investor of the opportunity to derive $1000 in income by selling it. Thus the investor is in effect investing this $1000 to retain machine A. This concept is crucial to understanding the cash flows stated in Table 9–7. The alternative is to acquire machine C at a cost of $12,500. This machine has a six-year life and no salvage value. Operating costs for machine C are $1200 per year. The minimum attractive rate of return is 8%.

Because the service lives of these two plans are different, we have to use the equivalent-annual-cost approach. Since the cash flows associated with the machine A/machine B alternative are uneven, it is more convenient to compute the present worth of these costs and then spread this amount uniformly over the 12-year total life of the project. Therefore we have

$$\begin{aligned} P_{A,B} &= \$1000 + \$25{,}000\,(P|F,8,2) + \$1000\,(P|A,8,2) \\ &\quad + \$800\,(P|A,8,10)\,(P|F,8,2) \\ &= \$1000 + \$25{,}000\,(0.8573) + \$1000(1.7833) \\ &\quad + \$800\,(6.7101)\,(0.8573) \\ &= \$1000 + \$21{,}433 + \$1783 + \$4602 = \$28{,}818 \end{aligned}$$

The last term in this expression deserves special mention. The $800 annual operating cost for machine B is expended over a 10-year period beginning at the end of year 2. The quantity $800 $(P|A,8,10)$ therefore represents the equivalent single sum at the end of year 2. Then we have to determine the present worth of this single sum

TABLE 9–7. CASH FLOWS IN EVALUATING REPLACEMENT ALTERNATIVES

	ALTERNATIVES A, B		*ALTERNATIVE C*	
Year	*Initial investment*	*Operating cost*	*Initial investment*	*Operating cost*
0	$ 1,000		$12,500	
1		$1,000		$1,200
2	25,000	1,000		1,200
3		800		1,200
4		800		1,200
5		800		1,200
6		800		1,200
7		800		
8		800		
9		800		
10		800		
11		800		
12		800		

at year 0. We can do so by multiplying by the factor $(P|F,8,2)$. The equivalent annual cost is then

$$\begin{aligned} A_{A,B} &= \$28{,}818\,(A|P,8,12) \\ &= \$28{,}818\,(0.1327) = \$3824 \end{aligned}$$

We can find the equivalent annual cost of the machine C alternative more directly, as follows.

$$\begin{aligned} A_C &= \$12{,}500\,(A|P,8,6) + \$1200 \\ &= \$12{,}500\,(0.2163) + \$1200 \\ &= \$2704 + \$1200 = \$3904 \end{aligned}$$

Thus we would decide to retain machine A for its remaining two-year life and then replace it with machine B, instead of replacing it immediately with machine C.

Exercise

9–12 The salvage value of a 5-year-old piece of test equipment is now $1800; it will remain at that value in future years. The annual maintenance cost for this equipment is $600, and it will remain at that value for the foreseeable future. We can replace this equipment with a unit costing $4000. The new unit will last 10 years and have a salvage value of $1000. Its maintenance costs are projected to be $250 per year. The available interest rate is 8%. Should we replace the existing equipment?

9–6 BREAK-EVEN AND MINIMUM-COST MODELS

An important part of the economic analysis of an engineering operation is to establish the least costly way of carrying out the operation. The cost equation is generally a function of the level of activity in the operation. There are two ways to analyze this level of activity. One is to establish the level of output or the size of a project that produces the minimum cost per unit of output. This type of analysis often uses the same techniques that we examined in Chapter 5. The second type of analysis is to determine the level of output at which one mode of operation changes to another. This level is called the break-even point.

BREAK-EVEN ANALYSIS

When the cost of two ways of carrying out an engineering operation is a function of the same decision variable, we can establish the value of this variable for which the two alternatives have the same cost. This value is called the *break-even point*. The first step in this analysis is to express the total cost of each alternative as a function of the decision variable:

$$C_1 = f_1(x), \qquad C_2 = f_2(x)$$

where C_1 is the total cost for alternative 1 and C_2 is the total cost for alternative 2, and each equals the sum of the fixed costs and variable costs for their respective environments. The next step is to find x such that C_1 is equal to C_2. We can do this by mathematically manipulating the expression

$$f_1(x) = f_2(x) \tag{9–20}$$

or by analyzing graphs of these functions.

One of the important problems to which we can apply break-even analysis is the so-called make-or-buy problem. Suppose, for example, that a manufacturer of electronic calculators must procure a plastic carrying case for the calculators. The manufacturer can buy these cases for \$2 each or fabricate them in-house for a fixed cost of \$100,000 per year and a variable cost of \$0.75 per case. The cost of the "buy" option is

$$C_1 = \$2x$$

where x is the total number of cases bought each year. The cost of the "make" option is

$$C_2 = \$100{,}000 + \$0.75x$$

Equating these expressions and solving for x gives

$$\$2x = \$100{,}000 + \$0.75x, \qquad \$1.25x = \$100{,}000$$
$$x = 80{,}000 \text{ units}$$

Figure 9–7 gives the graphical representation of these two cost functions; it shows that the break-even point is 80,000 units. This means that the manufacturer must

FIGURE 9–7 A break-even analysis of the make or buy option

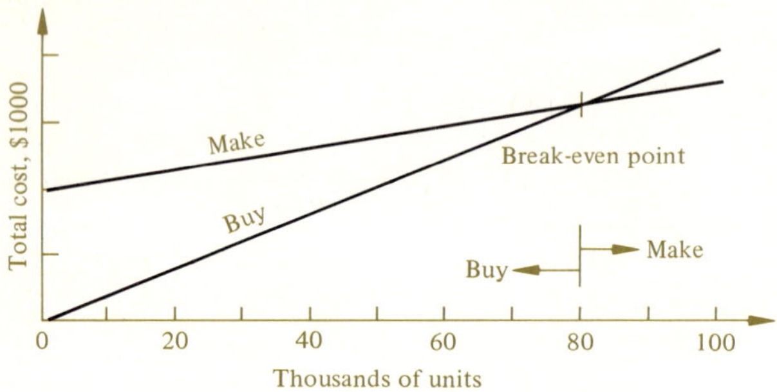

produce more than 80,000 units a year for the "make" option to be worthwhile. If the company anticipates producing only 65,000 units, the decision would be to buy the cases. If they estimate they'll need 95,000 units, they would manufacture the cases in-house.

MINIMUM-COST ANALYSIS

Sometimes the total cost of an engineering operation is a function of one component that increases as output increases, a second component that decreases as output increases, and a third that is fixed. In this case, the mathematical model has the form

$$C = rx + \frac{s}{x} + t \qquad (9\text{–}21)$$

where C = total cost of the activity, x = decision variable, and r, s, t = constant coefficients. In (9–21), the total cost C is a dependent variable, and x is an independent or decision variable. The rx term states that cost increases directly as x, while the s/x term shows that cost decreases as x increases. Thus the rx and s/x terms represent variable costs. The t term is the fixed costs.

One approach to finding the value of x that minimizes C is to compute C for various values of x, plot the results on a graph, and determine the minimum cost directly from the graph. A second and more direct approach is to apply the derivative procedure described in Chapter 5. We do this by taking the derivative of C with respect to x, equating the result to zero, and solving for x. For the total-cost model stated in Equation (9–21), this is

$$\frac{dC}{dx} = r - \frac{s}{x^2} = 0 \qquad \text{which yields} \qquad x = \sqrt{\frac{s}{r}}$$

For example, let's suppose that we want to determine the quantity Q of an item that we should buy to minimize the total annual cost of maintaining the item in

FIGURE 9–8 A minimum cost analysis

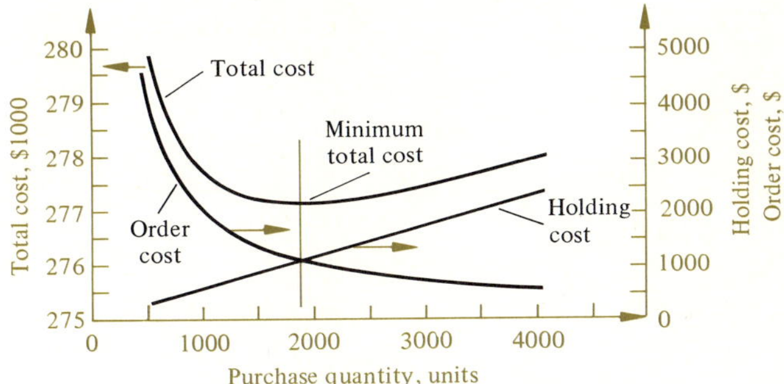

inventory. The annual demand for this item is 100,000 units. The cost of purchasing each unit is \$2.75. The order cost (the cost of placing an order) is \$20. The holding cost (which is a function of the space required to store a unit, the taxes, insurance, and interest on the capital required to buy the unit) is \$1.20 per unit per year. Then the model relating the total annual cost to these quantities is

$$C = \$1.20\,\frac{Q}{2} + \$20\,\frac{100{,}000}{Q} + \$2.75(100{,}000)$$

The first term states that the holding cost is a function of the average number of units held in inventory, which is $Q/2$. The second term gives the cost of placing $(100{,}000)/Q$ orders per year. The third term is the fixed cost of purchasing 100,000 units per year. Figure 9–8 shows a graphical solution of this problem. The minimum total annual cost corresponds to an order quantity Q of 1826 units. The more direct mathematical solution is as follows.

$$C = \$0.60Q + \frac{\$2{,}000{,}000}{Q} + \$275{,}000, \qquad \frac{dC}{dQ} = \$0.60 - \frac{2{,}000{,}000}{Q^2} = 0$$

$$\begin{aligned} Q &= \sqrt{\frac{2{,}000{,}000}{0.60}} \\ &= \sqrt{3{,}333{,}333} = 1825.74 \\ Q &= 1826 \text{ units per order} \end{aligned}$$

Exercises

9–13 An electric motor drives a pump that fills storage tanks in a refinery. It is estimated that the motor requires 420 horsepower-hours each day for 365 days. The following motors can be leased to perform this function.

Size (hp)	*Annual leasing charge*	*Operating cost* (per hp-hr)
20	$180	$0.0056
40	200	0.0052
75	210	0.0048
100	220	0.0045
150	240	0.0036
225	260	0.0030

Graphically show the total annual cost as a function of the size of the motor. Select the size that affords the minimum annual operating cost.

9–14 An engineering firm can purchase a small digital computer for $40,000. The firm estimates that the computer will last 5 years and have a salvage value of $4000 at the end of that time. Operating expenses are $80 per day, including maintenance and labor. As an alternative, they can buy computer time for $175 per day. At an interest rate of 10%, how many days per year must the firm operate the computer to justify buying it?

9–7 SUMMARY

This chapter has examined a number of the economic models that engineers need. These are important because engineering—the application and implementation of scientific principles—is practiced in a socioeconomic domain. That is why engineering students must satisfy minimum levels of course work in humanities and social studies. Further, engineer-in-training and professional engineering examinations in almost all states attest to the importance that practicing engineers attach to the basic principles of economic analysis. Economic problems constitute about 10% of these examinations.

Your early exposure to these economic concepts and models should enable you to adopt an economic framework for more advanced work in engineering. For instance, if you are examining two different truss designs in engineering mechanics, you may find it worthwhile to ask: How do the costs of these two designs compare?

PROBLEMS

P9–1 A tool-and-die shop has 40 horsepower of connected electrical load. It purchases electricity according to the following rates.

Kilowatt-hours per horse-power of connected load	0–60	61–120	120–360	Over 360
Cost per kilowatt-hour ($)	0.03	0.02	0.012	0.010

The shop presently uses 4000 kilowatt–hours per month.

(a) Calculate the monthly electric bill for this tool-and-die shop.

(b) The shop operator has an opportunity to acquire additional business that will require him to use an additional 1500 kilowatt–hours per month. At what rate should the operator value the cost of this additional electrical energy?

(c) The operator is considering the installation of new machines that will increase the connected load by 20%. Because these new machines will be used only for highly specialized jobs, they will add only 200 kilowatt–hours per month to the electrical usage. In considering the economic merit of purchasing these new machines, at what rate should the operator value this additional electrical energy?

P9–2 An engineering consulting firm is considering the construction of a new office. It can buy a corner lot for $80,000. The minimum office area that will satisfy the needs of the firm is 2800 ft². The maximum office area that can be constructed on the lot and satisfy all easement requirements is 4200 ft². Construction costs are $40 per ft².

(a) If office area is the decision variable, what is the fixed cost of this project?

(b) What is the variable cost of this project?

(c) Develop an equation that relates the total cost to the fixed and variable costs of the office construction.

P9–3 The owner of a frozen-food warehouse is considering an addition of 20,000 ft². This addition will have 300 linear ft of walls 28 ft high. The insulation of these walls can be either 6 inches or 9 inches thick. Using the additional 3 inches of insulation thickness will add $124,000 to the cost of constructing the addition, but will save 4.5 tons of refrigeration. Each ton of refrigeration costs $2200 to install and $960 per year in electrical energy and maintenance cost.

(a) What is the net difference in construction cost of using the 9-inch insulation?

(b) At an interest rate of 8%, how long will it take to recover the net cost of installing the 9-inch insulation?

P9–4 A firm is considering buying one of two machines. The minimum attractive rate of return is 10%; which of these two machines should the firm buy?

	Machine A	*Machine B*
Purchase cost	\$2200	\$7000
Annual benefit	980	1250
Salvage value	350	900
Useful life	6 years	12 years

P9–5 An asset has a first cost of \$18,000, an estimated useful life of 12 years, and an estimated salvage value of \$1800. Using the straight-line method of depreciation, find (a) the annual depreciation charge, (b) the book value after 9 years, and (c) the annual depreciation rate expressed as a percentage of first cost.

P9–6 An item of conveyor equipment for bridging a railroad track at an automotive plant costs \$40,000 installed, has a useful life of 8 years, and a salvage value of \$2000. Use the double-declining-balance method of depreciation (in which the annual depreciation rate is $2/n$). Find (a) the depreciation charge in the 4th year, (b) the book value at the end of the 6th year, and (c) the total amount depreciated at the end of the useful life of the conveyor.

P9–7 Solve Problem 9–6, using the sum-of-the-years'-digits method of depreciation.

P9–8 Certain projects last such a long time that their useful lives are considered infinite. If such projects have annual operating and maintenance costs "in perpetuity," the present worth of these perpetual costs, called the *capitalized cost,* is given by $P = A/i$.

(a) Using the concept of the limit from calculus, show how Equation (9–11) yields the capitalized cost.

(b) The annual maintenance cost of a steel trestle is \$2400. At an interest rate of 8%, what is the capitalized cost of these annual maintenance expenses?

(c) A university is seeking endowed professorships; each costs \$48,000 per year. How much must a benefactor donate to the university to have an endowed professorship named in his or her honor, given that the yield on this amount is 6% per annum?

P9–9 The initial cost of a proposed dam is \$100,000,000. The present worth of all future maintenance costs is \$40,000,000. The present worth of all costs paid by public users is estimated to be \$30,000,000. The present worth of all benefits is \$225,000,000. Show three ways to compute the benefit/cost ratio for the proposed project. Is the project worthwhile under each of these alternative approaches?

P9–10 The chief engineer in a manufacturing firm assigns a new engineer the task of developing a compound interest table (such as those shown in Appendix B) for the interest rate 13%, the firm's minimum attractive rate of return. The table is to have the six factors shown in Appendix B, and also values of these factors for each period for $1 \leq n \leq 30$, and for each five periods for $30 \leq n \leq 100$. The headings in the table must be labeled accordingly. Develop and execute a FORTRAN program that will satisfy the chief engineer's needs.

P9–11 The annual demand for an item that the Ajax Widget Company can either purchase or produce is 4000 units. Costs associated with these alternatives are as follows.

	Purchase	*Produce*
Item cost per unit	\$4.10	\$3.60
Order cost per purchase	\$25.00	—
Set-up cost per production run	—	\$250.00
Holding cost per unit per year	\$0.75	\$0.75
Production rate per year, units	—	24,000

Should Ajax purchase or produce the item? What is the most economical quantity to purchase or produce at a time?

CHAPTER TEN
SYSTEMS AND FEEDBACK

Throughout this text, we have consistently woven the theme of a system, primarily by example. In Chapter 1, we viewed the Earth as a system and discussed it in qualitative terms. Any system is characterized by a set of inputs and a set of outputs. Further, the relationships between inputs and outputs are essential to any useful analysis of systems.

To illustrate these concepts, let us consider the simple deterministic model that constitutes Newton's second law.

$$F = Ma$$

Figure 10-1 shows two sketches of a system governed by this equation. The first shows the physical components explicitly, while the second represents the system (which is just the mass M) as a block with one arrow entering from the left (input, the force F) and one arrow leaving to the right (the output, the acceleration a). The second sketch is called a *block diagram* because inputs and outputs are shown as labels on arrows entering and leaving the block that represents the system. Such diagrams are useful abstractions of physical systems.

10–1 EXAMPLES OF SYSTEMS

A *system* is any collection of things intended to perform a function or functions. Inputs represent applied stimuli, and outputs represent responses. In general, inputs are any identifiable physical entities that have observable effects on the behavior of the system, while outputs constitute measures of the system's response.

The common vending machine is a useful example of a simple system. Figure 10–2 is a block diagram showing the inputs and outputs of the machine from the viewpoint of a prospective purchaser. Identifiable inputs (things affecting subsequent behavior) include the amount of money deposited and the electricity it

FIGURE 10–1 System described by $F = Ma$: (a) pictorial view, (b) block diagram with input F and output a

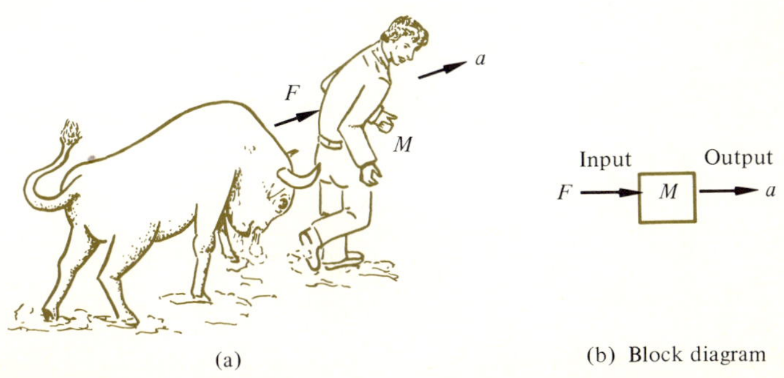

FIGURE 10–2 A block diagram of a vending machine—user's viewpoint

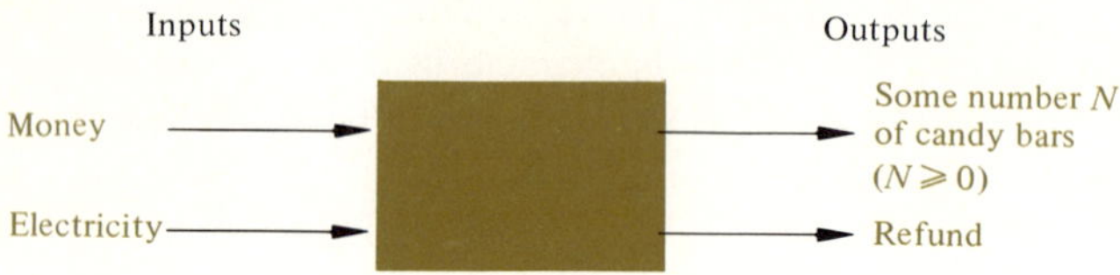

consumes for lighting to make the machine appear an attractive place to spend money. The outputs (responses to inputs) consist of some number N of candy bars and, in some cases, a refund.

Next consider the vending machine from the viewpoint of the person who services it. We need another block diagram (Figure 10–3), since this individual will see candy as an input and money as an output. The classification of physical processes into inputs and outputs thus depends on the perspective of the user. For a consumer, the vending machine absorbs money and produces candy bars. The individual who services the machine sees the consumption of candy bars and the output of money. Thus selecting inputs and outputs for a system depends on the physical processes as well as the perspective from which one considers the system.

As another example, consider the economy of the country. Since this is a complicated, imperfectly understood system, we shall consider only some of the more obvious inputs and outputs. Figure 10–4 shows a simplified block diagram; the input and output arrows labeled "etc." are measures of collective ignorance about the system.

The systems-engineering approach has also been applied to model the administration of criminal justice, beginning with the commission of a crime and ending with a prison term. Over the last decade, systems concepts have been successfully applied to an increasingly wide range of problems. This rapid evolution has generated fields of engineering that did not exist in the 1950s. Engineers now work in such areas as urban transportation and city planning.

Many systems are quite complex. If we hope to make any analytical progress with the systems concept, we must limit the discussion to simple systems with well-defined input and output variables and simple deterministic relationships between these variables. In Section 10–2, we shall develop algebraic models for such systems.

FIGURE 10–3 A block diagram of a vending machine—seller's viewpoint

FIGURE 10–4 A block diagram of the national economy

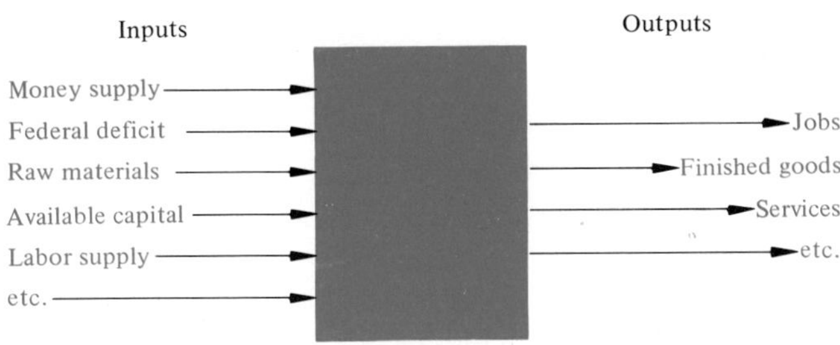

10–2 BASIC BUILDING BLOCKS OF SYSTEMS

A record player provides a suitable example for our study of systems. Its inputs are electricity and a record, and the output is the sound produced. Figure 10–5 shows this in a block diagram.

This simple block diagram represents many physical processes that we shall discuss briefly. A record contains spiral grooves that have rough surfaces produced by an impression from a master recording. Physics has produced techniques to "record" audible sounds in the analog form of surface roughness. When the record plays, this surface roughness deflects the stylus (commonly a diamond point) as the record revolves. The stylus is attached to a block of piezoelectric material that produces an electrical voltage when mechanically stressed or deformed. As the stylus moves through the grooves, its motion deforms the piezoelectric material, and an electrical voltage results. This voltage is a faithful electrical analog of the sound originally recorded. Since humans cannot hear electrical voltages, loudspeakers are used to convert electrical signals into audible sound.

Figure 10–6 shows a very simple sound system, in which the loudspeaker is connected directly to the stylus. Such systems are not manufactured and sold for a very simple reason—the sound produced in the loudspeaker would be inaudible. The electrical voltage produced by the motion of the stylus in the record grooves is much too small to drive a loudspeaker directly. The electrical voltage has to be increased to produce audible sound. To fully appreciate the methods used, we must examine the mathematical means commonly used to represent audible sound and electrical voltages.

FIGURE 10–5 A block diagram of a phonograph

FIGURE 10–6 Major elements of a simple phonograph

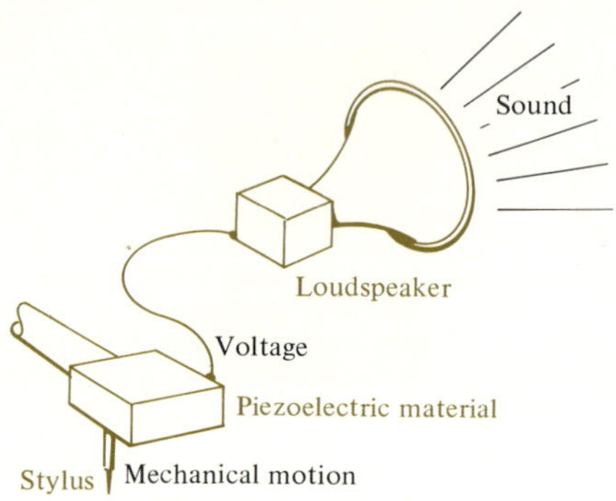

Let's start by examining pure tones produced by tuning forks, organ pipes, and horizontal piano strings. Sound is nothing more than a variation in pressure transmitted through the air. When someone speaks, his or her vocal cords vibrate and radiate zones of high and low pressure through the larynx and into the atmosphere. When these waves of pressure strike the ears of another person, they set the eardrums in motion. This motion is eventually transmitted to the brain as the sensation of sound.

We can describe sound at any point by changes in air pressure as a function of time. We can represent a particular sound by a plot of pressure (p) versus time (t), as Figure 10–7 illustrates. Note that as time increases, the pressure relaxes to the value p_0 that symbolizes the atmospheric pressure present in the absence of sound. (Sound also travels through liquids and solids as pressure waves.)

When a pressure detector is a fixed distance from a tuning fork that has been vibrating for several seconds, the equation of pressure versus time is

$$p(t) = p_0 + A \sin(\omega t + \phi) \tag{10–1}$$

FIGURE 10–7 Sound pressure p versus time t. As the sound dies away, the pressure p relaxes to the atmospheric pressure p_0.

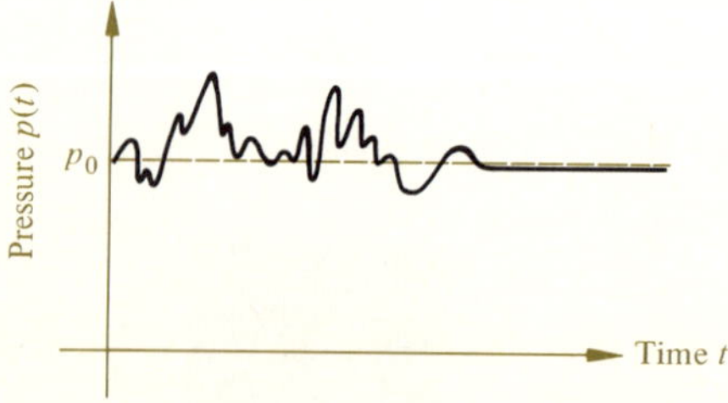

In this equation, A is a constant that depends on how hard the tuning fork was initially struck. The constant ϕ depends on the distance between the detector and the tuning fork, as well as on the time at which the fork was initially set in motion. The third new symbol in this equation is ω, the radian frequency. Physically ω depends on the size and mass of the tuning fork. It measures the rapidity with which the pressure waves fluctuate.

We may rewrite Equation (10–1) as

$$p(t) - p_0 = A \sin(\omega t + \phi) \tag{10–2}$$

The right-hand side of Equation (10–2) is a sinusoid. The three quantities that characterize it are the amplitude A, the radian frequency ω, and the phase ϕ.

Figure 10–8 illustrates the physical meaning of these quantities. The sinusoid is bounded by the values A and $-A$. The amplitude is the largest value, either positive or negative, attained by the function. The phase is the time difference between the pressure wave and that represented by $A \sin(\omega t)$. Changing the phase from $\phi = \pi/4$ radians to $\phi = 0$ radians simply shifts the entire curve in the direction of increasing time, as Figure 10–8 illustrates.

The period T of a sinusoid is the total time required for the function to complete one cycle. A cycle is a part of the function that begins at zero, increases to its maximum, drops back to zero, decreases to its minimum, and finally climbs back to zero. From Figure 10–8 and Equation (10–2), you can see that the following equation gives the period.

$$T = \frac{2\pi}{\omega} \tag{10–3}$$

The frequency f, cycles per unit of time, is the reciprocal of T.

$$f = T^{-1} = \frac{\omega}{2\pi} \tag{10–4}$$

FIGURE 10–8 Effect of phase shift on a sine wave. Both sinusoids have the same amplitude and frequency.

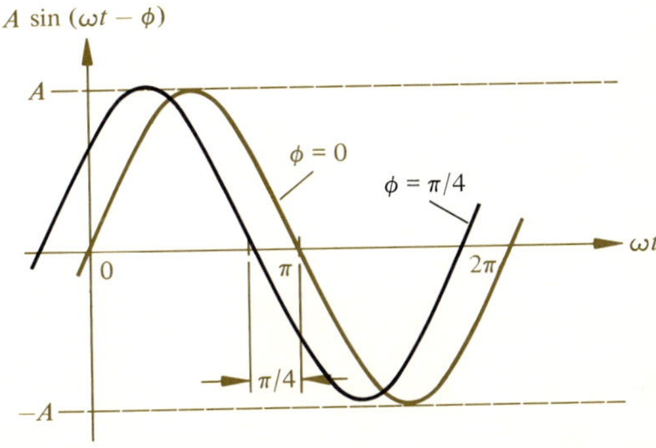

This equation shows that ω is a measure of how rapidly the function alternates between positive and negative extremes along the time axis. As the radian frequency is increased (by using a small, lighter tuning fork, for example), the sinusoid wiggles more rapidly. That is, the frequency f increases and the period T required for the function to complete one cycle decreases.

Frequency is directly related to the pitch with which humans hear sound. As frequency increases, the perceived pitch is higher. The human ear in a young person with normal hearing responds to sounds with frequencies in the range

$$25 \text{ Hz} < f < 18000 \text{ Hz}$$

Hz is an abbreviation for hertz, the SI unit of frequency. One hertz equals one cycle per second.

Sinusoids are very important in engineering and the physical sciences. If we know how a system responds to a sinusoidal input, we can determine its response to almost any other input as long as the system is well behaved (engineers say that the system must be linear). Fortunately a wide range of physical systems can be accurately modeled as linear. The output of a linear system excited by several inputs is equal to the sum of the outputs produced by each input acting alone.

To fulfill our intention of developing simple algebraic models to describe physical systems involving one input and one output, we have considered a record player and have identified the nature of acoustical and electrical signals. In particular, we have expressed such signals as functions of time. Sinusoids are useful for this purpose, since they accurately model pure musical tones.

Now we shall represent the electrical signal present at the output of a phonographic stylus as $v(t)$, where v denotes that this quantity is an electrical voltage. Recall that this voltage is too small to be fed directly to a loudspeaker. Engineers overcome this problem with electrical devices called amplifiers; the block diagram in Figure 10–9 accurately models these. The input to the device is the electrical signal $v(t)$, and the output is another signal $v_o(t)$. The following deterministic model relates these two entities.

$$v_o(t) = Gv(t) \qquad (10\text{–}5)$$

In this equation, the quantity G is called the gain. It is normally a constant greater than 1. The input and output signals have exactly the same shape when they are

FIGURE 10–9 Block diagram of an amplifier of gain G.

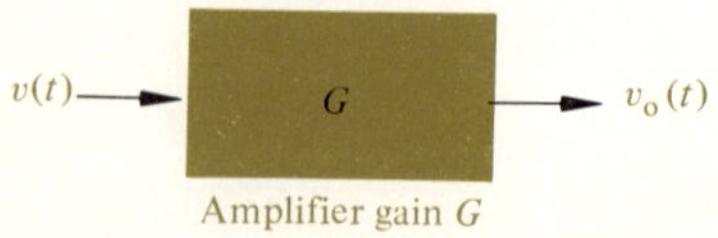

$v(t)$ → G → $v_o(t)$

Amplifier gain G

plotted versus time, but the output signal will have a greater amplitude. In mathematical terms (if we assume, for simplicity, that the phonograph record produces only pure tones), we may write the input signal in the form

$$v(t) = A \sin(\omega t + \phi) \tag{10–6}$$

Using Equation (10–5), we can write the output signal as

$$v_o(t) = Gv(t) = GA \sin(\omega t + \phi) \tag{10–7}$$

These two signals, plotted in Figure 10–10, have the same functional dependence on time. If both were played through loudspeakers, each would produce exactly the same pitch, but the amplifier output would produce a much louder sound than the input. The greater electrical signal available at the amplifier output drives the speaker with greater amplitude, and a louder sound results. The amplifier is the element missing from the simple system shown in Figure 10–6. Manufacturers of quality stereo equipment take great pains to extend the frequency response of their equipment so that the range is wide enough to be compatible with the hearing capabilities of the human ear. On an inexpensive tape recorder, high-frequency instruments (violin, cymbals, etc.) sound "washed out" because of the inferior frequency response of the amplifiers used. The devices simply cannot respond to high-frequency tones, and the model

$$v_o(t) = Gv(t)$$

is not valid for input signals of high frequency.

FIGURE 10–10 Amplifier input and output voltages versus time

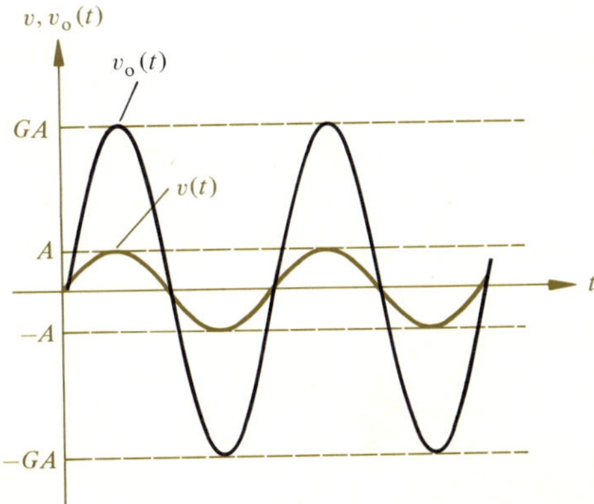

FIGURE 10–11 Model of a summer. The output equals the difference between the two inputs.

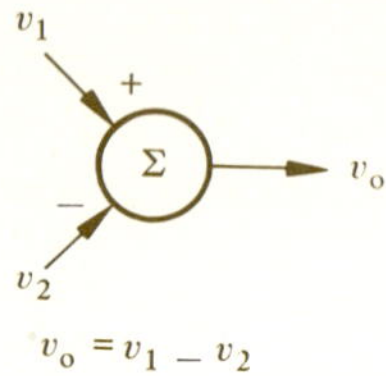

The amplifier gain G depends on the electrical components and interconnections used in construction. Any electrical device used for communications contains at least one amplifier. Examples include radios, televisions, tape recorders, and the telephone system. The amplifier is the first building block of the set that we shall use to algebraically analyze simple systems.

To expand this set of basic elements in a natural fashion, let us consider a typical home heating system. For the moment, let's focus on the thermostat, which is normally located in a central room. This device has two scales; one reads the actual room temperature and the other is set to the desired temperature. The thermostat compares the actual temperature with the desired temperature, and uses the difference to generate signals that control the furnace. When the room's temperature is above the desired temperature, the thermostat stops the flow of fuel to the furnace. When it is below, the thermostat directs the furnace to burn fuel. Thus the thermostat is generating a signal that depends on the difference between the actual and desired temperatures.

To analyze this and similar systems, we need a block diagram and associated model that behave as the thermostat does. Such a device is called a summer; it is characterized by the block diagram in Figure 10–11 and the following deterministic model.

$$v_o(t) = v_1(t) - v_2(t) \tag{10–8}$$

This device has two inputs ($v_1(t)$, $v_2(t)$), which are analogous to the two temperature settings on a thermostat; and it has one output. Note the algebraic signs near each of the input arrowheads. Other variants of summers are possible. Figure 10–12 shows these, along with the associated deterministic models.

FIGURE 10–12 Other forms of summers

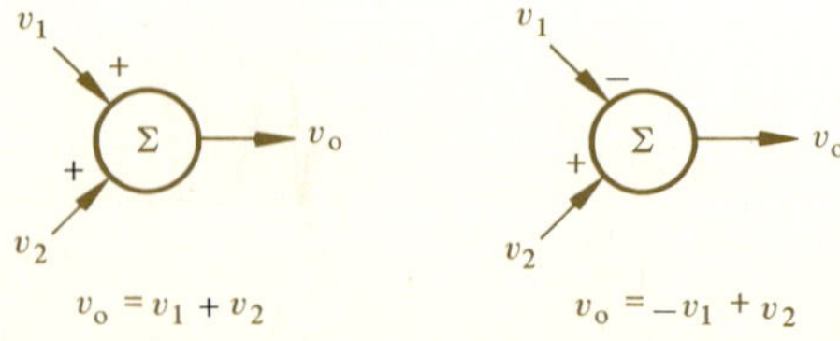

Aided by the concepts of an amplifier and a summer, we may now develop the concept of feedback, one of the most significant ideas in all system theory.

10–3 FEEDBACK

Imagine that you are standing at the end of a long bar and the bartender slides a foam-covered mug of cold beer along the bar toward you. How do you intercept the moving mug before it falls on the floor? You extend your hand and judge the distance between the moving mug and hand. You then begin to slowly move your hand away from the approaching mug in order to cushion its collision. If you do it properly, you will manage to arrest the mug's motion without spilling any of the precious contents.

The outputs of this system are the hand's position and velocity, while the inputs are the position and velocity of the mug and the desired position and velocity of your hand. As the mug approaches, you intuitively calculate where your hand should be and how fast it should be moving. The difference between the actual values of these two quantities and the desired values constitute error signals that cause you to change the position and velocity of your hand. The goal is to constantly reduce this error and to position your hand so that you can best cushion the impact of the mug when it reaches your hand. Figure 10–13 uses the basic algebraic summer to represent this.

The summer generates the error or difference between the desired and actual positions and velocities. The error message is an input to the body, which acts through the brain, nerves, and muscles to produce the hand's actual position and velocity. One of the summer's inputs and the system's output are identical; the dashed line in Figure 10–13 indicates this explicitly. This system provides our first example of feedback. The output of this system is fed back and serves as an input.

Feedback may at first be a troubling concept. To clarify it, we again consider the case of a thermostat and the associated heating equipment in a home. The output of this system is simply the temperature in the house. To control this, the homeowner spends money for equipment and fuel. At the same time, the temperature in the house (the output) is an input to the home heating system through the thermostat. So, in a very familiar physical example, you can see how the same physical quantity serves as both an input and an output. Figure 10–14 shows a block diagram of the

FIGURE 10–13 Feedback in an attempt to match the position and velocity of a moving object

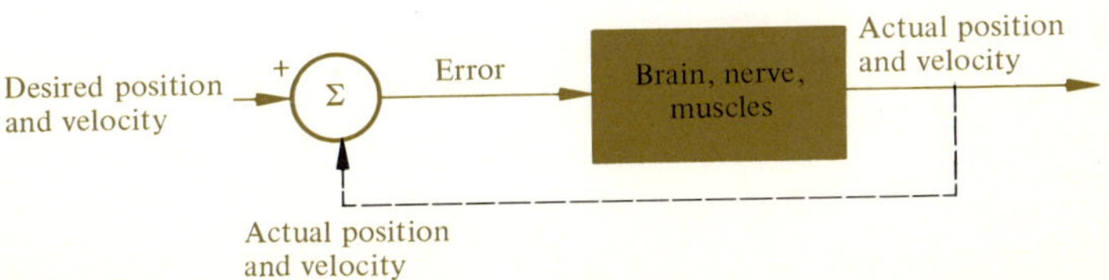

FIGURE 10–14 Block diagram of a home heating system. Using feedback, the summer produces a signal equal to the difference between the desired and actual temperatures.

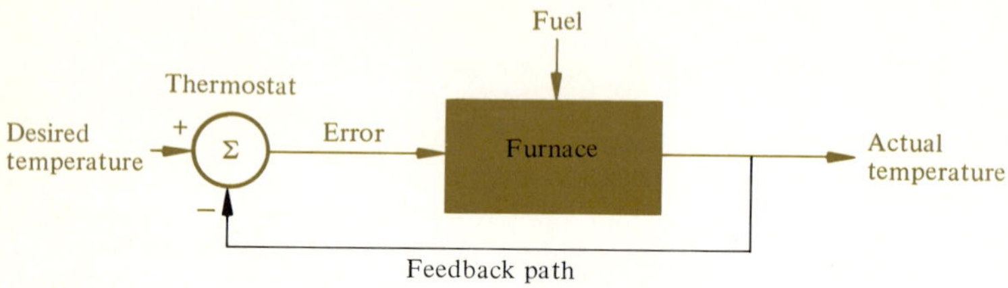

home heating system. In this example, placing a temperature-sensitive element inside the house physically supplies the feedback. In some other cases, the physical mechanism responsible for feedback is less obvious.

Feedback serves a control function. It uses error messages produced by comparing a desired output with an actual output so that a system can achieve a goal. You pick up objects in this way. You compare the position of your hand with the position of the object. Then through the system composed of your brain, nerves, muscles, and eyes, you act to reduce this error signal to zero so that your hand reaches the desired position. This is feedback. Many engineering design problems use feedback to control a process or device.

In addition to this control function, feedback also produces interesting behavior in simple systems. We shall examine a few of these examples in detail, but first we'll expand the concepts introduced in this chapter by discussing an example that involves manipulating systems composed of amplifiers and summers.

Consider the system shown in Figure 10–15. It is composed of two amplifiers—one with a gain of 100, the other with a gain G—and one summer. As always in analyzing a system, we are seeking a deterministic model linking the output $y(t)$ and the input $x(t)$.

Before we write any algebraic equations, we have to address a point that is often confusing. When you consider a block diagram like the one shown in Figure 10–15, do not think of the signals as present in a form that propagates around and around the feedback loop (like an echo). These signals are present in the steady state throughout the system. It follows that a signal can be associated with each connecting line between various blocks. The diagram appears again in Figure 10–16; it is

FIGURE 10–15 A two-amplifier system using feedback

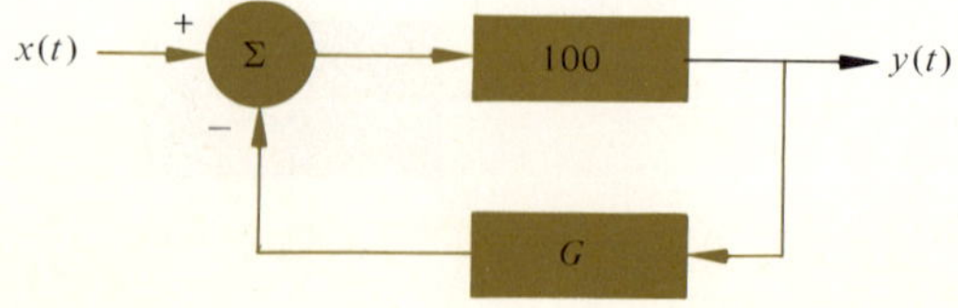

FIGURE 10–16 The input to the amplifier with gain G is equal to the system's output.

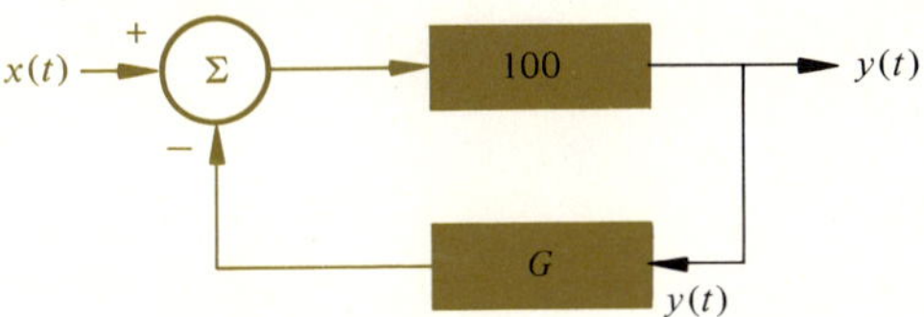

like Figure 10–15 except that we have identified the input to the amplifier with gain G as $y(t)$. Using Equation (10–5) with appropriate substitutions, we get the output of this amplifier.

$$Gy(t)$$

The block diagram appears again in Figure 10–17; this time we have explicitly identified the output of the amplifier. This output is, of course, one of the two inputs to the summer.

Applying the algebraic model given in Figure 10–11, we find that the output of the summer must be

$$x(t) - Gy(t)$$

However, the output of the summer is the input of the amplifier with gain 100. We redraw the relevant part of the block diagram in Figure 10–18. Applying Equation (10–5) again, we obtain an equation that involves the two amplifier gains, the system input $x(t)$, and the system output $y(t)$

$$y(t) = 100[x(t) - Gy(t)]$$

We can rearrange this equation to isolate the terms involving $y(t)$ on the left-hand side.

$$y(t) + 100Gy(t) = 100x(t)$$

Finally we can get the system output in terms of the input.

$$y(t) = \left(\frac{100}{1 + 100G}\right)x(t)$$

FIGURE 10–17 The output of the amplifier with gain G, $Gy(t)$.

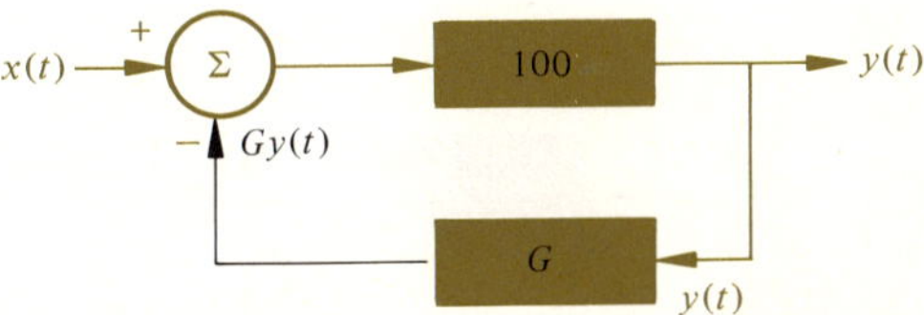

FIGURE 10–18 Part of the system shown in Figure 10–15

This represents a "solution" for the block diagram that first appeared in Figure 10–15. We have obtained the system output $y(t)$ in terms of the system input $x(t)$ and the two amplifier gains. The apparently complex system in Figure 10–15 behaves like a single amplifier; its gain is

$$\text{Gain} = \frac{100}{1 + 100G}$$

The block diagram in Figure 10–19 displays precisely the same relationship between input and output as does the three-component block diagram of Figure 10–15. This assertion is troublesome, since it indicates that systems containing feedback do not behave any differently from simple one-amplifier systems. Fortunately, the system composed of one summer and two amplifiers has a number of significant advantages over the simpler one-amplifier block diagram of Figure 10–19. We discuss these advantages in the next section.

Exercises

10–1 For $0 \leqslant t \leqslant 20$ sec, plot the following three sinusoids on a single sheet of graph paper. Use a calculator.

$$x(t) = 10 \sin\left(t - \frac{\pi}{4}\right), \qquad y(t) = 10 \sin(t), \qquad z(t) = 10 \sin\left(t + \frac{\pi}{4}\right)$$

$$0 \leqslant t \leqslant 20 \text{ sec.}$$

10–2 It is often convenient to use either an electronic calculator or a digital computer to prepare tables of values of sinusoids for different arguments. FORTRAN has library functions that aid in the evaluation of sinusoids. The following are examples.

Algebraic	FORTRAN
$y = \sin(x)$	Y = SIN(X)
$a = \cos(b)$	A = CØS(B)

FIGURE 10–19 A one-amplifier system "equivalent" to the system shown in Figure 10–15

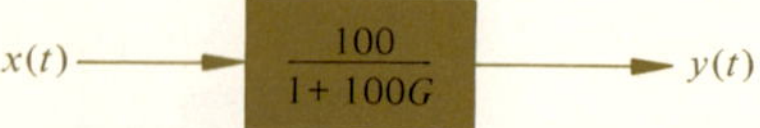

FIGURE 10–20 A two-amplifier system

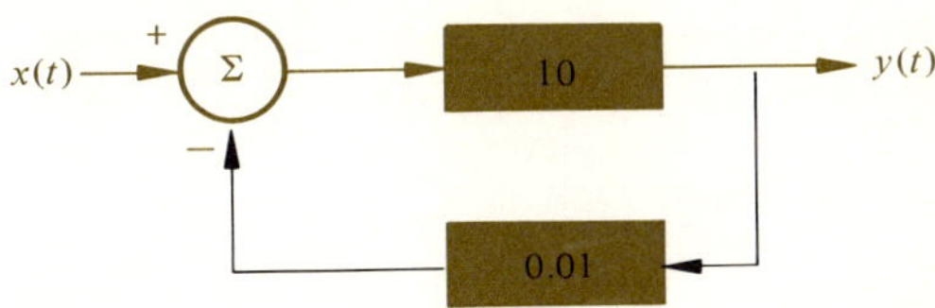

Write and execute a FORTRAN program to evaluate the following two sinusoids over the range $0 \leq t \leq 20$ sec.

$$x(t) = 1.22 \sin\left(t - \frac{\pi}{7}\right), \qquad y(t) = 2.87 \sin\left(t + \frac{3\pi}{8}\right)$$

10–3 Using an appropriately modified version of the program in Exercise 10–2, evaluate the following three sinusoids over the range $0 \leq t \leq 20$ sec.

$$a(t) = 10 \sin(t), \qquad b(t) = 10 \sin(4t) \qquad c(t) = 10 \sin(8t)$$

These results illustrate the effects of changing the radian frequency ω.

10–4 The output $v_o(t)$ and input $v(t)$ of an amplifier are given below. Plot both of these functions over the range $0 \leq t \leq 0.004$ sec, and determine the gain G of the amplifier used.

$$v_o(t) = 15 \sin\left(1800t + \frac{\pi}{4}\right), \qquad v(t) = 0.12 \sin\left(1800t + \frac{\pi}{4}\right)$$

10–5 Find an expression for the output $v_o(t)$ of the summer shown in Figure 10–11. It has the following inputs.

$$v_1(t) = 3 \sin(10t), \qquad v_2(t) = 2 \sin(10t)$$

10–6 Find an expression relating the output $y(t)$ to the input $x(t)$ for the system shown in Figure 10–20.

10–4 ADVANTAGES OF FEEDBACK

We can demonstrate one of the most striking advantages of feedback by examining the manufacture of two systems (one with feedback and the other without) for use as amplifiers in a complete phonograph. To do this, let's consider the block diagrams in Figure 10–21. We can readily find the overall gain of the first system, defined as the ratio of output to input, by using the techniques illustrated in analyzing the system of Figure 10–15.

FIGURE 10–21 Systems for sensitivity calculations

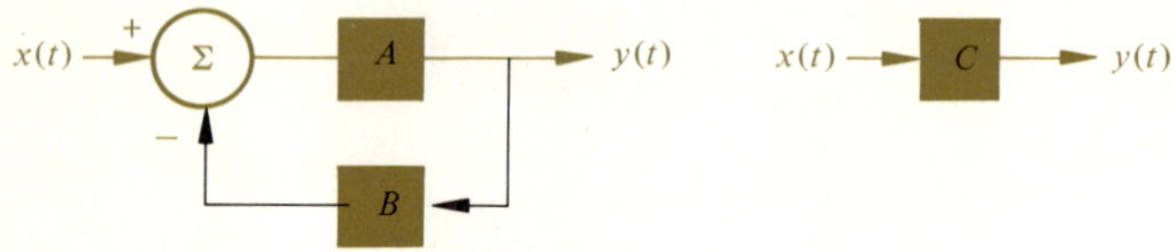

(a) System with feedback (b) System without feedback

$$\frac{y(t)}{x(t)} = \frac{A}{1 + AB} \tag{10–9}$$

The gain of the second system is, of course, C.

Consider the viewpoint of a manufacturer of both types of amplifiers. As a going concern, the company is naturally worried about the costs of the components used. Customers will pay only so much for the amplifiers, and the manufacturer has to keep costs below this level to ensure a profit and an adequate return on the stockholders' investment. The cost of components goes up as their accuracy increases. In simple terms, an amplifier with a gain accurate to within 1% will cost significantly more than an amplifier with a gain guaranteed to be accurate to within 10%. For example, imagine that we need an amplifier with a gain of 100. Amplifiers displaying gains between

$$90 \leq \text{gain} \leq 110 \qquad (10\% \text{ accuracy})$$

cost less than amplifiers with gains in the range

$$99 \leq \text{gain} \leq 101 \qquad (1\% \text{ accuracy}).$$

This should seem quite reasonable, since more accurate amplifiers require much more precisely made parts. We wish to investigate the effect of feedback on this issue.

In systems with one amplifier we can define a sensitivity (somewhat more simply than commonly appears in the engineering literature) by the expression

$$\frac{d(\text{system gain})}{d(\text{amplifier gain})} = \frac{dC}{dC} = 1.$$

The sensitivity is the ratio of a change in the total system gain to the change in the gain of the amplifier. This is a simple system (Figure 10–21(b)), and the numerical value obtained for the sensitivity is unity. The obvious physical interpretation is that a change of 10% in amplifier gain produces a 10% change in system gain. In the same way, a change of $x\%$ in amplifier gain produces a change of $x\%$ in system gain. In short, a sensitivity S of unity implies that the system gain changes in step with the amplifier gain. A simple system using an amplifier with a gain guaranteed only to 20% will have a 20% inaccuracy in its behavior. The sensitivities of the more complex system will display different results. Equation (10–9) gives the system gain.

$$\frac{y(t)}{x(t)} = \frac{A}{1 + AB}$$

Since there are two amplifiers, we can define two sensitivities.

$$\frac{\partial(\text{system gain})}{\partial A} \quad \text{and} \quad \frac{\partial(\text{system gain})}{\partial B}$$

Because we have an expression for the overall system gain in terms of the two amplifier gains A and B, we can calculate the two sensitivities by differentiating.

$$S_A = \frac{\partial(\text{system gain})}{\partial A} = \frac{\partial}{\partial A}\left(\frac{A}{1 + AB}\right)$$

$$= (1 + AB)^{-1} - AB(1 + AB)^{-2} = \left(\frac{1}{1 + AB}\right)^2$$

$$S_B = \frac{\partial(\text{system gain})}{\partial B} = \frac{\partial}{\partial B}\left(\frac{A}{1 + AB}\right)$$

$$= -\left(\frac{A}{1 + AB}\right)^2$$

To evaluate the three sensitivities, we select a set of values for A, B, and C such that the two systems have nearly the same system gain.

$$A = 10^4, \qquad B = 10^{-2}, \qquad C = 100$$

The gains of the two systems are essentially identical. To be sure, there is some small difference, but it is physically unimportant. Now we can evaluate the three sensitivities. Of course, the sensitivity of the single amplifier system is unity, but we list it below for completeness.

$$S_A = 0.000098, \qquad S_B = -0.98(10^4), \qquad S_C = 1$$

These results indicate some remarkable effects obtained through the use of feedback. Note that the overall sensitivity S_A to the large amplifier gain is essentially zero. Thus the manufacturer can use inexpensive components to construct this amplifier. The deviations in its gain A from a specified value will have no effect on the overall behavior of the system. The sensitivity S_A is so small for this case that variations in A are unimportant.

The sensitivity S_B has a value of 10^4. This means that a change of 1% in this amplifier gain will result in an enormous variation in the total behavior of the system. The saving feature is that the gain of this amplifier is less than unity; thus the amplifier can be produced, in most cases, rather inexpensively, even to high accuracy.

From this analysis we can draw the following conclusion. As long as the cost of a summer and a very precise amplifier with a gain less than unity is significantly less than the cost of an amplifier with gain much greater than unity, the more complex system is preferable. This is because the amplifier with gain A can be bought inexpensively, since its accuracy is unimportant.

10–5 FEEDBACK AND NOISE

In amplifying any signals, the engineer must be concerned with noise. It has many physical origins and is always present. Whether the noise is sufficient to corrupt some set of physical measurements depends on the cleverness of the individual designing the total system used to make these measurements. The importance of noise becomes painfully obvious in the laboratory. It can render the results of carelessly designed experiments meaningless.

We can explain the origins of noise by using different physical situations. Consider the transmission of a radio signal from one point to another. A transmitter and antenna radiate the electrical signal into the air, and it propagates according to a set of natural laws through the atmosphere until the receiving antenna picks it up. There is no way to control the environmental conditions that this signal experiences as it propagates from transmitter to receiver. Random factors change the signal. The most vivid example is lightning. If lightning strikes near the path of propagation, this catastrophic electrical event will distort the signal. In exactly the same way, the amount of humidity present in the air will affect (although usually to a very small degree) the rate at which the signal is dissipated as it travels.

The random motion of the atoms that compose all solids imposes lower limits on the experimental voltages and currents that one can hope to measure in the laboratory. This motion, which persists down to zero degrees Kelvin, produces small voltages and currents that corrupt the measured values.

Let's return for a moment to the problem in Section 10–4. The electrical signal produced by the motion of a diamond stylus in the grooves of a record may mix with other electrical signals present in the room in which the record player is being used. For example, improperly filtered CB radio transmitters generate signals that propagate through the air and are "picked up" by the system.

To illustrate the positive aspects of feedback, we shall now construct an algebraic model in which noise is explicitly included. We can do this by comparing the performance of the two systems of Figure 10–21 with respect to noise. We have redrawn these models in Figure 10–22, using the numerical values introduced in the latter part of Section 10–4.

FIGURE 10–22 Numerical values of gains for sensitivity comparisons

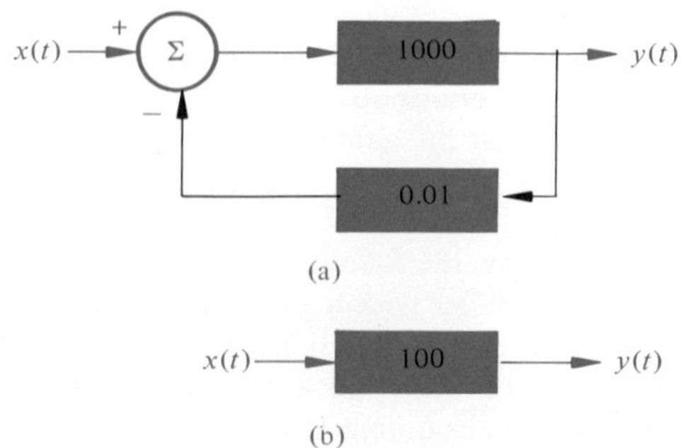

The function $n(t)$ will represent the noise signal. Thus the sum of the stylus signal and the noise gives the input.

$$x(t) + n(t)$$

The output of the system of Figure 10–22(b) follows immediately from the basic deterministic model for an amplifier.

$$y(t) = 100x(t) + 100n(t)$$

This demonstrates that noise shows up dramatically in the output of the amplifier, with its magnitude increased by a factor of 100, certainly not a desirable situation. Any noise that manages to intrude into the system's input will be amplified and heard. The simple one-amplifier phonograph does not offer any opportunity to improve the system.

Now we shall analyze the more complex system of Figure 10–22(a). A few minutes with a paper and pencil will show that exactly the same output results with noise present.

$$y(t) = 100x(t) + 100n(t)$$

In this situation, however, there is a way to improve the characteristics of the noise rejection. To see this, consider a modification of the phonograph. Split the amplifier with gain 10,000 into two pieces (one with gain D and the other with gain $10{,}000/D$). Figure 10–23 shows the block diagram for this modified system. Amplifier D is a well-shielded "preamplifier"; the noise is able to enter the system only after this device.

Application of the algebraic models and some manipulation yields the following equation for the system's output $y(t)$ in terms of the input $x(t)$ and the noise $n(t)$.

$$y(t) = 100x(t) + \frac{100}{D}n(t)$$

The amplifier gain D is available for design purposes. By choosing its value greater than 100, we can make the noise present in the output of the system less than the noise present at the input. As a concrete example, we choose $D = 1000$, with the following result:

$$y(t) = 100x(t) + 0.1n(t)$$

FIGURE 10–23 A system that uses feedback to reduce the effects of noise

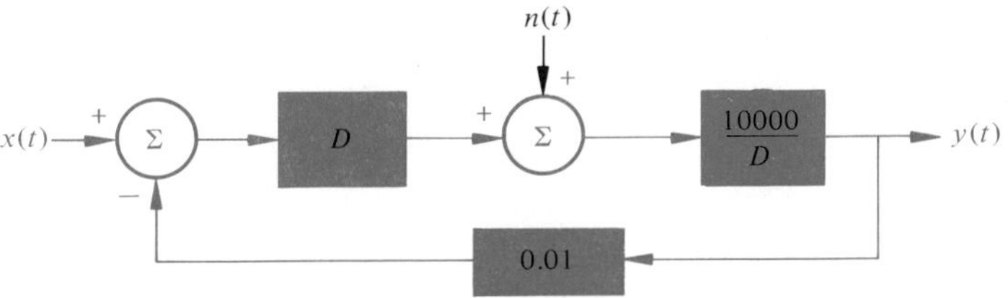

The output fed to the loudspeaker consists of two terms. The first is the desired output: it is 100 times the input. The second term depends on the inevitable noise, but the amplitude of this noise has been decreased by a factor of 10. Compare this $y(t)$ with the output for the one-amplifier system.

This excursion into noise has been rather brief. We have not begun to scratch the surface of the many facets of this concept. It has been enough for our purposes to look at one example of using a feedback path to reduce the undesired effects of noise. Through sophisticated mathematical models, engineers have developed many ways of decreasing the effects of noise in systems.

In the next section, we shall add one additional element to the basic set of system building blocks. This will widen the class of problems that we can analyze.

10–6 THE INTEGRATOR

A chemical engineer is responsible for a large chemical storage tank with both inlet and outlet valves, as sketched in Figure 10–24. The function $f(t)$ gives the rate in liters/hour at which chemical enters the tank. The function $e(t)$ gives the rate in liters/hour at which it leaves through the outlet valve. The tank has a total capacity of 1000 liters. At some initial time $t = 0$, the tank contains 400 liters. Because of a

FIGURE 10–24 A storage tank with inlet and outlet valves. The net flow of fluid into the tank is $f(t) - e(t)$.

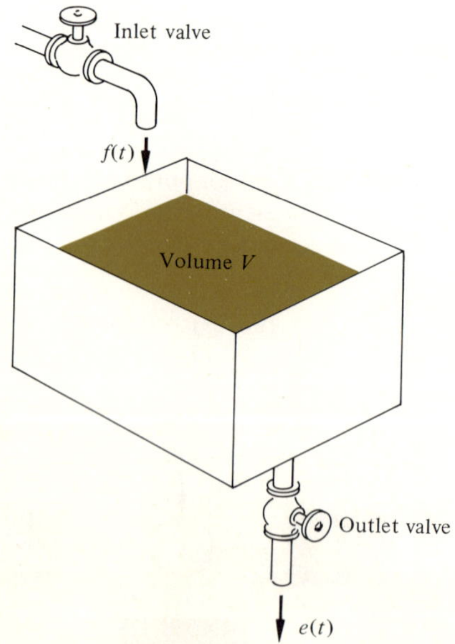

computer error, the two valves are set so that the functions $f(t)$ and $e(t)$ have the following forms.

$$f(t) = t^2 \quad \text{liters/hour,} \qquad e(t) = t \quad \text{liters/hour}$$

where t is measured in hours.

The problem is first to draw a block diagram that describes the operation of the system consisting of the tank and its two valves and then to use this diagram to determine the overflow time. We know the rates at which chemicals flow into and out of the tank, but we need an expression that gives the total volume at any time.

We can most easily explain the new concept by considering a simplified example. We shall consider only a flow rate and extract from this the fluid volume that has collected because of the flow. Let us write this rate as a general function $g(t)$, with units of volume per time. The new algebraic-system model has the flow rate as an input and the total volume of the fluid as an output. Such an element is called an *integrator;* see the block diagram of Figure 10–25. The deterministic model associated with this element is

$$V(t) = V(0) + \int_0^t g(t)\,dt. \tag{10–10}$$

The second input is necessary to account for any initial volume $V(0)$ of fluid that is present at $t = 0$. We can now solve this problem by using the integrator as the essential element, since the net input is

$$f(t) - e(t)$$

Figure 10–26 shows the block diagram. The following equation holds for $V(t)$.

$$V(t) = 400 + \int_0^t (t^2 - t)\,dt$$

since $f(t) = t^2$ and $e(t) = t$.

To obtain a solution, we can use the computer program developed in Chapter 7 to perform numerical integration. Or, since we have chosen relatively simple functions

FIGURE 10–25 Block diagram of an integrator

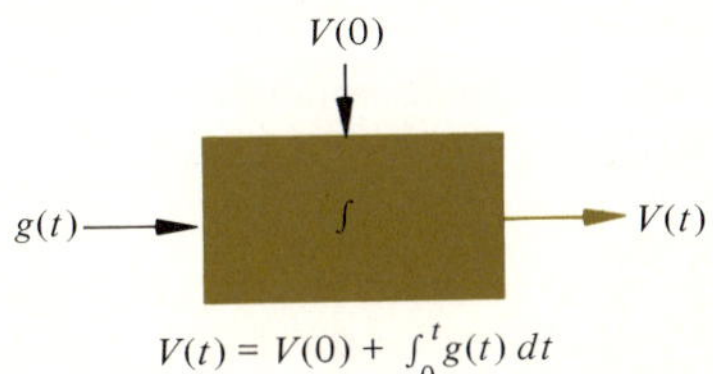

FIGURE 10–26 Evaluation of stored liquid volume using an integrator

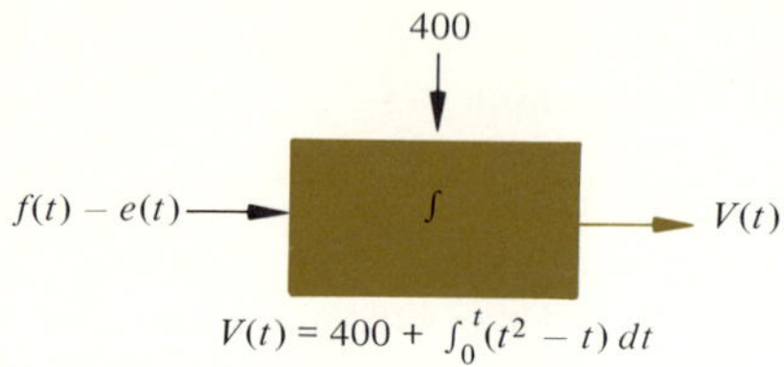

for the input, we can alternatively carry out the integration using the closed-form solutions available from integral calculus. This results in

$$V(t) = 400 + \frac{t^3}{3} - \frac{t^2}{2}$$

Denoting the time at which the tank overflows as t_0, we may write

$$1000 = 400 + \frac{t_0^3}{3} - \frac{t_0^2}{2}$$

We can readily cast this equation into a form that enables us to use the Newton–Raphson algorithm discussed in Chapter 5.

$$\frac{t_0^3}{3} - \frac{t_0^2}{2} - 600 = 0$$

The approximate solution is 12.7 hours.

The key insight in applying the integrator model is to identify the input as the difference between the filling and emptying rates. This should be acceptable on physical grounds, since the difference between these two quantities is the net rate at which the tank is filling.

This problem does not illustrate the real power of the integrator. We used it simply to raise, in as basic a way as possible, the need for this element. We can demonstrate the scope of the integrator more completely by returning to the topics from elementary mechanics that we first introduced in Section 7–9. We shall now analyze a problem using a block diagram.

The problem involves the trajectory of an Apollo space vehicle when it reenters the atmosphere. We measure the position of the vehicle in feet by three distances—x_1, x_2, and x_3—in a fixed-reference coordinate system whose origin is the center of the Earth. Figure 10–27 shows this. We find the total distance x from the center of the Earth by using simple trigonometric concepts. This distance is

$$x = (x_1^2 + x_2^2 + x_3^2)^{1/2}. \qquad (10\text{–}11)$$

The radius R of the Earth is approximately $R = 6372$ km. Since altitude is the distance above the surface, we can write the altitude of the vehicle as

$$\text{Altitude} = x - 6372 \text{ km}$$

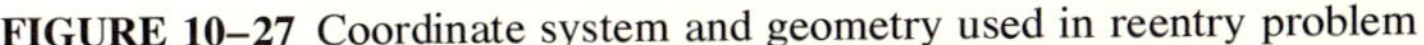

FIGURE 10–27 Coordinate system and geometry used in reentry problem

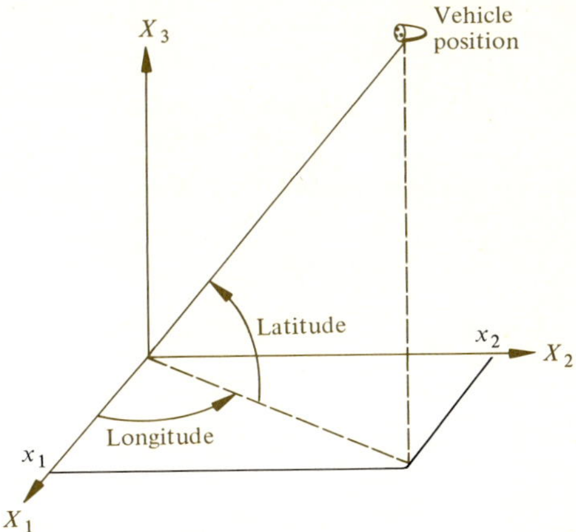

By referring to Figure 10–27, we can find the latitude and longitude of the vehicle in degrees.

$$\text{Latitude} = \frac{180}{\pi}\sin^{-1}\left(\frac{x_3}{x}\right), \qquad \text{Longitude} = \frac{180}{\pi}\tan^{-1}\left(\frac{x_2}{x_1}\right)$$

Next we define a set of three velocities v_1, v_2, and v_3 as the vehicle velocities in the three reference directions X_1, X_2, and X_3. In terms of these quantities, the total speed of the vehicle is

$$v = (v_1^2 + v_2^2 + v_3^2)^{1/2}$$

Using the basic integrator concept, we can relate the position $x_1(t)$ to the velocity v_1 by the following deterministic model and the block diagram of Figure 10–28.

$$x_1(t) = x_1(0) + \int_0^t v_1(t)\,dt$$

FIGURE 10–28 Change in position as the integral of velocity

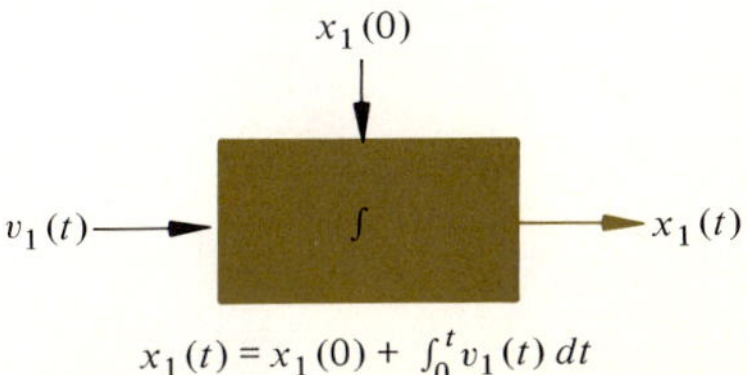

We can write similar equations relating positions and velocities in the other two dimensions.

$$x_2(t) = x_2(0) + \int_0^t v_2(t)\,dt, \qquad x_3(t) = x_3(0) + \int_0^t v_3(t)\,dt$$

Further, we can define a set of three accelerations a_1, a_2, and a_3, and we can relate the velocities in each of the three directions to the respective accelerations by means of integrator-type equations. To do this we have to take the initial values of velocity into account.

$$v_1(t) = v_1(0) + \int_0^t a_1(t)\,dt, \qquad v_2(t) = v_2(0) + \int_0^t a_2(t)\,dt,$$

$$v_3(t) = v_3(0) + \int_0^t a_3(t)\,dt$$

Now we treat the acceleration in a given direction as one input, along with the initial velocity and position. Then Figure 10–29 shows the relationship between vehicle position and acceleration. This figure uses two integrators. We could draw similar figures for the quantities in the other two reference directions.

With this set of block diagrams and an understanding of the principles we have discussed, we can find the reentry trajectory. To do so, we need to have the accelerations as functions of time and a set of six initial conditions (three initial positions and three initial velocities). Problems at the end of this chapter provide the opportunity to apply these techniques.

As a last concept, we consider a vehicle in which control of the acceleration during flight is possible. We can then introduce a feedback path into the block diagram to enable us to compare the position $x_1(t)$ and the desired position $x_{10}(t)$. A summer generates the difference between the actual and desired motion, and with this difference, we can control the acceleration. Figure 10–30 shows the modified block diagram. Of course, we must also control x_2 and x_3.

The details of the block labeled control in Figure 10–30 are too advanced for us to consider, but this example shows another use of the basic integrator. The integral of a quantity has an important meaning in many physical situations, and an integrator will normally appear in the block diagrams of such systems.

FIGURE 10–29 Evaluation of position using acceleration, initial velocity, and initial position as inputs

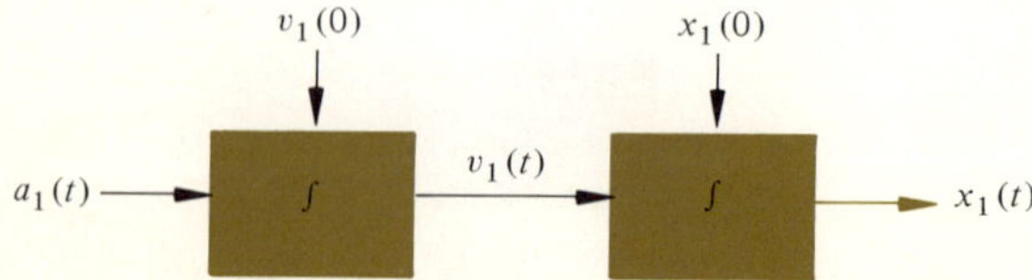

FIGURE 10–30 A block diagram of a basic control system

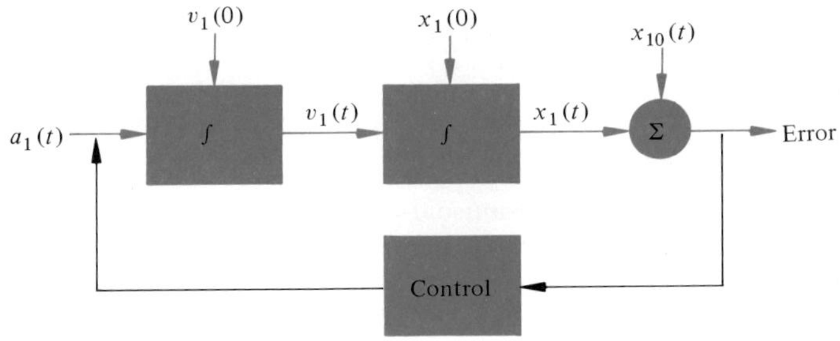

10–7 ANALOG COMPUTERS

The digital computer performs all calculations by manipulating numbers and instructions in binary form. In contrast, analog computers use voltages that vary continuously over the operating range. The voltages are used as analogs of other physical quantities. Section 2–1 discussed this concept in the measurement of chemical volumes in storage tanks. We summed five voltages using an amplifier. The output voltage, reflecting the total chemical volume, could have been read out on an oscilloscope, an *x-y* plotter, a strip-chart recorder, a voltmeter, or in digital form following an analog-to-digital conversion.

The analog computer is particularly well suited for solving differential equations. It is also ideal for simulation. The analog computer consists of units identical to the basic block diagrams: summers, amplifiers, and integrators. Equipment to generate trigonometric functions and exponentials useful as system inputs is also available.

Although the analog computer has advantages in certain classes of dynamic, control problems and in simulation, the digital computer has greater utility for solving engineering problems. Thus we shall not discuss analog computers in detail in this book. But we advise you to become familiar with them in later courses.

10–8 SUMMARY

This completes our brief survey of systems and feedback. We have used the input–output concept to describe an interacting collection of physical objects intended to perform some function. We have defined the basic building blocks and presented several examples of their use. We have also examined the important topic of feedback by using examples from different areas.

PROBLEMS

P10–1 Plot the following three sinusoids on a single sheet of graph paper for $0 \leq t \leq 2$ sec.

$$v_1(t) = 2 \sin\left(10t - \frac{2\pi}{3}\right), \qquad v_2(t) = 2 \sin(10t)$$

$$v_3(t) = 2 \sin\left(10t + \frac{2\pi}{3}\right)$$

You may calculate the values of the functions using either an electronic calculator or a digital computer.

P10–2 Plot the following three sinusoids on a single sheet of graph paper and include at least two cycles of each.

$$v_1(t) = 2 \sin\left(10t + \frac{\pi}{4}\right), \qquad v_2(t) = 2 \sin\left(40t + \frac{\pi}{4}\right),$$

$$v_3(t) = 2 \sin\left(100t + \frac{\pi}{4}\right)$$

For $v_3(t)$, approximately how many cycles would be contained in the time interval $0 \leq t \leq 0.5$ sec?

P10–3 The summer shown in Figure 10–11 has the following inputs.

$$v_1(t) = 50 \sin(18t), \qquad v_2(t) = 19 \sin(18t)$$

Find an expression for the output $v_o(t)$.

P10–4 The summer shown in Figure 10–11 has the following inputs.

$$v_1(t) = 50 \sin(18t), \qquad v_2(t) = 40 \sin(10t)$$

Find an expression for the output $v_o(t)$, and use a computer program to generate values of both inputs and the output over the range $0 \leq t \leq 1$ sec. Plot all three functions on a single sheet of graph paper.

P10–5 Briefly discuss the use of feedback in each of the following.
(a) A refrigerator
(b) Filling a glass with water
(c) Approaching a traffic light while driving an automobile
(d) An oven in which a pizza is being heated

P10–6 For the system shown in Figure P10–1, find the output $v_o(t)$, given

$$v_i(t) = 10 \sin\left(18t + \frac{\pi}{4}\right), \qquad G = 100$$

P10–7 For the system shown in Figure P10–1, find the gain G, given

$$v_i(t) = 10 \sin(11t), \qquad v_o(t) = 8 \sin(11t)$$

FIGURE P10–1

FIGURE P10–2

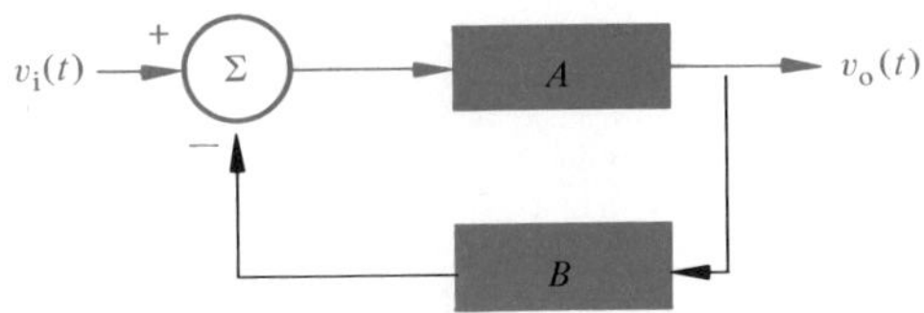

P10–8 What is the input $v_i(t)$ of the system shown in Figure P10–1, if the gain and the output are

$$G = 1414, \qquad v_o(t) = 2 \sin (1000t)$$

P10–9 For the system shown in Figure P10–2, what is the gain of the feedback amplifier B? The amplitude, output, and input are

$$A = 10, \qquad v_o = 20 \sin (1500t), \qquad v_i = 4 \sin (1500t)$$

P10–10 Find the output $v_o(t)$ for the system shown in Figure P10–2.

$$A = 1000, \qquad B = 0.1, \qquad v_i = 5 \sin (1200t)$$

P10–11 Find the output $v_o(t)$ of the system shown in Figure P10–3, given

$$C = 10000, \qquad D = 0.1, \qquad v_i(t) = 126 \sin (240t)$$

P10–12 Find the amplifier gain C of the system shown in Figure P10–3, given

$$v_o(t) = 1000 \sin (100t), \qquad v_i(t) = 10 \sin (100t), \qquad D = 0.001$$

P10–13 For the system shown in Figure P10–4, find an expression for the overall gain

$$\frac{v_o(t)}{v_i(t)}$$

in terms of the two amplifier gains (E and F).

P10–14 Find the output of the system shown in Figure P10–4, given

$$E = 10000, \qquad F = 0.1, \qquad v_i = 5 \sin (1100t)$$

FIGURE P10–3

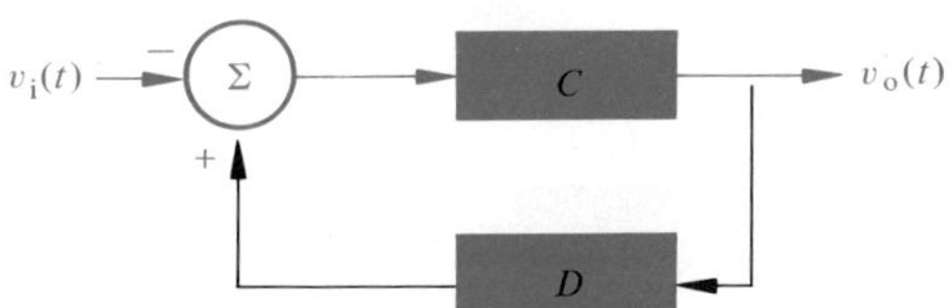

FIGURE P10–4

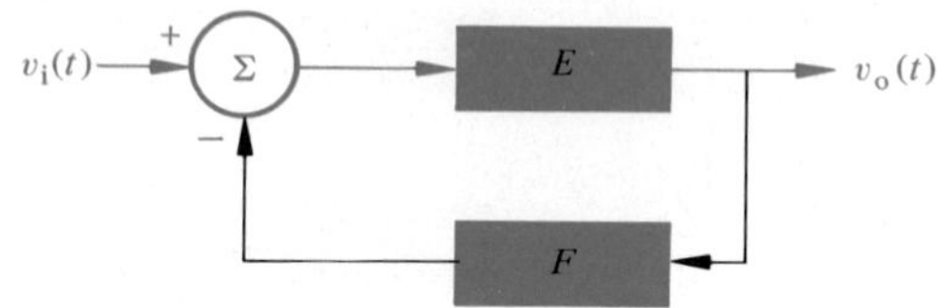

P10–15 Find the output of the system shown in Figure P10–4, given

$$E = 99, \qquad F = 0.01, \qquad v_i = 5 \sin (6400t)$$

P10–16 For the system shown in Figure P10–4, with $E = 100$, what value of F will make the overall gain of the system infinite? What is the physical meaning of an infinite system gain?

P10–17 For the two-amplifier and one-summer configuration of Figure 10–21(a), evaluate the system sensitivities S_A and S_B for the following gains.

$$A = 10^5, \qquad B = 10^{-1}$$

P10–18 Consider the block diagram of an amplifier system shown in Figure P10–5.
(a) Calculate the system gain $v_o(t)/v_i(t)$ and the sensitivities S_A, S_B, and S_C with respect to the individual gains A, B, and C.
(b) Evaluate the system gain and three sensitivities for the following conditions.

$$A = 10^4, \qquad B = 0.1, \qquad C = 0.1$$

(c) The amplifier system shown in Figure 10–21(a) resulted in a system gain

$$\frac{v_o(t)}{v_i(t)} = \frac{A}{1 + AB}$$

Calculate the system gain and sensitivities for

$$A = 10^4, \qquad B = 10^{-2}$$

FIGURE P10–5

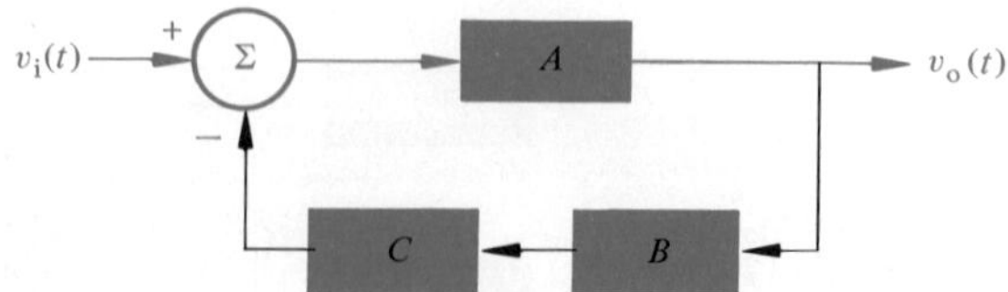

(d) For the two systems of parts (b) and (c), compare the system gains and sensitivities. Which system is better from the standpoint of sensitivity?

P10–19 The temperature T of a fluid is fluctuating sinusoidally

$$T(t) = A \sin(\omega t) + A$$

We wish to measure this temperature by immersing a thermometer in the fluid, but the thermometer has a finite capacity for heat, and cannot measure the correct instantaneous temperature. Its reading always lags behind the actual fluid temperature. We know that the thermometer reading follows the relationship

$$\text{Reading} = \frac{A \sin(\omega t + \phi)}{\sqrt{1 + \omega^2 \tau^2}} + A$$

where ϕ is the phase lag in radians, $\phi = -\tan^{-1}(\omega\tau)$, and τ is the "time constant" of the thermometer.

(a) Draw a block diagram of this system.

(b) For $A = 300$ K and $\omega = 0.2\ \text{sec}^{-1}$, plot the fluid temperature and the thermometer reading for $0 \leqslant t \leqslant 60$ sec, with (i) $\tau = 5$ sec, (ii) $\tau = 2$ sec.

(c) Which is the better thermometer? Why?

P10–20 Consider a system with noise present and with a block diagram identical to that shown in Figure 10–23. If we choose $D = 10{,}000$, find the system output $y(t)$ in terms of the system input $x(t)$ and the noise $n(t)$.

Find the ratio of the term present in the output due to noise to the noise itself. We symbolize this ratio as NR.

Engineers find it convenient to consider a system's ability to reject noise in terms of a quantity known as noise attenuation:

$$\text{Attenuation} = -20 \log_{10}(NR) \qquad \text{[decibels]}$$

Although this is a dimensionless quantity, it is customary to attach the units decibels to the attenuation. For the system of Figure 10–23, calculate the noise attenuation in decibels.

P10–21 The solution of this problem rests on a thorough understanding of the material presented in Section 10–6. Here we give the initial position and velocity, each in three components, for a vehicle reentering the Earth's atmosphere. In addition, we give the measured accelerations at discrete time intervals. Calculate the position of the vehicle as a function of time. Use a suitable FORTRAN program that performs numerical integration to calculate the velocities from the acceleration data and the initial velocities. Once you have found the velocities as functions of time, you can calculate the positions, since the initial positions and velocities are as follows.

$x_1(0) = 6.465(10^3)$ km $\quad v_1(0) = -0.314$ km/sec
$x_2(0) = 0 \quad v_2(0) = 78.57$ km/sec
$x_3(0) = 0 \quad v_3(0) = 0$

The accelerations are as follows.

Time (sec)	a_1 (ft/sec^2)	a_2 (ft/sec^2)	a_3 (ft/sec^2)
0	−31.5	−2.50	−0.60
5	−31.5	−3.30	−0.75
10	−31.5	−4.21	−0.93
15	−31.55	−5.41	−1.18
20	−31.6	−6.63	−1.45
25	−31.55	−8.40	−1.83
30	−31.5	−10.13	−2.26
35	−31.3	−13.00	−2.86
40	−31.2	−15.26	−3.49
45	−31.0	−18.80	−4.43
50	−30.5	−22.89	−5.39
55	−29.8	−28.73	−7.00
60	−29.2	−34.28	−8.26
65	−27.5	−42.17	−10.37
70	−26.6	−51.25	−12.48
75	−24.0	−64.20	−15.45
80	−22.0	−76.13	−18.43
85	−18.0	−92.27	−22.31
90	−14.8	−111.37	−26.20
95	−12.4	−130.63	−28.30
100	−10.2	−149.72	−30.41

Convert the accelerations to metric units before you carry out the necessary integrations.

APPENDIX A

TABLES OF RANDOM NUMBERS

TABLE A-1. TWO-DIGIT RANDOM NUMBERS

33	46	21	54	73	95	55	19	64	53
85	74	03	68	25	77	78	52	83	71
27	80	44	92	93	75	97	31	58	11
82	04	70	29	84	88	62	43	37	81
91	63	01	79	05	61	65	49	42	89
40	86	28	09	34	66	32	35	67	30
20	87	36	14	07	90	23	69	48	08
96	57	02	51	15	45	76	94	59	38
13	50	22	10	00	47	06	72	11	16
26	24	56	99	12	41	39	17	18	98

67	61	02	40	62	49	84	01	92	86
31	51	69	00	18	90	17	39	24	30
07	12	54	59	79	11	99	35	63	73
93	46	75	82	57	47	09	74	80	23
27	91	50	21	77	19	72	94	65	14
42	37	98	05	16	52	43	83	48	87
13	36	60	71	38	96	76	64	78	03
06	25	15	04	29	81	70	34	45	88
97	32	68	08	95	44	26	11	66	41
33	20	56	28	11	53	85	22	10	89

63	23	12	08	81	50	19	00	75	88
96	28	17	91	39	55	22	80	27	84
06	16	87	21	85	59	38	33	99	46
64	05	69	10	79	25	77	41	73	11
49	40	36	01	24	29	35	56	37	60
66	98	78	43	72	02	82	42	57	52
58	51	20	54	68	15	92	94	11	31
30	48	03	89	34	44	97	47	67	07
26	86	61	04	09	90	74	65	83	71
76	62	11	18	95	70	93	53	11	45

97	23	05	66	94	09	53	79	39	63
73	13	58	74	70	91	61	88	15	46
04	51	11	25	48	03	29	26	77	10
71	47	57	62	01	90	67	85	96	34
87	56	17	75	42	22	93	80	35	40
14	68	16	82	45	49	20	69	89	06
50	95	65	30	07	02	60	37	54	00
92	32	76	27	44	18	43	59	72	28
64	83	36	38	33	99	84	78	55	08
24	11	21	52	31	41	86	19	81	11

24	88	52	65	63	38	02	16	18	03
26	84	08	95	74	29	49	77	61	21
17	94	81	78	82	36	20	06	05	14
80	96	86	40	32	22	72	33	51	50
97	66	31	87	89	90	47	45	73	83
69	13	58	28	85	59	12	35	07	62
56	42	55	39	98	10	01	11	54	48
11	30	19	91	79	23	75	43	37	71
41	60	67	25	00	15	99	09	68	70
34	76	92	04	93	11	27	57	53	44

TABLE A-2. FOUR-DIGIT RANDOM NUMBERS

2372	3681	1176	4372	6090	7633	1425	0300
9416	4236	1114	9001	4420	5952	6374	5118
3780	7065	8810	9714	9441	9656	3415	4023
3846	7310	9690	2796	9004	5302	2216	6022
6631	6026	6924	7748	4616	8404	9326	0761
1073	6027	0947	5884	7216	0784	0206	4620
6310	6725	4001	3921	7959	2908	6256	1804
4962	3978	9657	2579	9000	1235	6853	0441
1409	4932	7352	0166	5269	0559	6377	3672
5078	7864	1919	1181	0257	1352	6246	5749
8721	1027	8116	9900	6794	2109	1943	3123
1693	2493	0161	8975	2836	6690	5051	0543
8241	4998	6260	3024	2239	6666	0283	2148
0778	5784	8140	7228	0545	8662	7513	7557
8170	1444	5580	0922	5760	6701	8806	2971
9011	7765	5940	6193	4140	9545	0454	7258
9909	4573	8697	1466	0967	3043	9999	3053
8768	5574	4973	1107	6333	7471	8274	2848
3066	3202	2065	4010	5921	9879	6426	2088
3135	8458	2979	2193	6787	1433	7953	5263
0447	5758	0961	4391	9758	5730	7004	0894
2772	9027	9654	7122	6283	4047	8174	3067
5275	4495	9934	9592	8586	5632	6964	1536
6983	8515	8683	5906	7733	3691	2986	5138
4401	0601	4445	1699	0629	8931	8365	0252
6665	8166	9452	3664	7355	1599	3838	9081
0388	1045	3216	0336	3515	8505	9839	2936
9502	1033	1122	7872	7578	5061	2604	0515
9102	6417	8023	0832	3226	2308	5252	1186
0289	1499	6835	7961	6693	8947	3892	3266
5016	1138	2128	2969	9109	8374	8704	7304
5928	0272	8720	0314	3848	0705	8044	4362
6211	8456	5273	5981	8867	9814	9530	9294
0432	9389	2887	4052	5360	6139	9032	9382
5447	8690	3555	3562	9819	7299	5862	9922
4435	2107	3173	0517	4986	5711	9828	8017
4091	8833	2619	6657	6815	1420	7631	3446
2442	4079	2938	1356	2132	1035	7462	5898
8675	9408	8818	8672	3114	1081	8898	4102
4976	3379	5934	5637	0857	4845	1802	7651
0124	2329	3301	9289	6470	5661	6173	6533
4084	6154	0608	8700	7174	5184	6980	5664
1604	9094	0569	2011	7389	6680	4015	4416
0802	5507	6263	8460	4833	3299	6733	1154
6767	0657	3482	5417	1613	1361	4095	2759
0148	6492	8066	0412	0318	8638	9411	9167
0748	2425	8259	8167	5115	0188	2896	6123
1118	2041	2628	7844	3854	3577	5168	9258
9479	3996	9107	9117	3184	7488	6716	3345
9070	0759	4365	9804	9979	2084	3130	0468

APPENDIX B
COMPOUND INTEREST TABLES

FORMULAS FOR CALCULATING COMPOUND INTEREST FACTORS

Single-payment compound-amount factor $(F\|P,i,n)$	$[(1+i)^n]$
Single-payment present-worth factor $(P\|F,i,n)$	$\left[\frac{1}{(1+i)^n}\right]$
Equal-payment-series compound-amount factor $(F\|A,i,n)$	$\left[\frac{(1+i)^n-1}{i}\right]$
Equal-payment-series sinking-fund factor $(A\|F,i,n)$	$\left[\frac{i}{(1+i)^n-1}\right]$
Equal-payment-series present-worth factor $(P\|A,i,n)$	$\left[\frac{(1+i)^n-1}{i(1+i)^n}\right]$
Equal-payment-series capital-recovery factor $(A\|P,i,n)$	$\left[\frac{i(1+i)^n}{(1+i)^n-1}\right]$

TABLE B-1. ½% COMPOUND INTEREST TABLES

	SINGLE PAYMENT		*EQUAL-PAYMENT SERIES*			
Number of periods (n)	*Compound-amount factor* (To find $F\|P,i,n$)	*Present-worth factor* (To find $P\|F,i,n$)	*Compound-amount factor* (To find $F\|A,i,n$)	*Sinking-fund factor* (To find $A\|F,i,n$)	*Present-worth factor* (To find $P\|A,i,n$)	*Capital-recovery factor* (To find $A\|P,i,n$)
1	1.0050	0.9950	1.0000	1.0000	0.9950	1.0050
2	1.0100	0.9901	2.0050	0.4988	1.9851	0.5038
3	1.0151	0.9851	3.0150	0.3317	2.9702	0.3367
4	1.0202	0.9802	4.0301	0.2481	3.9505	0.2531
5	1.0253	0.9754	5.0503	0.1980	4.9259	0.2030
6	1.0304	0.9705	6.0755	0.1646	5.8964	0.1696
7	1.0355	0.9657	7.1059	0.1407	6.8621	0.1457
8	1.0407	0.9609	8.1414	0.1228	7.8230	0.1278
9	1.0459	0.9561	9.1821	0.1089	8.7791	0.1139
10	1.0511	0.9513	10.2280	0.0978	9.7304	0.1028
11	1.0564	0.9466	11.2792	0.0887	10.6770	0.0937
12	1.0617	0.9419	12.3356	0.0811	11.6189	0.0861
13	1.0670	0.9372	13.3972	0.0746	12.5562	0.0796
14	1.0723	0.9326	14.4642	0.0691	13.4887	0.0741
15	1.0777	0.9279	15.5365	0.0644	14.4166	0.0694
16	1.0831	0.9233	16.6142	0.0602	15.3399	0.0652
17	1.0885	0.9187	17.6973	0.0565	16.2586	0.0615
18	1.0939	0.9141	18.7858	0.0532	17.1728	0.0582
19	1.0994	0.9096	19.8797	0.0503	18.0824	0.0553
20	1.1049	0.9051	20.9791	0.0477	18.9874	0.0527
21	1.1104	0.9006	22.0840	0.0453	19.8880	0.0503
22	1.1160	0.8961	23.1944	0.0431	20.7841	0.048
23	1.1216	0.8916	24.3104	0.0411	21.6757	0.0461
24	1.1272	0.8872	25.4320	0.0393	22.5629	0.0443
25	1.1328	0.8828	26.5591	0.0377	23.4456	0.0427
26	1.1385	0.8784	27.6919	0.0361	24.3240	0.0411
27	1.1442	0.8740	28.8304	0.0347	25.1980	0.0397
28	1.1499	0.8697	29.9745	0.0334	26.0677	0.0384
29	1.1556	0.8653	31.1244	0.0321	26.9330	0.0371
30	1.1614	0.8610	32.2800	0.0310	27.7941	0.0360
31	1.1672	0.8567	33.4414	0.0299	28.6508	0.0349
32	1.1730	0.8525	34.6086	0.0289	29.5033	0.0339
33	1.1789	0.8482	35.7817	0.0279	30.3515	0.0329
34	1.1848	0.8440	36.9606	0.0271	31.1955	0.0321
35	1.1907	0.8398	38.1454	0.0262	32.0354	0.0312
40	1.2208	0.8191	44.1588	0.0226	36.1722	0.0276
45	1.2516	0.7990	50.3242	0.0199	40.2072	0.0249
50	1.2832	0.7793	56.6452	0.0177	44.1428	0.0227
55	1.3156	0.7601	63.1258	0.0158	47.9814	0.0208
60	1.3489	0.7414	69.7700	0.0143	51.7256	0.0193
65	1.3829	0.7231	76.5821	0.0131	55.3775	0.0181
70	1.4178	0.7053	83.5661	0.0120	58.9394	0.0170
75	1.4536	0.6879	90.7265	0.0110	62.4136	0.0160
80	1.4903	0.6710	98.0677	0.0102	65.8023	0.0152
85	1.5280	0.6545	105.5943	0.0095	69.1075	0.0145
90	1.5666	0.6383	113.3109	0.0088	72.3313	0.0138
95	1.6061	0.6226	121.2224	0.0082	75.4757	0.0132
100	1.6467	0.6073	129.3337	0.0077	78.5426	0.0127

TABLE B-2. 1% COMPOUND INTEREST TABLES

	SINGLE PAYMENT		*EQUAL-PAYMENT SERIES*			
Number of periods (n)	*Compound-amount factor* (To find $F\|P,i,n$)	*Present-worth factor* (To find $P\|F,i,n$)	*Compound-amount factor* (To find $F\|A,i,n$)	*Sinking-fund factor* (To find $A\|F,i,n$)	*Present-worth factor* (To find $P\|A,i,n$)	*Capital-recovery factor* (To find $A\|P,i,n$)
1	1.0100	0.9901	1.0000	1.0000	0.9901	1.0100
2	1.0201	0.9803	2.0100	0.4975	1.9704	0.5075
3	1.0303	0.9706	3.0301	0.3300	2.9410	0.3400
4	1.0406	0.9610	4.0604	0.2463	3.9020	0.2563
5	1.0510	0.9515	5.1010	0.1960	4.8534	0.2060
6	1.0615	0.9420	6.1520	0.1625	5.7955	0.1725
7	1.0721	0.9327	7.2135	0.1386	6.7282	0.1486
8	1.0829	0.9235	8.2857	0.1207	7.6517	0.1307
9	1.0937	0.9143	9.3685	0.1067	8.5660	0.1167
10	1.1046	0.9053	10.4622	0.0956	9.4713	0.1056
11	1.1157	0.8963	11.5668	0.0865	10.3676	0.0965
12	1.1268	0.8874	12.6825	0.0788	11.2551	0.0888
13	1.1381	0.8787	13.8093	0.0724	12.1337	0.0824
14	1.1495	0.8700	14.9474	0.0669	13.0037	0.0769
15	1.1610	0.8613	16.0969	0.0621	13.8651	0.0721
16	1.1726	0.8528	17.2579	0.0579	14.7179	0.0679
17	1.1843	0.8444	18.4304	0.0543	15.5623	0.0643
18	1.1961	0.8360	19.6147	0.0510	16.3983	0.0610
19	1.2081	0.8277	20.8109	0.0481	17.2260	0.0581
20	1.2202	0.8195	22.0190	0.0454	18.0456	0.0554
21	1.2324	0.8114	23.2392	0.0430	18.8570	0.0530
22	1.2447	0.8034	24.4716	0.0409	19.6604	0.0509
23	1.2572	0.7954	25.7163	0.0389	20.4558	0.0489
24	1.2697	0.7876	26.9735	0.0371	21.2434	0.0471
25	1.2824	0.7798	28.2432	0.0354	22.0232	0.0454
26	1.2953	0.7720	29.5256	0.0339	22.7952	0.0439
27	1.3082	0.7644	30.8209	0.0324	23.5596	0.0424
28	1.3213	0.7568	32.1291	0.0311	24.3164	0.0411
29	1.3345	0.7493	33.4504	0.0299	25.0658	0.0399
30	1.3478	0.7419	34.7849	0.0287	25.8077	0.0387
31	1.3613	0.7346	36.1327	0.0277	26.5423	0.0377
32	1.3749	0.7273	37.4941	0.0267	27.2696	0.0367
33	1.3887	0.7201	38.8690	0.0257	27.9897	0.0357
34	1.4026	0.7130	40.2577	0.0248	28.7027	0.0348
35	1.4166	0.7059	41.6603	0.0240	29.4086	0.0340
40	1.4889	0.6717	48.8864	0.0205	32.8347	0.0305
45	1.5648	0.6391	56.4811	0.0177	36.0945	0.0277
50	1.6446	0.6080	64.4632	0.0155	39.1961	0.0255
55	1.7285	0.5785	72.8525	0.0137	42.1472	0.0237
60	1.8167	0.5504	81.6697	0.0122	44.9550	0.0222
65	1.9094	0.5237	90.9366	0.0110	47.6266	0.0210
70	2.0068	0.4983	100.6763	0.0099	50.1685	0.0199
75	2.1091	0.4741	110.9128	0.0090	52.5871	0.0190
80	2.2167	0.4511	121.6715	0.0082	54.8882	0.0182
85	2.3298	0.4292	132.9790	0.0075	57.0777	0.0175
90	2.4486	0.4084	144.8633	0.0069	59.1609	0.0169
95	2.5735	0.3886	157.3538	0.0064	61.1430	0.0164
100	2.7048	0.3697	170.4814	0.0059	63.0289	0.0159

TABLE B-3. 2% COMPOUND INTEREST TABLES

	SINGLE PAYMENT		*EQUAL-PAYMENT SERIES*			
Number of periods (n)	*Compound-amount factor* (To find $F\|P,i,n$)	*Present-worth factor* (To find $P\|F,i,n$)	*Compound-amount factor* (To find $F\|A,i,n$)	*Sinking-fund factor* (To find $A\|F,i,n$)	*Present-worth factor* (To find $P\|A,i,n$)	*Capital-recovery factor* (To find $A\|P,i,n$)
1	1.0200	0.9804	1.0000	1.0000	0.9804	1.0200
2	1.0404	0.9612	2.0200	0.4950	1.9416	0.5150
3	1.0612	0.9423	3.0604	0.3268	2.8839	0.3468
4	1.0824	0.9238	4.1216	0.2426	3.8077	0.2626
5	1.1041	0.9057	5.2040	0.1922	4.7135	0.2122
6	1.1262	0.8880	6.3081	0.1585	5.6014	0.1785
7	1.1487	0.8706	7.4343	0.1345	6.4720	0.1545
8	1.1717	0.8535	8.5830	0.1165	7.3255	0.1365
9	1.1951	0.8368	9.7546	0.1025	8.1622	0.1225
10	1.2190	0.8203	10.9497	0.0913	8.9826	0.1113
11	1.2434	0.8043	12.1687	0.0822	9.7868	0.1022
12	1.2682	0.7885	13.4121	0.0746	10.5753	0.0946
13	1.2936	0.7730	14.6803	0.0681	11.3484	0.0881
14	1.3195	0.7579	15.9739	0.0626	12.1062	0.0826
15	1.3459	0.7430	17.2934	0.0578	12.8493	0.0778
16	1.3728	0.7284	18.6393	0.0537	13.5777	0.0737
17	1.4002	0.7142	20.0121	0.0500	14.2919	0.0700
18	1.4282	0.7002	21.4123	0.0467	14.9920	0.0667
19	1.4568	0.6864	22.8406	0.0438	15.6785	0.0638
20	1.4859	0.6730	24.2974	0.0412	16.3514	0.0612
21	1.5157	0.6598	25.7833	0.0388	17.0112	0.0588
22	1.5460	0.6468	27.2990	0.0366	17.6580	0.0566
23	1.5769	0.6342	28.8450	0.0347	18.2922	0.0547
24	1.6084	0.6217	30.4219	0.0329	18.9139	0.0529
25	1.6406	0.6095	32.0303	0.0312	19.5235	0.0512
26	1.6734	0.5976	33.6709	0.0297	20.1210	0.0497
27	1.7069	0.5859	35.3443	0.0283	20.7069	0.0483
28	1.7410	0.5744	37.0512	0.0270	21.2813	0.0470
29	1.7758	0.5631	38.7922	0.0258	21.8444	0.0458
30	1.8114	0.5521	40.5681	0.0246	22.3965	0.0446
31	1.8476	0.5412	42.3794	0.0236	22.9377	0.0436
32	1.8845	0.5306	44.2270	0.0226	23.4683	0.0426
33	1.9222	0.5202	46.1116	0.0217	23.9886	0.0417
34	1.9607	0.5100	48.0338	0.0208	24.4986	0.0408
35	1.9999	0.5000	49.9945	0.0200	24.9986	0.0400
40	2.2080	0.4529	60.4020	0.0166	27.3555	0.0366
45	2.4379	0.4102	71.8927	0.0139	29.4902	0.0339
50	2.6916	0.3715	84.5794	0.0118	31.4236	0.0318
55	2.9717	0.3365	98.5865	0.0101	33.1748	0.0301
60	3.2810	0.3048	114.0515	0.0088	34.7609	0.0288
65	3.6225	0.2761	131.1262	0.0076	36.1975	0.0276
70	3.9996	0.2500	149.9779	0.0067	37.4986	0.0267
75	4.4158	0.2265	170.7918	0.0059	38.6771	0.0259
80	4.8754	0.2051	193.7720	0.0052	39.7445	0.0252
85	5.3829	0.1858	219.1439	0.0046	40.7113	0.0246
90	5.9431	0.1683	247.1567	0.0040	41.5869	0.0240
95	6.5617	0.1524	278.0850	0.0036	42.3800	0.0236
100	7.2446	0.1380	312.2323	0.0032	43.0984	0.0232

TABLE B-4. 3% COMPOUND INTEREST TABLES

	SINGLE PAYMENT		*EQUAL-PAYMENT SERIES*			
Number of periods (n)	*Compound-amount factor* (To find $F\|P,i,n$)	*Present-worth factor* (To find $P\|F,i,n$)	*Compound-amount factor* (To find $F\|A,i,n$)	*Sinking-fund factor* (To find $A\|F,i,n$)	*Present-worth factor* (To find $P\|A,i,n$)	*Capital-recovery factor* (To find $A\|P,i,n$)
1	1.0300	0.9709	1.0000	1.0000	0.9709	1.0300
2	1.0609	0.9426	2.0300	0.4926	1.9135	0.5226
3	1.0927	0.9151	3.0909	0.3235	2.8286	0.3535
4	1.1255	0.8885	4.1836	0.2390	3.7171	0.2690
5	1.1593	0.8626	5.3091	0.1884	4.5797	0.2184
6	1.1941	0.8375	6.4684	0.1546	5.4172	0.1846
7	1.2299	0.8131	7.6625	0.1305	6.2303	0.1605
8	1.2668	0.7894	8.8923	0.1125	7.0197	0.1425
9	1.3048	0.7664	10.1591	0.0984	7.7861	0.1284
10	1.3439	0.7441	11.4639	0.0872	8.5302	0.1172
11	1.3842	0.7224	12.8078	0.0781	9.2526	0.1081
12	1.4258	0.7014	14.1920	0.0705	9.9540	0.1005
13	1.4685	0.6810	15.6178	0.0640	10.6350	0.0940
14	1.5126	0.6611	17.0863	0.0585	11.2961	0.0885
15	1.5580	0.6419	18.5989	0.0538	11.9379	0.0838
16	1.6047	0.6232	20.1569	0.0496	12.5611	0.0796
17	1.6528	0.6050	21.7616	0.0460	13.1661	0.0760
18	1.7024	0.5874	23.4144	0.0427	13.7535	0.0727
19	1.7535	0.5703	25.1169	0.0398	14.3238	0.0698
20	1.8061	0.5537	26.8704	0.0372	14.8775	0.0672
21	1.8603	0.5375	28.6765	0.0349	15.4150	0.0649
22	1.9161	0.5219	30.5368	0.0327	15.9369	0.0627
23	1.9736	0.5067	32.4529	0.0308	16.4436	0.0608
24	2.0328	0.4919	34.4265	0.0290	16.9355	0.0590
25	2.0938	0.4776	36.4593	0.0274	17.4131	0.0574
26	2.1566	0.4637	38.5530	0.0259	17.8768	0.0559
27	2.2213	0.4502	40.7096	0.0246	18.3270	0.0546
28	2.2879	0.4371	42.9309	0.0233	18.7641	0.0533
29	2.3566	0.4243	45.2189	0.0221	19.1885	0.0521
30	2.4273	0.4120	47.5754	0.0210	19.6004	0.0510
31	2.5001	0.4000	50.0027	0.0200	20.0004	0.0500
32	2.5751	0.3883	52.5028	0.0190	20.3888	0.0490
33	2.6523	0.3770	55.0778	0.0182	20.7658	0.0482
34	2.7319	0.3660	57.7302	0.0173	21.1318	0.0473
35	2.8139	0.3554	60.4621	0.0165	21.4872	0.0465
40	3.2620	0.3066	75.4013	0.0133	23.1148	0.0433
45	3.7816	0.2644	92.7199	0.0108	24.5187	0.0408
50	4.3839	0.2281	112.7969	0.0089	25.7298	0.0389
55	5.0821	0.1968	136.0716	0.0073	26.7744	0.0373
60	5.8916	0.1697	163.0534	0.0061	27.6756	0.0361
65	6.8300	0.1464	194.3328	0.0051	28.4529	0.0351
70	7.9178	0.1263	230.5941	0.0043	29.1234	0.0343
75	9.1789	0.1089	272.6309	0.0037	29.7018	0.0337
80	10.6409	0.0940	321.3630	0.0031	30.2008	0.0331
85	12.3357	0.0811	377.8570	0.0026	30.6312	0.0326
90	14.3005	0.0699	443.3489	0.0023	31.0024	0.0323
95	16.5782	0.0603	519.2720	0.0019	31.3227	0.0319
100	19.2186	0.0520	607.2877	0.0016	31.5989	0.0316

TABLE B-5. 4% COMPOUND INTEREST TABLES

	SINGLE PAYMENT		*EQUAL-PAYMENT SERIES*			
Number of periods (n)	*Compound-amount factor* (To find $F\|P,i,n$)	*Present-worth factor* (To find $P\|F,i,n$)	*Compound-amount factor* (To find $F\|A,i,n$)	*Sinking-fund factor* (To find $A\|F,i,n$)	*Present-worth factor* (To find $P\|A,i,n$)	*Capital-recovery factor* (To find $A\|P,i,n$)
1	1.0400	0.9615	1.0000	1.0000	0.9615	1.0400
2	1.0816	0.9246	2.0400	0.4902	1.8861	0.5302
3	1.1249	0.8890	3.1216	0.3203	2.7751	0.3603
4	1.1699	0.8548	4.2465	0.2355	3.6299	0.2755
5	1.2167	0.8219	5.4163	0.1846	4.4518	0.2246
6	1.2653	0.7903	6.6330	0.1508	5.2421	0.1908
7	1.3159	0.7599	7.8983	0.1266	6.0021	0.1666
8	1.3686	0.7307	9.2142	0.1085	6.7327	0.1485
9	1.4233	0.7026	10.5828	0.0945	7.4353	0.1345
10	1.4802	0.6756	12.0061	0.0833	8.1109	0.1233
11	1.5395	0.6496	13.4864	0.0741	8.7605	0.1141
12	1.6010	0.6246	15.0258	0.0666	9.3851	0.1066
13	1.6651	0.6006	16.6268	0.0601	9.9856	0.1001
14	1.7317	0.5775	18.2919	0.0547	10.5631	0.0947
15	1.8009	0.5553	20.0236	0.0499	11.1184	0.0899
16	1.8730	0.5339	21.8245	0.0458	11.6523	0.0858
17	1.9479	0.5134	23.6975	0.0422	12.1657	0.0822
18	2.0258	0.4936	25.6454	0.0390	12.6593	0.0790
19	2.1068	0.4746	27.6712	0.0361	13.1339	0.0761
20	2.1911	0.4564	29.7781	0.0336	13.5903	0.0736
21	2.2788	0.4388	31.9692	0.0313	14.0292	0.0713
22	2.3699	0.4220	34.2480	0.0292	14.4511	0.0692
23	2.4647	0.4057	36.6179	0.0273	14.8568	0.0673
24	2.5633	0.3901	39.0826	0.0256	15.2470	0.0656
25	2.6658	0.3751	41.6459	0.0240	15.6221	0.0640
26	2.7725	0.3607	44.3117	0.0226	15.9828	0.0626
27	2.8834	0.3468	47.0842	0.0212	16.3296	0.0612
28	2.9987	0.3335	49.9676	0.0200	16.6631	0.0600
29	3.1187	0.3207	52.9663	0.0189	16.9837	0.0589
30	3.2434	0.3083	56.0849	0.0178	17.2920	0.0578
31	3.3731	0.2965	59.3283	0.0169	17.5885	0.0569
32	3.5081	0.2851	62.7015	0.0159	17.8736	0.0559
33	3.6484	0.2741	66.2095	0.0151	18.1476	0.0551
34	3.7943	0.2636	69.8579	0.0143	18.4112	0.0543
35	3.9461	0.2534	73.6522	0.0136	18.6646	0.0536
40	4.8010	0.2083	95.0255	0.0105	19.7928	0.0505
45	5.8412	0.1712	121.0294	0.0083	20.7200	0.0483
50	7.1067	0.1407	152.6671	0.0066	21.4822	0.0466
55	8.6464	0.1157	191.1592	0.0052	22.1086	0.0452
60	10.5196	0.0951	237.9907	0.0042	22.6235	0.0442
65	12.7987	0.0781	294.9684	0.0034	23.0467	0.0434
70	15.5716	0.0642	364.2905	0.0027	23.3945	0.0427
75	18.9453	0.0528	448.6314	0.0022	23.6804	0.0422
80	23.0498	0.0434	551.2450	0.0018	23.9154	0.0418
85	28.0436	0.0357	676.0901	0.0015	24.1085	0.0415
90	34.1193	0.0293	827.9833	0.0012	24.2673	0.0412
95	41.5114	0.0241	1012.7846	0.0010	24.3978	0.0410
100	50.5049	0.0198	1237.6237	0.0008	24.5050	0.0408

TABLE B-6. 5% COMPOUND INTEREST TABLES

	SINGLE PAYMENT		*EQUAL-PAYMENT SERIES*			
Number of periods (n)	*Compound-amount factor* (To find $F\|P,i,n$)	*Present-worth factor* (To find $P\|F,i,n$)	*Compound-amount factor* (To find $F\|A,i,n$)	*Sinking-fund factor* (To find $A\|F,i,n$)	*Present-worth factor* (To find $P\|A,i,n$)	*Capital-recovery factor* (To find $A\|P,i,n$)
1	1.0500	0.9524	1.0000	1.0000	0.9524	1.0500
2	1.1025	0.9070	2.0500	0.4878	1.8594	0.5378
3	1.1576	0.8638	3.1525	0.3172	2.7232	0.3672
4	1.2155	0.8227	4.3101	0.2320	3.5460	0.2820
5	1.2763	0.7835	5.5256	0.1810	4.3295	0.2310
6	1.3401	0.7462	6.8019	0.1470	5.0757	0.1970
7	1.4071	0.7107	8.1420	0.1228	5.7864	0.1728
8	1.4775	0.6768	9.5491	0.1047	6.4632	0.1547
9	1.5513	0.6446	11.0266	0.0907	7.1078	0.1407
10	1.6289	0.6139	12.5779	0.0795	7.7217	0.1295
11	1.7103	0.5847	14.2068	0.0704	8.3064	0.1204
12	1.7959	0.5568	15.9171	0.0628	8.8633	0.1128
13	1.8856	0.5303	17.7130	0.0565	9.3936	0.1065
14	1.9799	0.5051	19.5986	0.0510	9.8986	0.1010
15	2.0789	0.4810	21.5786	0.0463	10.3797	0.0963
16	2.1829	0.4581	23.6575	0.0423	10.8378	0.0923
17	2.2920	0.4363	25.8404	0.0387	11.2741	0.0887
18	2.4066	0.4155	28.1324	0.0355	11.6896	0.0855
19	2.5270	0.3957	30.5390	0.0327	12.0853	0.0827
20	2.6533	0.3769	33.0660	0.0302	12.4622	0.0802
21	2.7860	0.3589	35.7193	0.0280	12.8212	0.0780
22	2.9253	0.3418	38.5052	0.0260	13.1630	0.0760
23	3.0715	0.3256	41.4305	0.0241	13.4886	0.0741
24	3.2251	0.3101	44.5020	0.0225	13.7986	0.0725
25	3.3864	0.2953	47.7271	0.0210	14.0939	0.0710
26	3.5557	0.2812	51.1135	0.0196	14.3752	0.0696
27	3.7335	0.2678	54.6691	0.0183	14.6430	0.0683
28	3.9201	0.2551	58.4026	0.0171	14.8981	0.0671
29	4.1161	0.2429	62.3227	0.0160	15.1411	0.0660
30	4.3219	0.2314	66.4388	0.0151	15.3725	0.0651
31	4.5380	0.2204	70.7608	0.0141	15.5928	0.0641
32	4.7649	0.2099	75.2988	0.0133	15.8027	0.0633
33	5.0032	0.1999	80.0638	0.0125	16.0025	0.0625
34	5.2533	0.1904	85.0670	0.0118	16.1929	0.0618
35	5.5160	0.1813	90.3203	0.0111	16.3742	0.0611
40	7.0400	0.1420	120.7998	0.0083	17.1591	0.0583
45	8.9850	0.1113	159.7002	0.0063	17.7741	0.0563
50	11.4674	0.0872	209.3480	0.0048	18.2559	0.0548
55	14.6356	0.0683	272.7126	0.0037	18.6335	0.0537
60	18.6792	0.0535	353.5837	0.0028	18.9293	0.0528
65	23.8399	0.0419	456.7980	0.0022	19.1611	0.0522
70	30.4264	0.0329	588.5285	0.0017	19.3427	0.0517
75	38.8327	0.0258	756.6537	0.0013	19.4850	0.0513
80	49.5614	0.0202	971.2288	0.0010	19.5965	0.0510
85	63.2544	0.0158	1245.0871	0.0008	19.6838	0.0508
90	80.7304	0.0124	1594.6073	0.0006	19.7523	0.0506
95	103.0347	0.0097	2040.6935	0.0005	19.8059	0.0505
100	131.5013	0.0076	2610.0252	0.0004	19.8479	0.0504

TABLE B-7. 6% COMPOUND INTEREST TABLES

	SINGLE PAYMENT		*EQUAL-PAYMENT SERIES*			
Number of periods (n)	*Compound-amount factor* (To find $F\|P,i,n$)	*Present-worth factor* (To find $P\|F,i,n$)	*Compound-amount factor* (To find $F\|A,i,n$)	*Sinking-fund factor* (To find $A\|F,i,n$)	*Present-worth factor* (To find $P\|A,i,n$)	*Capital-recovery factor* (To find $A\|P,i,n$)
1	1.0600	0.9434	1.0000	1.0000	0.9434	1.0600
2	1.1236	0.8900	2.0600	0.4854	1.8334	0.5454
3	1.1910	0.8396	3.1836	0.3141	2.6730	0.3741
4	1.2625	0.7921	4.3746	0.2286	3.4651	0.2886
5	1.3382	0.7473	5.6371	0.1774	4.2124	0.2374
6	1.4185	0.7050	6.9753	0.1434	4.9173	0.2034
7	1.5036	0.6651	8.3938	0.1191	5.5824	0.1791
8	1.5938	0.6274	9.8975	0.1010	6.2098	0.1610
9	1.6895	0.5919	11.4913	0.0870	6.8017	0.1470
10	1.7908	0.5584	13.1808	0.0759	7.3601	0.1359
11	1.8983	0.5268	14.9716	0.0668	7.8869	0.1268
12	2.0122	0.4970	16.8699	0.0593	8.3838	0.1193
13	2.1329	0.4688	18.8821	0.0530	8.8527	0.1130
14	2.2609	0.4423	21.0151	0.0476	9.2950	0.1076
15	2.3966	0.4173	23.2760	0.0430	9.7122	0.1030
16	2.5404	0.3936	25.6725	0.0390	10.1059	0.0990
17	2.6928	0.3714	28.2129	0.0354	10.4773	0.0954
18	2.8543	0.3503	30.9057	0.0324	10.8276	0.0924
19	3.0256	0.3305	33.7600	0.0296	11.1581	0.0896
20	3.2071	0.3118	36.7856	0.0272	11.4699	0.0872
21	3.3996	0.2942	39.9927	0.0250	11.7641	0.0850
22	3.6035	0.2775	43.3923	0.0230	12.0416	0.0830
23	3.8197	0.2618	46.9958	0.0213	12.3034	0.0813
24	4.0489	0.2470	50.8156	0.0197	12.5504	0.0797
25	4.2919	0.2330	54.8645	0.0182	12.7834	0.0782
26	4.5494	0.2198	59.1564	0.0169	13.0032	0.0769
27	4.8223	0.2074	63.7058	0.0157	13.2105	0.0757
28	5.1117	0.1956	68.5281	0.0146	13.4062	0.0746
29	5.4184	0.1846	73.6398	0.0136	13.5907	0.0736
30	5.7435	0.1741	79.0582	0.0126	13.7648	0.0726
31	6.0881	0.1643	84.8017	0.0118	13.9291	0.0718
32	6.4534	0.1550	90.8898	0.0110	14.0840	0.0710
33	6.8406	0.1462	97.3432	0.0103	14.2302	0.0703
34	7.2510	0.1379	104.1838	0.0096	14.3681	0.0696
35	7.6861	0.1301	111.4348	0.0090	14.4982	0.0690
40	10.2857	0.0972	154.7620	0.0065	15.0463	0.0665
45	13.7646	0.0727	212.7435	0.0047	15.4558	0.0647
50	18.4202	0.0543	290.3359	0.0034	15.7619	0.0634
55	24.6503	0.0406	394.1720	0.0025	15.9905	0.0625
60	32.9877	0.0303	533.1282	0.0019	16.1614	0.0619
65	44.1450	0.0227	719.0829	0.0014	16.2891	0.0614
70	59.0759	0.0169	967.9322	0.0010	16.3845	0.0610
75	79.0569	0.0126	1300.9487	0.0008	16.4558	0.0608
80	105.7960	0.0095	1746.5999	0.0006	16.5091	0.0606
85	141.5789	0.0071	2342.9817	0.0004	16.5489	0.0604
90	189.4645	0.0053	3141.0752	0.0003	16.5787	0.0603
95	253.5463	0.0039	4209.1042	0.0002	16.6009	0.0602
100	339.3021	0.0029	5638.3681	0.0002	16.6175	0.0602

TABLE B-8. 7% COMPOUND INTEREST TABLES

	SINGLE PAYMENT		*EQUAL-PAYMENT SERIES*			
Number of periods (n)	*Compound-amount factor* (To find $F\|P,i,n$)	*Present-worth factor* (To find $P\|F,i,n$)	*Compound-amount factor* (To find $F\|A,i,n$)	*Sinking-fund factor* (To find $A\|F,i,n$)	*Present-worth factor* (To find $P\|A,i,n$)	*Capital-recovery factor* (To find $A\|P,i,n$)
1	1.0700	0.9346	1.0000	1.0000	0.9346	1.0700
2	1.1449	0.8734	2.0700	0.4831	1.8080	0.5531
3	1.2250	0.8163	3.2149	0.3111	2.6243	0.3811
4	1.3108	0.7629	4.4399	0.2252	3.3872	0.2952
5	1.4026	0.7130	5.7507	0.1739	4.1002	0.2439
6	1.5007	0.6663	7.1533	0.1398	4.7665	0.2098
7	1.6058	0.6227	8.6540	0.1156	5.3893	0.1856
8	1.7182	0.5820	10.2598	0.0975	5.9713	0.1675
9	1.8385	0.5439	11.9780	0.0835	6.5152	0.1535
10	1.9672	0.5083	13.8164	0.0724	7.0236	0.1424
11	2.1049	0.4751	15.7836	0.0634	7.4987	0.1334
12	2.2522	0.4440	17.8885	0.0559	7.9427	0.1259
13	2.4098	0.4150	20.1406	0.0497	8.3577	0.1197
14	2.5785	0.3878	22.5505	0.0443	8.7455	0.1143
15	2.7590	0.3624	25.1290	0.0398	9.1079	0.1098
16	2.9522	0.3387	27.8881	0.0359	9.4466	0.1059
17	3.1588	0.3166	30.8402	0.0324	9.7632	0.1024
18	3.3799	0.2959	33.9990	0.0294	10.0591	0.0994
19	3.6165	0.2765	37.3790	0.0268	10.3356	0.0968
20	3.8697	0.2584	40.9955	0.0244	10.5940	0.0944
21	4.1406	0.2415	44.8652	0.0223	10.8355	0.0923
22	4.4304	0.2257	49.0057	0.0204	11.0612	0.0904
23	4.7405	0.2109	53.4361	0.0187	11.2722	0.0887
24	5.0724	0.1971	58.1767	0.0172	11.4693	0.0872
25	5.4274	0.1842	63.2490	0.0158	11.6536	0.0858
26	5.8074	0.1722	68.6765	0.0146	11.8258	0.0846
27	6.2139	0.1609	74.4838	0.0134	11.9867	0.0834
28	6.6488	0.1504	80.6977	0.0124	12.1371	0.0824
29	7.1143	0.1406	87.3465	0.0114	12.2777	0.0814
30	7.6123	0.1314	94.4608	0.0106	12.4090	0.0806
31	8.1451	0.1228	102.0730	0.0098	12.5318	0.0798
32	8.7153	0.1147	110.2182	0.0091	12.6466	0.0791
33	9.3253	0.1072	118.9334	0.0084	12.7538	0.0784
34	9.9781	0.1002	128.2588	0.0078	12.8540	0.0778
35	10.6766	0.0937	138.2369	0.0072	12.9477	0.0772
40	14.9745	0.0668	199.6351	0.0050	13.3317	0.0750
45	21.0025	0.0476	285.7493	0.0035	13.6055	0.0735
50	29.4570	0.0339	406.5289	0.0025	13.8007	0.0725
55	41.3150	0.0242	575.9286	0.0017	13.9399	0.0717
60	57.9464	0.0173	813.5204	0.0012	14.0392	0.0712
65	81.2729	0.0123	1146.7552	0.0009	14.1099	0.0709
70	113.9894	0.0088	1614.1342	0.0006	14.1604	0.0706
75	159.8760	0.0063	2269.6574	0.0004	14.1964	0.0704
80	224.2344	0.0045	3189.0627	0.0003	14.2220	0.0703
85	314.5003	0.0032	4478.5761	0.0002	14.2403	0.0702
90	441.1030	0.0023	6287.1854	0.0002	14.2533	0.0702
95	618.6697	0.0016	8823.8535	0.0001	14.2626	0.0701
100	867.7163	0.0012	12381.6618	0.0001	14.2693	0.0701

TABLE B-9. 8% COMPOUND INTEREST TABLES

	SINGLE PAYMENT		*EQUAL-PAYMENT SERIES*			
Number of periods (n)	*Compound-amount factor* (To find $F\|P,i,n$)	*Present-worth factor* (To find $P\|F,i,n$)	*Compound-amount factor* (To find $F\|A,i,n$)	*Sinking-fund factor* (To find $A\|F,i,n$)	*Present-worth factor* (To find $P\|A,i,n$)	*Capital-recovery factor* (To find $A\|P,i,n$)
1	1.0800	0.9259	1.0000	1.0000	0.9259	1.0800
2	1.1664	0.8573	2.0800	0.4808	1.7833	0.5608
3	1.2597	0.7938	3.2464	0.3080	2.5771	0.3880
4	1.3605	0.7350	4.5061	0.2219	3.3121	0.3019
5	1.4693	0.6806	5.8666	0.1705	3.9927	0.2505
6	1.5869	0.6302	7.3359	0.1363	4.6229	0.2163
7	1.7138	0.5835	8.9228	0.1121	5.2064	0.1921
8	1.8509	0.5403	10.6366	0.0940	5.7466	0.1740
9	1.9990	0.5002	12.4876	0.0801	6.2469	0.1601
10	2.1589	0.4632	14.4866	0.0690	6.7101	0.1490
11	2.3316	0.4289	16.6455	0.0601	7.1390	0.1401
12	2.5182	0.3971	18.9771	0.0527	7.5361	0.1327
13	2.7196	0.3677	21.4953	0.0465	7.9038	0.1265
14	2.9372	0.3405	24.2149	0.0413	8.2442	0.1213
15	3.1722	0.3152	27.1521	0.0368	8.5595	0.1168
16	3.4259	0.2919	30.3243	0.0330	8.8514	0.1130
17	3.7000	0.2703	33.7502	0.0296	9.1216	0.1096
18	3.9960	0.2502	37.4502	0.0267	9.3719	0.1067
19	4.3157	0.2317	41.4463	0.0241	9.6036	0.1041
20	4.6610	0.2145	45.7620	0.0219	9.8181	0.1019
21	5.0338	0.1987	50.4229	0.0198	10.0168	0.0998
22	5.4365	0.1839	55.4568	0.0180	10.2007	0.0980
23	5.8715	0.1703	60.8933	0.0164	10.3711	0.0964
24	6.3412	0.1577	66.7648	0.0150	10.5288	0.0950
25	6.8485	0.1460	73.1059	0.0137	10.6748	0.0937
26	7.3964	0.1352	79.9544	0.0125	10.8100	0.0925
27	7.9881	0.1252	87.3508	0.0114	10.9352	0.0914
28	8.6271	0.1159	95.3388	0.0105	11.0511	0.0905
29	9.3173	0.1073	103.9659	0.0096	11.1584	0.0896
30	10.0627	0.0994	113.2832	0.0088	11.2578	0.0888
31	10.8677	0.0920	123.3459	0.0081	11.3498	0.0881
32	11.7371	0.0852	134.2135	0.0075	11.4350	0.0875
33	12.6760	0.0789	145.9506	0.0069	11.5139	0.0869
34	13.6901	0.0730	158.6267	0.0063	11.5869	0.0863
35	14.7853	0.0676	172.3168	0.0058	11.6546	0.0858
40	21.7245	0.0460	259.0565	0.0039	11.9246	0.0839
45	31.9204	0.0313	386.5056	0.0026	12.1084	0.0826
50	46.9016	0.0213	573.7702	0.0017	12.2335	0.0817
55	68.9139	0.0145	848.9232	0.0012	12.3186	0.0812
60	101.2571	0.0099	1253.2133	0.0008	12.3766	0.0808
65	148.7798	0.0067	1847.2481	0.0005	12.4160	0.0805
70	218.6064	0.0046	2720.0801	0.0004	12.4428	0.0804
75	321.2045	0.0031	4002.5566	0.0002	12.4611	0.0802
80	471.9548	0.0021	5886.9354	0.0002	12.4735	0.0802
85	693.4565	0.0014	8655.7061	0.0001	12.4820	0.0801
90	1018.9151	0.0010	12723.9386	0.0001	12.4877	0.0801
95	1497.1205	0.0007	18701.5069	0.0001	12.4917	0.0801
100	2199.7613	0.0005	27484.5157	0.0000	12.4943	0.0800

TABLE B-10. 9% COMPOUND INTEREST TABLES

	SINGLE PAYMENT		*EQUAL-PAYMENT SERIES*			
Number of periods (*n*)	*Compound-amount factor* (To find $F\|P,i,n$)	*Present-worth factor* (To find $P\|F,i,n$)	*Compound-amount factor* (To find $F\|A,i,n$)	*Sinking-fund factor* (To find $A\|F,i,n$)	*Present-worth factor* (To find $P\|A,i,n$)	*Capital-recovery factor* (To find $A\|P,i,n$)
1	1.0900	0.9174	1.0000	1.0000	0.9174	1.0900
2	1.1881	0.8417	2.0900	0.4785	1.7591	0.5685
3	1.2950	0.7722	3.2781	0.3051	2.5313	0.3951
4	1.4116	0.7084	4.5731	0.2187	3.2397	0.3087
5	1.5386	0.6499	5.9847	0.1671	3.8897	0.2571
6	1.6771	0.5963	7.5233	0.1329	4.4859	0.2229
7	1.8280	0.5470	9.2004	0.1087	5.0330	0.1987
8	1.9926	0.5019	11.0285	0.0907	5.5348	0.1807
9	2.1719	0.4604	13.0210	0.0768	5.9952	0.1668
10	2.3674	0.4224	15.1929	0.0658	6.4177	0.1558
11	2.5804	0.3875	17.5603	0.0569	6.8052	0.1469
12	2.8127	0.3555	20.1407	0.0497	7.1607	0.1397
13	3.0658	0.3262	22.9534	0.0436	7.4869	0.1336
14	3.3417	0.2992	26.0192	0.0384	7.7862	0.1284
15	3.6425	0.2745	29.3609	0.0341	8.0607	0.1241
16	3.9703	0.2519	33.0034	0.0303	8.3126	0.1203
17	4.3276	0.2311	36.9737	0.0270	8.5436	0.1170
18	4.7171	0.2120	41.3013	0.0242	8.7556	0.1142
19	5.1417	0.1945	46.0185	0.0217	8.9501	0.1117
20	5.6044	0.1784	51.1601	0.0195	9.1285	0.1095
21	6.1088	0.1637	56.7645	0.0176	9.2922	0.1076
22	6.6586	0.1502	62.8733	0.0159	9.4424	0.1059
23	7.2579	0.1378	69.5319	0.0144	9.5802	0.1044
24	7.9111	0.1264	76.7898	0.0130	9.7066	0.1030
25	8.6231	0.1160	84.7009	0.0118	9.8226	0.1018
26	9.3992	0.1064	93.3240	0.0107	9.9290	0.1007
27	10.2451	0.0976	102.7231	0.0097	10.0266	0.0997
28	11.1671	0.0895	112.9682	0.0089	10.1161	0.0989
29	12.1722	0.0822	124.1354	0.0081	10.1983	0.0981
30	13.2677	0.0754	136.3075	0.0073	10.2737	0.0973
31	14.4618	0.0691	149.5752	0.0067	10.3428	0.0967
32	15.7633	0.0634	164.0370	0.0061	10.4062	0.0961
33	17.1820	0.0582	179.8003	0.0056	10.4644	0.0956
34	18.7284	0.0534	196.9823	0.0051	10.5178	0.0951
35	20.4140	0.0490	215.7108	0.0046	10.5668	0.0946
40	31.4094	0.0318	337.8824	0.0030	10.7574	0.0930
45	48.3273	0.0207	525.8587	0.0019	10.8812	0.0919
50	74.3575	0.0134	815.0836	0.0012	10.9617	0.0912
55	114.4083	0.0087	1260.0918	0.0008	11.0140	0.0908
60	176.0313	0.0057	1944.7921	0.0005	11.0480	0.0905
65	270.8460	0.0037	2998.2885	0.0003	11.0701	0.0903
70	416 7301	0.0024	4619.2232	0.0002	11.0844	0.0902
75	641.1909	0.0016	7113.2321	0.0001	11.0938	0.0901
80	986.5517	0.0010	10950.5741	0.0001	11.0998	0.0901
85	1517.9320	0.0007	16854.8003	0.0001	11.1038	0.0901
90	2335.5266	0.0004	25939.1842	0.0000	11.1064	0.0900
95	3593.4971	0.0003	39916.6350	0.0000	11.1080	0.0900
100	5529.0408	0.0002	61422.6755	0.0000	11.1091	0.0900

TABLE B-11. 10% COMPOUND INTEREST TABLES

	SINGLE PAYMENT		*EQUAL-PAYMENT SERIES*			
Number of periods (n)	*Compound-amount factor* (To find $F\|P,i,n$)	*Present-worth factor* (To find $P\|F,i,n$)	*Compound-amount factor* (To find $F\|A,i,n$)	*Sinking-fund factor* (To find $A\|F,i,n$)	*Present-worth factor* (To find $P\|A,i,n$)	*Capital-recovery factor* (To find $A\|P,i,n$)
1	1.1000	0.9091	1.0000	1.0000	0.9091	1.1000
2	1.2100	0.8264	2.1000	0.4762	1.7355	0.5762
3	1.3310	0.7513	3.3100	0.3021	2.4869	0.4021
4	1.4641	0.6830	4.6410	0.2155	3.1699	0.3155
5	1.6105	0.6209	6.1051	0.1638	3.7908	0.2638
6	1.7716	0.5645	7.7156	0.1296	4.3553	0.2296
7	1.9487	0.5132	9.4872	0.1054	4.8684	0.2054
8	2.1436	0.4665	11.4359	0.0874	5.3349	0.1874
9	2.3579	0.4241	13.5795	0.0736	5.7590	0.1736
10	2.5937	0.3855	15.9374	0.0627	6.1446	0.1627
11	2.8531	0.3505	18.5312	0.0540	6.4951	0.1540
12	3.1384	0.3186	21.3843	0.0468	6.8137	0.1468
13	3.4523	0.2897	24.5227	0.0408	7.1034	0.1408
14	3.7975	0.2633	27.9750	0.0357	7.3667	0.1357
15	4.1772	0.2394	31.7725	0.0315	7.6061	0.1315
16	4.5950	0.2176	35.9497	0.0278	7.8237	0.1278
17	5.0545	0.1978	40.5447	0.0247	8.0216	0.1247
18	5.5599	0.1799	45.5992	0.0219	8.2014	0.1219
19	6.1159	0.1635	51.1591	0.0195	8.3649	0.1195
20	6.7275	0.1486	57.2750	0.0175	8.5136	0.1175
21	7.4002	0.1351	64.0025	0.0156	8.6487	0.1156
22	8.1403	0.1228	71.4027	0.0140	8.7715	0.1140
23	8.9543	0.1117	79.5430	0.0126	8.8832	0.1126
24	9.8497	0.1015	88.4973	0.0113	8.9847	0.1113
25	10.8347	0.0923	98.3471	0.0102	9.0770	0.1102
26	11.9182	0.0839	109.1818	0.0092	9.1609	0.1092
27	13.1100	0.0763	121.0999	0.0083	9.2372	0.1083
28	14.4210	0.0693	134.2099	0.0075	9.3066	0.1075
29	15.8631	0.0630	148.6309	0.0067	9.3696	0.1067
30	17.4494	0.0573	164.4940	0.0061	9.4269	0.1061
31	19.1943	0.0521	181.9434	0.0055	9.4790	0.1055
32	21.1138	0.0474	201.1378	0.0050	9.5264	0.1050
33	23.2252	0.0431	222.2515	0.0045	9.5694	0.1045
34	25.5477	0.0391	245.4767	0.0041	9.6086	0.1041
35	28.1024	0.0356	271.0244	0.0037	9.6442	0.1037
40	45.2593	0.0221	442.5926	0.0023	9.7791	0.1023
45	72.8905	0.0137	718.9048	0.0014	9.8628	0.1014
50	117.3909	0.0085	1163.9085	0.0009	9.9148	0.1009
55	189.0591	0.0053	1880.5914	0.0005	9.9471	0.1005
60	304.4816	0.0033	3034.8164	0.0003	9.9672	0.1003
65	490.3707	0.0020	4893.7073	0.0002	9.9796	0.1002
70	789.7470	0.0013	7887.4696	0.0001	9.9873	0.1001
75	1271.8954	0.0008	12708.9537	0.0001	9.9921	0.1001
80	2048.4002	0.0005	20474.0021	0.0000	9.9951	0.1000
85	3298.9690	0.0003	32979.6903	0.0000	9.9970	0.1000
90	5313.0226	0.0002	53120.2261	0.0000	9.9981	0.1000
95	8556.6760	0.0001	85556.7605	0.0000	9.9988	0.1000

TABLE B-12. 12% COMPOUND INTEREST TABLES

	SINGLE PAYMENT		*EQUAL-PAYMENT SERIES*			
Number of periods (*n*)	*Compound-amount factor* (To find $F\|P,i,n$)	*Present-worth factor* (To find $P\|F,i,n$)	*Compound-amount factor* (To find $F\|A,i,n$)	*Sinking-fund factor* (To find $A\|F,i,n$)	*Present-worth factor* (To find $P\|A,i,n$)	*Capital-recovery factor* (To find $A\|P,i,n$)
1	1.1200	0.8929	1.0000	1.0000	0.8929	1.1200
2	1.2544	0.7972	2.1200	0.4717	1.6901	0.5917
3	1.4049	0.7118	3.3744	0.2963	2.4018	0.4163
4	1.5735	0.6355	4.7793	0.2092	3.0373	0.3292
5	1.7623	0.5674	6.3528	0.1574	3.6048	0.2774
6	1.9738	0.5066	8.1152	0.1232	4.1114	0.2432
7	2.2107	0.4523	10.0890	0.0991	4.5638	0.2191
8	2.4760	0.4039	12.2997	0.0813	4.9676	0.2013
9	2.7731	0.3606	14.7757	0.0677	5.3282	0.1877
10	3.1058	0.3220	17.5487	0.0570	5.6502	0.1770
11	3.4785	0.2875	20.6546	0.0484	5.9377	0.1684
12	3.8960	0.2567	24.1331	0.0414	6.1944	0.1614
13	4.3635	0.2292	28.0291	0.0357	6.4235	0.1557
14	4.8871	0.2046	32.3926	0.0309	6.6282	0.1509
15	5.4736	0.1827	37.2797	0.0268	6.8109	0.1468
16	6.1304	0.1631	42.7533	0.0234	6.9740	0.1434
17	6.8660	0.1456	48.8837	0.0205	7.1196	0.1405
18	7.6900	0.1300	55.7497	0.0179	7.2497	0.1379
19	8.6128	0.1161	63.4397	0.0158	7.3658	0.1358
20	9.6463	0.1037	72.0524	0.0139	7.4694	0.1339
21	10.8038	0.0926	81.6987	0.0122	7.5620	0.1322
22	12.1003	0.0826	92.5026	0.0108	7.6446	0.1308
23	13.5523	0.0738	104.6029	0.0096	7.7184	0.1296
24	15.1786	0.0659	118.1552	0.0085	7.7843	0.1285
25	17.0001	0.0588	133.3339	0.0075	7.8431	0.1275
26	19.0401	0.0525	150.3339	0.0067	7.8957	0.1267
27	21.3249	0.0469	169.3740	0.0059	7.9426	0.1259
28	23.8839	0.0419	190.6989	0.0052	7.9844	0.1252
29	26.7499	0.0374	214.5828	0.0047	8.0218	0.1247
30	29.9599	0.0334	241.3327	0.0041	8.0552	0.1241
31	33.5551	0.0298	271.2926	0.0037	8.0850	0.1237
32	37.5817	0.0266	304.8477	0.0033	8.1116	0.1233
33	42.0915	0.0238	342.4294	0.0029	8.1354	0.1229
34	47.1425	0.0212	384.5210	0.0026	8.1566	0.1226
35	52.7996	0.0189	431.6635	0.0023	8.1755	0.1223
40	93.0510	0.0107	767.0914	0.0013	8.2438	0.1213
45	163.9876	0.0061	1358.2300	0.0007	8.2825	0.1207
50	289.0022	0.0035	2400.0182	0.0004	8.3045	0.1204
55	509.3206	0.0020	4236.0050	0.0002	8.3170	0.1202
60	897.5969	0.0011	7471.6411	0.0001	8.3240	0.1201
65	1581.8725	0.0006	13173.9374	0.0001	8.3281	0.1201
70	2787.7998	0.0004	23223.3319	0.0000	8.3303	0.1200
75	4913.0558	0.0002	40933.7987	0.0000	8.3316	0.1200

TABLE B-13. 15% COMPOUND INTEREST TABLES

	SINGLE PAYMENT		*EQUAL-PAYMENT SERIES*			
Number of periods (n)	*Compound-amount factor* (To find $F\|P,i,n$)	*Present-worth factor* (To find $P\|F,i,n$)	*Compound-amount factor* (To find $F\|A,i,n$)	*Sinking-fund factor* (To find $A\|F,i,n$)	*Present-worth factor* (To find $P\|A,i,n$)	*Capital-recovery factor* (To find $A\|P,i,n$)
1	1.1500	0.8696	1.0000	1.0000	0.8696	1.1500
2	1.3225	0.7561	2.1500	0.4651	1.6257	0.6151
3	1.5209	0.6575	3.4725	0.2880	2.2832	0.4380
4	1.7490	0.5718	4.9934	0.2003	2.8550	0.3503
5	2.0114	0.4972	6.7424	0.1483	3.3522	0.2983
6	2.3131	0.4323	8.7537	0.1142	3.7845	0.2642
7	2.6600	0.3759	11.0668	0.0904	4.1604	0.2404
8	3.0590	0.3269	13.7268	0.0729	4.4873	0.2229
9	3.5179	0.2843	16.7858	0.0596	4.7716	0.2096
10	4.0456	0.2472	20.3037	0.0493	5.0188	0.1993
11	4.6524	0.2149	24.3493	0.0411	5.2337	0.1911
12	5.3503	0.1869	29.0017	0.0345	5.4206	0.1845
13	6.1528	0.1625	34.3519	0.0291	5.5831	0.1791
14	7.0757	0.1413	40.5047	0.0247	5.7245	0.1747
15	8.1371	0.1229	47.5804	0.0210	5.8474	0.1710
16	9.3576	0.1069	55.7175	0.0179	5.9542	0.1679
17	10.7613	0.0929	65.0751	0.0154	6.0472	0.1654
18	12.3755	0.0808	75.8364	0.0132	6.1280	0.1632
19	14.2318	0.0703	88.2118	0.0113	6.1982	0.1613
20	16.3665	0.0611	102.4436	0.0098	6.2593	0.1598
21	18.8215	0.0531	118.8101	0.0084	6.3125	0.1584
22	21.6447	0.0462	137.6316	0.0073	6.3587	0.1573
23	24.8915	0.0402	159.2764	0.0063	6.3988	0.1563
24	28.6252	0.0349	184.1678	0.0054	6.4338	0.1554
25	32.9190	0.0304	212.7930	0.0047	6.4641	0.1547
26	37.8568	0.0264	245.7120	0.0041	6.4906	0.1541
27	43.5353	0.0230	283.5688	0.0035	6.5135	0.1535
28	50.0656	0.0200	327.1041	0.0031	6.5335	0.1531
29	57.5755	0.0174	377.1697	0.0027	6.5509	0.1527
30	66.2118	0.0151	434.7451	0.0023	6.5660	0.1523
31	76.1435	0.0131	500.9569	0.0020	6.5791	0.1520
32	87.5651	0.0114	577.1005	0.0017	6.5905	0.1517
33	100.6998	0.0099	664.6655	0.0015	6.6005	0.1515
34	115.8048	0.0086	765.3654	0.0013	6.6091	0.1513
35	133.1755	0.0075	881.1702	0.0011	6.6166	0.1511
40	267.8635	0.0037	1779.0903	0.0006	6.6418	0.1506
45	538.7693	0.0019	3585.1285	0.0003	6.6543	0.1503
50	1083.6574	0.0009	7217.7163	0.0001	6.6605	0.1501
55	2179.6222	0.0005	14524.1479	0.0001	6.6636	0.1501
60	4383.9987	0.0002	29219.9916	0.0000	6.6651	0.1500
65	8817.7874	0.0001	58778.5826	0.0000	6.6659	0.1500
70	17735.7200	0.0001	**********	0.0000	6.6663	0.1500
75	35672.8680	0.0000	**********	0.0000	6.6665	0.1500

TABLE B-14. 20% COMPOUND INTEREST TABLES

	SINGLE PAYMENT		*EQUAL-PAYMENT SERIES*			
Number of periods (n)	*Compound-amount factor* (To find $F\|P,i,n$)	*Present-worth factor* (To find $P\|F,i,n$)	*Compound-amount factor* (To find $F\|A,i,n$)	*Sinking-fund factor* (To find $A\|F,i,n$)	*Present-worth factor* (To find $P\|A,i,n$)	*Capital-recovery factor* (To find $A\|P,i,n$)
1	1.2000	0.8333	1.0000	1.0000	0.8333	1.2000
2	1.4400	0.6944	2.2000	0.4545	1.5278	0.6545
3	1.7280	0.5787	3.6400	0.2747	2.1065	0.4747
4	2.0736	0.4823	5.3680	0.1863	2.5887	0.3863
5	2.4883	0.4019	7.4416	0.1344	2.9906	0.3344
6	2.9860	0.3349	9.9299	0.1007	3.3255	0.3007
7	3.5832	0.2791	12.9159	0.0774	3.6046	0.2774
8	4.2998	0.2326	16.4991	0.0606	3.8372	0.2606
9	5.1598	0.1938	20.7989	0.0481	4.0310	0.2481
10	6.1917	0.1615	25.9587	0.0385	4.1925	0.2385
11	7.4301	0.1346	32.1504	0.0311	4.3271	0.2311
12	8.9161	0.1122	39.5805	0.0253	4.4392	0.2253
13	10.6993	0.0935	48.4966	0.0206	4.5327	0.2206
14	12.8392	0.0779	59.1959	0.0169	4.6106	0.2169
15	15.4070	0.0649	72.0351	0.0139	4.6755	0.2139
16	18.4884	0.0541	87.4421	0.0114	4.7296	0.2114
17	22.1861	0.0451	105.9306	0.0094	4.7746	0.2094
18	26.6233	0.0376	128.1167	0.0078	4.8122	0.2078
19	31.9480	0.0313	154.7400	0.0065	4.8435	0.2065
20	38.3376	0.0261	186.6880	0.0054	4.8696	0.2054
21	46.0051	0.0217	225.0256	0.0044	4.8913	0.2044
22	55.2061	0.0181	271.0307	0.0037	4.9094	0.2037
23	66.2474	0.0151	326.2369	0.0031	4.9245	0.2031
24	79.4968	0.0126	392.4842	0.0025	4.9371	0.2025
25	95.3962	0.0105	471.9811	0.0021	4.9476	0.2021
26	114.4755	0.0087	567.3773	0.0018	4.9563	0.2018
27	137.3706	0.0073	681.8528	0.0015	4.9636	0.2015
28	164.8447	0.0061	819.2233	0.0012	4.9697	0.2012
29	197.8136	0.0051	984.0680	0.0010	4.9747	0.2010
30	237.3763	0.0042	1181.8816	0.0008	4.9789	0.2008
31	284.8516	0.0035	1419.2579	0.0007	4.9824	0.2007
32	341.8219	0.0029	1704.1095	0.0006	4.9854	0.2006
33	410.1863	0.0024	2045.9314	0.0005	4.9878	0.2005
34	492.2235	0.0020	2456.1176	0.0004	4.9898	0.2004
35	590.6682	0.0017	2948.3411	0.0003	4.9915	0.2003
40	1469.7716	0.0007	7343.8578	0.0001	4.9966	0.2001
45	3657.2620	0.0003	18281.3099	0.0001	4.9986	0.2001
50	9100.4382	0.0001	45497.1908	0.0000	4.9995	0.2000

INDEX

ABCDEFGHIJ–H–7987